Pvt. Ltd.

Handbook of Techniques in Biotechnology

RASHMI RANADE

University Department of Life Sciences
University of Mumbai, Mumbai – 400 032

&

Prof. SANJAY DESHMUKH

University Department of Life Sciences
University of Mumbai, Mumbai – 400 032

2013

Studium Press (India) Pvt. Ltd.

Handbook of
Techniques in Biotechnology

ISBN: 978-93-80012-55-1

Published by:

Studium Press (India) Pvt. Ltd.
4735/22, 2nd Floor, Prakash Deep Building
(Near Delhi Medical Association),
Ansari Road, Darya Ganj, New Delhi-110 002
Tel.: 91-11-43240200-15 (15 lines), Fax: 91-11-43240215;
E-mail: studiumpress@gmail.com

Printed at:

Akash Press, New Delhi 110 020

About the Authors

Rashmi Ranade

Ms. Rashmi Ranade received her BSc (Agriculture) degree from Dr B. S. Konkan Krishi Vidyapeeth, Dapoli and MSc (Biotechnology) from Dr Y. S. P. University of Horticulture and Forestry, Solan; both with distinction. She has done extensive research in plant biotechnology and biotic stress resistance in plants. She has also worked at National Research Centre on Plant Biotechnology (NRCPB), IARI, New Delhi. She has teaching experience as a Guest Faculty in Dept. of Life Sciences, University of Mumbai, Kalina, Santacruz for conducting lectures of MSc Life Sciences (Environmental Biotechnology). Her research interests are plant tissue culture, plant improvement through biotechnological methods and molecular characterization.

Sanjay Deshmukh

Dr. Sanjay Deshmukh received his MSc (Botany) and PhD (Ecology) from The Institute of Science, Mumbai. He became LEAD International Fellow (Leadership for Environment and Development), London in 2006. Dr. Sanjay has gained a unique experience of working with various Research/ Corporate/ non-Govt. and Govt. Academic Establishments of national & international repute encompassing a period of over 2 decades. He worked with Dr Swaminathan in the reputed M.S. Swaminathan Research Foundation (MSSRF), Chennai for 6 years. He has worked as Director in various NGOs engaged in conservation of ecosystems and environment. He is currently working as Head, Professor of Life Sciences, at Dept. of Life Sciences, University of Mumbai. He has received numerous national and international awards for his contribution to life sciences and education. Dr. Sanjay has held several important academic positions within the University system, one of them being Chairman of Board of Studies in Life Sciences. Dr. Sanjay's significant research contribution is in areas such as community-based mangrove management, conservation, development and economic evaluation of coastal biodiversity, sacred grove conservation, bamboo resources development as well as ecological assessment of developmental projects. He has authored many research papers in reputed scientific journals.

PREFACE

Biotechnology is the most upcoming subject today and thousands of students graduate and post graduate in biotechnology every year. Being a subject of applied science, biotechnology inevitably brings every student to a stage where they should have knowledge and some experience in experimental techniques. Students have to perform a number of techniques in practicals as well as while doing research. Today, numerous students, teachers and research fellows employ a number of techniques while conducting research in biotechnology. Biotechnology is applied to various economic sectors like agriculture, pharmaceutical industries, livestock management, chemical industries, environmental management, human disease research, cancer research etc.

The book, 'Handbook of Techniques in Biotechnology' includes techniques, which one has to perform in lab and any graduate or postgraduate should know about them. The discipline of biotechnology accommodates principles and techniques of microbiology, biochemistry, molecular biology, as well as other branches of biology and uses them primarily for genetic studies or genetic manipulations. So the book aims at saving the valuable time of a researcher usually which is lost in acquiring basic information regarding the techniques. This book may serve as a textbook for both undergraduate and postgraduate courses in biotechnology and molecular biology. It can also be used as laboratory handbook in research laboratories.

Each chapter of this book contains a major theory part, which explains the principle and working of the technique, and its applications. Secondly, a generalized protocol is included wherever possible, which provides useful information for actually performing the techniques in lab.

The chapters in this book are framed in such a way that the reader not only understands the basic principle behind each technique, but also has some idea about how they are actually performed. New techniques are coming up every now and then; so students feel left behind as they don't know how the technique is performed. Here the chapters give clear-cut idea about the techniques and their variations.

The chapters, which contain protocols for the experimental technique, have details of buffer solutions, chemicals and other ingredients required in the process. Nevertheless it is assumed that the reader has some basic knowledge of instruments like micropipettes, centrifuge, spectrophotometer, hot water bath or gel casting tray and how to use them. The most important and unique feature of this book is 'troubleshooting'. It includes tips and cautions to avoid commonly made errors while performing the techniques.

Although a number of books have been written on biotechnology, genetics and life sciences, they are mostly theory oriented and even if they mention any techniques, the material provided, is not sufficient to execute them. There are also very few books, which explain both theory and protocols associated with techniques in biotechnology. This makes it difficult for student, teacher or researcher to properly understand various research methods and to apply them. Therefore to fill this void, we have made an attempt to explain basic concepts of techniques in biotechnology, combined with protocols for their execution and troubleshooting tips to avoid errors while performing them.

For teachers, this book may serve as a small handbook, which rounds up all the major techniques taught today in biotech courses. The techniques given in this book are not just for performing in lab. They comprise an important part of the theory syllabus of biotech courses in most of the UGC-recognized universities in the country.

The subject of this book is an integral part of many biotechnology courses taught all over India. This book will be useful for biotechnology undergraduates, postgraduates, teachers and researchers alike. The contents of the book also make an important part of the syllabus for various entrance exams like UPSC, UGC-NET, GATE, CSIR-JRF, SRF, DBT-JRF, JNU, ICAR-JRF, SRF, ICMR, AIIMS, BHU, IGNOU etc. The readership for this subject is quite big and growing each year.

We expect that this book will help students and researchers while applying biotechnological methods. We welcome any suggestions/criticism for further improvements, since there always remains scope for the same.

We take this opportunity to thank many of our students and colleagues who gave us new ideas and inspired us in the process of writing this book. We also want to appreciate ready assistance provided by J. N. Library of Mumbai University and British Council Library, Mumbai; that provided necessary literature sources.

We would like to express our appreciation for the generosity and patient support given by Dr J. N. Govil, Vice President (Publications) & Managing Editor, of Studium Press (India) and his staff who have strived hard to publish this book speedily without compromising the quality of the work.

We want to thank almighty for His grace and blessings, without which this extensive book project was not possible. In the end, we want to express our gratitude towards our family members for their love and support that stayed with us throughout this project.

Table of Contents

1

Electrophoresis

1.1 INTRODUCTION

Electrophoresis is a technique that is used to separate and sometimes purify macromolecules, especially proteins and nucleic acids that differ in size, charge or conformation. Today, it is one of the most widely used techniques in biochemistry and molecular biology.

Gel electrophoresis is a separation technique commonly used in modern laboratories to separate macromolecules such as DNA or proteins. As the name suggests, gel electrophoresis utilizes an electrical current to separate target molecules based on size, charge and other physical properties, through a porous gel matrix. In this technique, charged molecules are separated by the use of an electric field. Electrophoresis is a molecular separation technique that involves the use of high voltage electric current for inducing the movement of charged molecules such as proteins, DNA, nucleic acids in a support medium. The movement of charged molecules is called mobility. The mobility of molecules is towards the opposite charge for a negatively charged particle, for instance, a protein molecule with a negative charge moves towards the positive pole of the support medium.

Manipulating and analyzing nucleic acids and proteins are fundamental processes in the field of genetics, proteomics and molecular biology. The separation of complex mixtures of DNA into fragments of different sizes by electrophoresis became a well-established technique by the early 1970's. The basic electrophoretic principle was discovered by Arne Tiselius, which was called 'moving boundary electrophoresis' in free solution. Using this technique, he separated human serum into its four major components- α-globulin, β-globulin, γ-globulin and albumin. The Nobel Prize in Chemistry in 1948 was awarded to Arne Tiselius for his research on electrophoresis and adsorption analysis, especially for his discoveries concerning the complex nature of the serum proteins. His investigations have resulted in the development of specialized apparatus for methodology and technique of electrophoresis, which is in use today.

Later in 1984 Schwartz and Cantor described pulsed field gel electrophoresis (PFGE), introducing a new way to separate DNA. In particular, PFGE resolved extremely large DNA for the first time, raising the upper size limit of DNA separation in agarose gel from 30-50 kbp to over 10,000 kbp.

Electrophoresis is probably the main technique for molecular separation in today's cell biology laboratory. Because it is such a powerful technique, and yet reasonably easy and inexpensive, it has become a common technique. In spite of the many physical arrangements for the apparatus, and regardless of the medium through which molecules are allowed to migrate, all electrophoretic separations depend upon the charge distribution of the molecules being separated.

Electrophoresis can be one-dimensional having one plane of separation or two-dimensional. One-dimensional electrophoresis is used for most routine protein and nucleic acid separations. Two-dimensional separation of proteins is used for fingerprinting and can be extremely accurate in resolving all of the proteins present within a cell.

The example of one-dimensional electrophoresis is agarose gel electrophoresis, polyacrylamide gel electrophoresis (native or SDS-PAGE) and isoelectric focusing (IEF). The example of two-dimensional electrophoresis is 2D-PAGE.

1.2 PRINCIPLE AND THEORY OF ELECTROPHORESIS

The fundamental requirement for electrophoresis is that the molecules under study must have either net positive or negative charge. Most biological macromolecules carry electric charge and thus move in response to the electric field applied. Electrophoresis is a method used in clinical and research laboratories for separating molecules according to their molecular sizes. In this technique, electric current is passed through a medium that contains the mixture of molecules. Each kind of molecule travels through the medium at a different rate, depending on its electrical charge and molecular size. Separation of the molecules occurs based on these differences.

In case of nucleic acids, phosphate groups present in the sugar-phosphate backbone are negatively charged at neutral pH. In case of proteins, the net charge depends upon the amino acid composition of the polypeptide chain. Glutamic acid and aspartic acid are negatively charged while lysine and arginine are positively charged. The charge on histidine is weakly positive. The proteins are called zwitter ions, which means a molecule having both positive and negative charges. The net charge on the proteins at particular pH is determined by its content of acidic and basic amino acids and by their degree of ionization at the particular pH. Thus proteins carry either net negative or positive charge depending on the pH.

Therefore, many important biological macromolecules like amino acids, peptides, carbohydrates, nucleotides and proteins exist in the solution as electrically charged species. Under the influence of electric current, these charged particles migrate to either cathode or anode, depending on the nature of their charge.

Proteins and nucleic acids are electrophoresed within a matrix or gel. Most commonly, the gel is cast in the shape of a thin slab, with wells for loading the sample. The gel is immersed within an electrophoresis buffer that provides ions to carry a current and to maintain the pH at a relatively constant value.

The electrophoretic mobility is defined as the velocity of a particle per unit field strength. The choice of applied voltage and separation path length determine the field strength. The field strength and time of electrophoretic run are important parameters in optimizing electrophoretic separation. The conductive medium of electric charge is usually a buffer, also referred to as run buffer. The mixture of analytes is introduced into the medium containing the run buffer and electric field is applied.

The electrophoretic separations are usually performed in agarose or polyacrylamide gel. The gel has a sieving effect so that movement of larger compounds is retarded more than smaller compounds. This means that in gel electrophoresis, two compounds of the same charge but different sizes are separated according to sizes. Agarose is a type of polysaccharide extracted from seaweed. It is typically used at concentrations of 0.5 to 2%. The higher the agarose concentration, stiffer is the gel. Agarose gels are extremely easy to prepare; you simply mix agarose powder with buffer solution, melt it by heating, and pour the gel. Agarose gels have a large range of separation, but relatively low resolving power. By varying the concentration of agarose, fragments of DNA from about 200 to 50,000 bp can be separated using standard electrophoretic techniques.

Polyacrylamide is a cross-linked polymer of acrylamide. The length of the polymer chains depends upon the concentration of acrylamide used, which is typically between 3.5 and 20%. Polyacrylamide gels are significantly more difficult to prepare than agarose gels. Because oxygen inhibits the polymerization process, they must be poured between glass plates.

In the electrophoresis of proteins, pH of the conductive medium has a profound influence on the protein mobility in gel. If the operative pH is the same as pI of the protein, it will not migrate during electrophoresis. At pH values below their isoelectric point (pI), proteins will move towards cathode and while at pH values above their pI, proteins will migrate towards anode.

1.2.1 Joule Heating

During the electrophoretic separation, electric energy is also converted into heat. This phenomenon is known as joule heating. This can result in harmful effects such as enzyme denaturation, increased diffusion of proteins and sometimes, damage to the instrument. This heating forms a temperature gradient across the gel. This leads to convective flows within the buffer, resulting in band broadening and loss of separation resolution.

To minimize joule heating, heat generation can be reduced by applying lower electric field and decreasing the conductivity of the buffer. But the application of lower separation voltage results in the increased gel run time and poorer separation efficiency. The choice of buffer strength is also crucial, as higher the ionic strength, greater the conductivity and greater is the heat generated.

In case of sodium dodecyl sulphate-polyacrylamide gel electrophoresis, which is shortly denoted as SDS-PAGE, the analytes are treated such that they all exhibit the same charge to size ratio. Therefore they are separated only by the difference in their sizes. In case of isoelectric focusing, analytes such as proteins are separated according to their isoelectric point (pI). A combination of two separation methods is two-dimensional gel electrophoresis (2D-PAGE) that allows separation of thousands of proteins at a time.

The electrophoresis equipment contains gel-casting tray, comb for making wells, electrophoresis unit and power supply. Sometimes external cooling apparatus that works by circulation of water around the electrophoretic apparatus is also used, especially in case of electrophoresis of enzymes.

Separation of biomolecules, separation of bands, unique banding patterns, and molecular weight marker help identify molecular weight of unknown DNA fragment or protein from electrophoresis and Rf values can be measured.

Zymogram is a type of banding pattern obtained by electrophoresis of isozymes.

1.3 APPLICATIONS OF ELECTROPHORESIS

The vast applications of electrophoresis are most evident in the health or medical industry, including antibiotic and vaccine analysis. Protein and DNA analysis are also important electrophoresis applications. Apart from allowing researchers to map and see the differences in the genetic sequences of various species, electrophoretic DNA analysis also provides a reliable tool in forensic investigations. Electrophoretic technique is basis of DNA fingerprinting.

Electrophoresis is a very useful separation method employed in a great variety of applications, including the separation and analysis of biological samples such as nucleic acids and proteins, and environmental pesticide or pollutant analysis. Electrophoresis comes in various different forms making it a very versatile and adaptable technique.

Electrophoresis is one way of analyzing DNA, deoxyribonucleic acid, which is the code that contains all the traits that are inherited from the parents. Through electrophoresis, specific DNA sequences can be analyzed, isolated and cloned. The analyzed DNA may be used in forensic investigations and paternity tests. The ribonucleic acid (RNA) molecules can be separated by using gel electrophoresis in the same way. This makes the basis of RNA analysis in various genetic researches.

Electrophoresis has advanced our understanding of the structure and function of proteins. These molecules are needed by our body cells and may be analyzed, for instance, by getting blood and urine samples. Then through electrophoresis, the amount of proteins in the blood or urine are measured and compared to established normal values to diagnose a disease.

The application of electrophoresis in antibiotic studies dates back to the 1950s. Further studies on electrophoresis have led to improved electrophoretic techniques for the development of new antibiotics. These drugs, such as penicillin, are among the widely prescribed drugs against bacterial infections. With electrophoresis, experts are not only able to synthesize new antibiotics but are also able to analyze which types of bacteria are antibiotic-resistant.

Vaccine analysis is one of the most important applications of electrophoresis. There are several vaccines that have to be purified, processed and analyzed through electrophoresis, such as the influenza vaccine, hepatitis vaccine and polio vaccine.

Electrophoretic separation of nucleic acids is the pre-requisite for hybridization procedures like southern and northern blotting. Electrophoresis technique is widely used in medical research, as well as proteins and proteomics research. In agriculture, electrophoresis along with other analysis techniques is used in variety identification, varietal purity measurement, identification of genetically modified organisms, confirmation of gene transfer in organisms, identification of somaclonal variation, and mutations, measurement of hybrid purity, selection of superior genotypes using genetic markers, identification of pure lines, quality control in hybrid seed production and tissue culture raised plants; all these processes are based on electrophoretic techniques.

Electrophoretic techniques form an important component of major projects like documentation of rare and endangered species, conservation of germplasm, and genome sequencing projects etc.

1.4 AGAROSE GEL ELECTROPHORESIS

1.4.1 Introduction

Agarose gel electrophoresis technique came in use around late 1970s. Agarose is a polysaccharide obtained from seaweed. Agarose gel electrophoresis is a technique that is used very often by scientists to separate biological macromolecules. The development of methodology for separating DNA molecules by the technique of electrophoresis in agarose gels was a major breakthrough in the field of molecular biology. Today agarose gels are routinely used to monitor the results of DNA extractions, recombinant plasmid constructs, polymerase chain reactions etc. It also forms the basis of blotting and hybridization techniques such as southern and northern hybridization. Agarose gel electrophoresis can be done in native gels or denaturing gels. Denaturing agarose gel electrophoresis is primarily done to analyze RNA molecules, but can also be used for DNA analysis.

Agarose gel electrophoresis is a simple and highly effective method for separating, identifying, and purifying 0.5 to 25 kb DNA fragments.

1.4.2 Principle and Working

The solid medium for electrophoresis can be formed into a gel within a tube or it can be layered into flat sheets. The horizontal slab gel electrophoresis is most commonly performed technique using agarose gel. Horizontal slab gels are usually poured on a glass or plastic tray. Electrophoresis is carried out with the gel placed in electrophoresis tank and submerged in the buffer. The resistance of the gel to the passage of electric current is almost same as that of the buffer and thus the current flows along the length of the gel. In the horizontal electrophoresis system, only one gel can be run at a time. In this system thin gels can be used which provide higher efficiency. Horizontal slab gel system is also easier to handle and clean.

The Horizontal agarose gel is comparatively easy to prepare. It is easy to load the samples and the gel can be stained either before or after the electrophoresis. Slab gels are designed with multiple lanes set up such that samples run in parallel. The size and number of the lanes can be varied and, since the samples run in the same medium, there is less likelihood of sample variation due to minor changes in the gel structure. Slab gels are unquestionably the technique of choice for any blot analyses and for autoradiographic analysis. Consequently, for laboratories performing routine nucleic acid analyses, and those employing antigenic controls, slab gels have become standard.

Agarose gel electrophoresis is a type of continuous system in which both the gel buffers and tank buffers used are the same. Agarose gels are used when large pores for analysis of molecules over 10 nm is to be done. For preparation of good quality agarose gels, fine quality agarose of molecular grade is used. Gel should be as transparent as possible to obtain highest band resolution. The pore size of the gel depends on the amount of agarose used. In general gels with pore size of 150 nm at 1% concentration of agarose to 500 nm at 0.16% agarose concentration are used. Agarose is dissolved by boiling in water and it forms gel on cooling. In order to separate molecules, there must be tiny slits/wells made in the agarose gel by placing a comb into the gelling mixture before it solidifies. The resulting wells can be filled with the particular substances to be separated.

Urea or formamide is used to denature DNA or RNA for denaturing gel electrophoresis. Electrophoresis of single stranded DNA occurs at denaturing conditions while electrophoresis of double stranded DNA occurs in native gel conditions.

1.4.3 Procedure of Agarose Gel Electrophoresis

The agarose gel is used invariably for separation of DNA or RNA. For visualization of bands, ethidium bromide dye is used. This dye can be added to the samples before loading on the gel or directly added in the gelling mixture. After dissolving agarose completely by boiling, the gelling mixture is allowed to cool down to 50°C. Then ethidium bromide is added and then gelling mixture is allowed to set.

Materials Required

- Gel casting tray
- Agarose 0.8%
- Ethidium bromide (10 mg/ml)
- Purified DNA samples

- Electrophoretic apparatus
- Electrophoresis buffer- various types of buffers are used for electrophoresis.
- Gel loading buffer

Bromophenol blue	0.25%
Xylene cyanol FF	0.25%
Glycerol in water	30%

Store at 4°C.

Procedure

1. Prepare sufficient electrophoresis buffer (1x) (Refer Table 1.1) to fill the electrophoresis tank and to prepare the gel.

Table 1.1: Different types of buffers used in the electrophoresis system

Electrophoresis buffers	*Composition (1x)*
TBE buffer (Tris borate buffer) pH=8.0	Tris base 89 mM Boric acid 89 mM EDTA 2 mM
TAE buffer (Tris acetate buffer) pH=7.6	Tris base 40 mM Acetic acid 20 mM EDTA 1 mM
TPE buffer (Tris phosphate buffer) pH=8.2	Tris base 40 mM Phosphate (NaH_2PO_4) 30 mM EDTA 1 mM

2. Take the gel-casting tray. Usually the gel-casting tray comes as rectangular glass/plastic box whose opposite narrow sides are open and having notches for placement of comb.
3. Seal the sides of the tray using adhesive tape to form a mould. Place the gel-casting tray on even horizontal surface.
4. Add agarose to electrophoresis buffer at 0.8% concentration, in a clean glass container. Heat the solution till all the agarose is completely dissolved.
5. Allow the melted agarose solution to cool down to 55-60°C. Add ethidium bromide (0.5 μg/ml) using micropipette and mix well.
6. Place the comb properly and firmly on the gel-casting tray. Pour the gelling mixture slowly into the mould and allow the gel to set. Take care to avoid trapping of bubbles while pouring the gel (Fig. 1.1).
7. After the gel is completely set, carefully remove the comb and the tape.

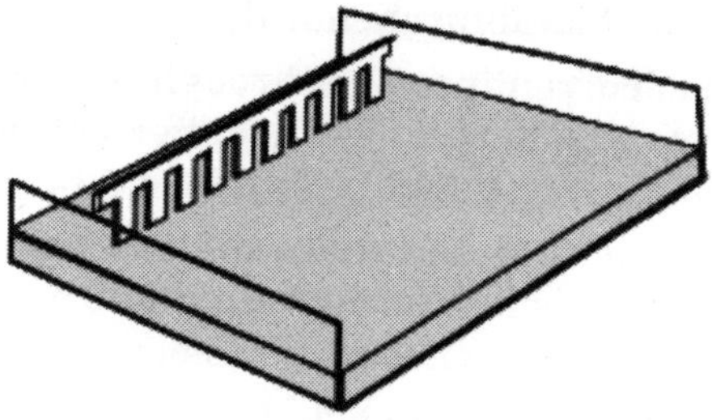

Fig. 1.1: Gel casting tray and comb

8. Carefully place the gel in the electrophoresis tank. Always place the gel in the proper orientation with wells-side towards negative terminal or anode.
9. Add enough amount of buffer to submerge the gel completely. Allow the gel to soak in the buffer for 10 minutes. Make sure all the wells are filled with buffer.
10. Mix DNA samples with loading buffer. Take about 8 µl of DNA and mix with 4 µl of loading mixture. Load the samples in the well carefully using micropipette. The glycerol present in the loading buffer helps the loading mixture to settle at the bottom of the well. Make sure there are no air bubbles in the wells.
11. Close the lid of the gel tank and switch on the electric current. The voltage of 5 V per cm length of the gel should be applied.
12. The bromophenol blue dye should migrate from the wells into the body of the gel. Run the gel until bromophenol blue migrates to the appropriate distance.
13. Switch off the electric current. Remove lid. Carefully remove the gel from electrophoresis tank.
14. Visualize the separated bands under UV transilluminator.

1.4.4 Troubleshooting Guide

1. Ethidium bromide is evaporated at high temperatures; it must never be added to the gelling mixture before dissolving the agarose. Boiling of mixture containing ethidium bromide will produce toxic fumes.
2. Always use micropipette to add ethidium bromide solution. Never use normal pipette where you have to suck the solution by mouth. Any accidental ingestion of ethidium bromide is extremely hazardous to health. Ethidium bromide is a potential carcinogenic agent.
3. The gel cast in the tray should not be uneven. It should be free from cracks and scratches.
4. The bubbles trapped in the gel adversely affect the separation of biomolecules. The perfectly cast gel should never contain any bubbles.

5. Wear gloves while handling chemicals.
6. The indication of current passing through the electrophoretic buffer is appearance of small bubbles in the buffer solution. Electrophoretic apparatus should never be left unattended especially for long hours.
7. The gels should be clear and free from any turbidity. Transparent gels facilitate better visualization of separated bands. For this reason, agarose of high purity or molecular biology grade should be used.
8. The same batch of the buffer should be used for both preparation of gel and electrophoresis tank buffer.
9. The gelling mixture should be poured in the mould when it is hot, and before the gel is starting to set. Once polymerization starts, the gel will not set properly.
10. To avoid mixing of the samples, always use new clean pipette tip to load each sample. Discard the tip after each use.

1.5 SDS-PAGE ELECTROPHORESIS

1.5.1 Introduction

Raymond and Weintraub first used the polyacrylamide gels for electrophoresis in 1959. In 1971, Laemmli introduced SDS electrophoresis.

PAGE is one of the most widely used electrophoresis techniques and separates proteins through a polyacrylamide gel matrix. Two types of PAGE can be carried out – native PAGE, in which electrophoresis is carried out under non-denaturing conditions and separation is based on the protein's charge and hydrodynamic size. Other type is SDS-PAGE, in which proteins are denatured prior to electrophoresis and separation is based on a protein's mass or molecular weight.

Protein separation by SDS-PAGE can be used to estimate relative molecular mass, to determine the relative abundance of major proteins in a sample, and to determine the distribution of proteins among fractions. The purity of protein samples can be assessed and the progress of a fractionation or purification procedure can be followed.

The most widely used polyacrylamide gel electrophoresis method was developed by Laemmli in 1970 using the denaturing (SDS) discontinuous method. This protocol relies on the presence of SDS (sodium dodecyl sulfate) and β-mercaptoethanol to denature the proteins, dissociate the proteins into subunits, and to coat them with negative charges (SDS).

This relatively uniform charge to mass ratio allows the proteins to migrate in an electric field and separate according to size. The Laemmli system uses buffers of different pH and composition to generate a voltage gradient and a discontinuous pH between the stacking and the resolving gel. A 4% acrylamide stacking gel (pH 6.8) is poured on top of a 10% acrylamide resolving gel (pH 8.8). The stacking gel with a large pore size serves to concentrate all of the proteins on top of the resolving gel. After entering the resolving gel, which has a smaller pore size, the proteins are separated according to relative molecular size.

1.5.2 Principle and Working

PAGE is usually performed as vertical gel electrophoresis. In vertical electrophoresis, multiple gels can be run at a time. Up to 3 mm thick gel is recommended for vertical systems. Very thin gels show higher sensitivity and are easier to stain. But too thin gel will present problems in sample loading.

Polyacrylamide gels are formed from the polymerization of two compounds, acrylamide and N, N-methylene-bisacrylamide. Bisacrylamide is a cross-linking agent for the gels. The polymerization is initiated by the addition of ammonium persulfate along with N,N,N,N,-tetramethylethylenediamine (TEMED). The gels are neutral, hydrophilic, and three-dimensional networks of long hydrocarbons cross-linked by methylene groups.

PAGE (Polyacrylamide Gel Electrophoresis) is an analytical method used to separate components of a protein mixture based on their size. The technique is based upon the principle that a charged molecule will migrate in an electric field towards an electrode with opposite charge. The general electrophoresis techniques cannot be used to determine the molecular weight of biological molecules because the mobility of a substance in the gel depends on both charge and size.

The separation of molecules within a gel is determined by the relative size of the pores formed within the gel. The pore size of a gel is determined by two factors, the total amount of acrylamide present (designated as %T) and the amount of cross-linker (%C). As the total amount of acrylamide increases, the pore size decreases. With cross-linking, 5%C gives the smallest pore size. Any increase or decrease in %C increases the pore size.

The PAGE gels are designated as percent solutions and will have two necessary parameters. The total acrylamide is given as % (w/v) of the acrylamide plus the bisacrylamide. Thus, a 7.5% T would indicate that there is a total of 7.5 g of acrylamide and bisacrylamide per 100 ml of gel.

Typically, proteins carry both positive and negatively charged amino acids as a result of this, proteins are zwitterionic compounds, *i.e.* carrying both positive and negative charge. To overcome this, the biological samples need to be treated so that they acquire uniform charge, so that the electrophoretic mobility depends primarily on size. For this reason, different protein molecules with different shapes and sizes, needs to be denatured with the help of detergent like SDS so that the proteins lose their secondary, tertiary or quaternary structure. When the detergent SDS is used with proteins, all of the proteins become negatively charged by their attachment to the SDS anions. The proteins being covered by SDS are negatively charged and when loaded onto a gel and placed in an electric field, migrate towards the anode. The proteins are separated by a molecular sieving effect based on size. After the visualization of bands by a staining technique, like Coomassie Blue or silver staining, the size of a protein can be calculated by comparing its migration distance with that of a known molecular weight marker.

Thus when proteins are separated on a polyacrylamide gel, with denaturation by SDS, the procedure is abbreviated as SDS-PAGE. This technique has become a standard means for molecular weight determination of proteins.

Proteins of molecular weights ranging from 10,000 to 1,000,000 Da may be separated with 7½% acrylamide gels, while proteins with higher molecular weights require lower acrylamide gel concentrations. The higher the gel concentration, the smaller the pore size of the gel and the better it will be able to separate smaller molecules. The percent gel to use depends on the molecular weight of the protein to be separated. Recommended gel concentrations are 5% for proteins ranging from 60,000 to 200,000 Daltons, 10% gels for a range of 16,000 to 70,000 Daltons and 15% gels for a range of 12,000 to 45,000 Daltons.

Relative mobility of proteins

PAGE is working upon the principle in which, the charged molecule will migrate towards the oppositely charged electrode, through highly cross-linked matrix. Separation occurs due to different rates of migration achieved by the magnitude of charge and frictional resistance related to the size.

$$R_f = ZE/F$$

Where,

Z = charge on the molecule

E = voltage applied

F = frictional resistance

Direction of movement is determined from Z is as follows,

If $Z < 0$, then → anode

If $Z > 0$, then → cathode

If $Z = 0$, then no movement

The atmospheric oxygen often creates hindrance in the polymerization of polyacrylamide gels. To minimize oxidation, gels are usually cast in vertically held glass plates separated by separating strips of uniform thickness, held together by screws. Gel is allowed to cool in the same mould and the whole mould along with gel is placed in the electrophoretic unit. In vertical electrophoresis, gel is never taken out of the mould or cassette.

The separated protein bands are usually detected by Coomassie brilliant Blue stain after electrophoresis. Coomassie dye is the most popular reagent for staining protein bands in electrophoretic gels. In acidic buffer conditions, Coomassie dye binds to basic and hydrophobic residues of proteins, changing from dull reddish-brown to intense blue. As with all staining methods, Coomassie dye reagents detect some proteins better than others based on their chemistry of action and differences in protein composition. Another popular method for detecting protein bands within a gel is silver staining, which deposits metallic silver onto the surface of a gel at the location of protein bands. Silver staining was first introduced as a general protein stain useful in polyacrylamide gel analysis in 1979 by Merrill and coworkers.

To obtain optimal resolution of proteins, a stacking gel is cast over the top of the resolving gel. Typically resolving gels are made in 6%, 8%, 10%, 12% or 15%. Stacking gel (5%) is poured on top of the resolving gel and a gel comb which forms the wells and defines the lanes, is placed. The percentage chosen depends on the size of the protein that one wishes to identify or probe in the sample. The smaller the known weight of target molecule, the higher is the percentage that should be used.

The stacking gel has a lower concentration of acrylamide (5%) for larger pore size, lower pH (6.8) and a different ionic content. This allows the proteins in a loaded sample to be concentrated into a tight band during the first few minutes of electrophoresis before entering the resolving portion of a gel. A stacking gel is not necessary when using a gradient gel, as the gradient itself performs this function.

Protein samples prepared for SDS-PAGE analysis are denatured by heating in the presence of a sample buffer containing 1% SDS with or without reducing agents such as 20 mM DTT (Dithiothreitol) and 2-mercaptoethanol. The protein sample is mixed with the sample buffer and boiled for 3-5 minutes, then cooled to room temperature before it is pipetted into the sample well of a gel. Loading buffers also contain glycerol so that they are heavier than water and sink neatly to the bottom of the buffer-submerged well when added to a gel.

To assess the relative molecular weight of a protein on a gel, protein molecular weight markers are run in the outer lanes of the gel for comparison.

TEMED (N,N,N',N'-tetramethylethylenediamine) is used for polymerization of polyacrylamide gels. TEMED stabilizes free radicals and improves polymerization. The rate of polymerization and the properties of the resulting gel depend on the concentrations of free radicals. Increasing the amount of free radicals results in increase in gel turbidity and a decrease in gel elasticity. Decreasing the amount free radical helps in better polymerization of gel. Ammonium persulfate (APS) is another agent used for the same reason. APS is a source of free radicals and is often used as an initiator for gel formation. An alternative source of free radicals is riboflavin, which generates free radicals in photochemical reaction. When riboflavin is used, gelling mixture needs to be illuminated with fluorescent light for a sufficient period of time.

1.5.3 Procedure of SDS-PAGE Electrophoresis

Materials required

- *1.5 M Tris (pH 8.8)*

  To make 500 ml of 1.5 M Tris (pH 8.8) stock solution, take 90.825 g Tris Base and add 400 ml distilled water. Adjust pH to 8.8 with con. HCl. Make the final volume to 500 ml using distilled water.

- *0.5 M Tris (pH 6.8)*

  To make 500 ml of 0.5 M Tris (pH 6.8) stock solution, take 30.275 g of Tris Base and add 400 ml distilled water. Adjust pH to 6.8. Make the final volume to 500 ml using distilled water.

- *30% Acrylamide*

  To prepare 30% acrylamide stock solution of 100 ml volume, take 29 g of acrylamide and 1 g of N,N'-methylenebisacrylamide in small amount of distilled water. Heat the solution to dissolve ingredients. Adjust final volume to 100 ml. Sterilize by filtering through Millipore filter. Store the solution at 4°C in dark.

- *10% SDS*

  To make 100 ml stock of 10% SDS, take 10 g of SDS (Sodium dodecyl sulfate) and add distilled water about 90 ml. Heat the solution to ~60°C to facilitate dissolving of SDS. At higher temperatures, SDS creates profuse foaming so do not sterilize in autoclave.

- TEMED is used undiluted.

- *10% APS*

  To prepare 10% APS stock solution of 10 ml volume, take 1 g of Ammonium persulfate. Add distilled water to make volume of 10 ml. This solution may be stored at 4°C for several weeks.

- *10X reservoir buffer* (Use 1x for running gel)

To make 10x reservoir buffer of 1 litre volume, take 30 g of Tris base and 144 g of Glycine. Make final volume 1 litre with distilled water. Add 5 ml/litre 20% SDS for electrophoresis.

- *Laemmli sample buffer (2x)*

  To make 2x Laemmli sample buffer of total volume of 16 ml, take distilled water 10.4 ml, 1.2 ml of 0.5M Tris (pH 6.8), 1.9 ml of Glycerol, 1.0 ml of 20% SDS, 0.5 ml of β-mercaptoethanol, and 1% bromophenol blue. Store aliquots at –20°C.

- *Fixative solution*

  Fixatives used to fix proteins to the gel are either 20% (w/v) trichloroacetic acid (TCA), or methanolic/ethanolic solutions of acetic acid.

- *Coomassie blue dye solution*

  Take 0.1% Coomassie Blue in 40% ethanol and 10% acetic acid.

  Destaining solution (50% methanol, 10% glacial acetic acid)

Procedure

Preparation of protein samples

Protein samples are diluted to 1:2 in Laemmli sample buffer and are boiled for 5 minutes. The β-mercaptoethanol in the sample buffer reduces the protein's disulfide bonds and the SDS denatures the proteins. The sample buffer contains glycerol to increase the density so that when the sample is loaded it sinks to the bottom of the well. Bromophenol blue dye is in the sample buffer to monitor the protein run during electrophoresis process.

Casting the SDS-PAGE gel

1. Assemble the vertical gel cassette apparatus. Lay the rectangular glass plate down first, then place two spacers of equal thickness along the opposite edges of the rectangular plate. Next, place the shorter glass plate on top of the spacers so that the bottom ends of the spacers and glass plates are aligned. At this point, the spacers should be sticking up about 5 mm above the long glass plate.
2. Loosen the four screws on the clamp assembly and stand it up so that the screws are facing away from you. Firmly grasp the glass plates with the spacers between them and gently slide it into the clamp assembly along the front face of the acrylic pressure plate. The longer glass plate should be against the acrylic pressure plate of the clamp assembly. Tighten the top two screws of the clamp assembly.
3. Place the clamp assembly into the alignment slot on the casting stand so that the clamp screws face away from you. Loosen the top two screws to allow the plates and spacers to settle against the casting

stand base. Gently tighten both pairs of screws. Transfer the clamp assembly to one of the casting slots in the casting stand.

4. Prepare the resolving gel mixture (Refer Table 1.2). It is important to add all the reagents except the TEMED to the gel. Once the TEMED and APS have been added to the polyacrylamide solution, it will polymerize in a few minutes, and hence they are added last. Allow the gel to polymerize.

Table 1.2: Various percentages of resolving gel mixture

Components	*6%*	*8%*	*10%*	*12%*	*15%*
Distilled water	2.6 ml	2.3 ml	1.9 ml	1.6 ml	1.1 ml
Tris 1.5 M (pH=8.8)	1.3 ml	1.3 ml	1.3 ml	1.3 ml	1.3 ml
Acrylamide solution 30%	1.0 ml	1.3 ml	1.7 ml	2.0 ml	2.5 ml
SDS 10%	50 µl	50 µl	50 µl	50 µl	50 µl
APS 10%	50 µl	50 µl	50 µl	50 µl	50 µl
TEMED	4 µl	3 µl	2 µl	2 µl	2 µl

In the given procedure, 10% resolving gel is used.

5. Pour the stacking gel directly on top of the polymerized resolving gel (Refer Table 1.3). Insert the comb gently and quickly before the gel polymerizes. Leave to polymerize until gelling mixture gets solid. Thereafter, remove the comb gently.

Table 1.3: Stacking gel mixture (5%)

Components	*5% gel*
Distilled water	1.4 ml
Tris 0.5 M (pH=8.8)	250 µl
Acrylamide solution 30%	330 µl
SDS 10%	20 µl
APS 10%	20 µl
TEMED	2 µl

6. Assemble the vertical gel electrophoresis apparatus.
7. Fill the inner chamber to the top with 1x Reservoir Buffer (with 0.1% SDS). Pour additional 1x Reservoir Buffer (with 0.1% SDS) into the outer chamber until the level is above the bottom of the glass plate.
8. Load the protein samples in the well, using the long gel-loading tips or a Hamilton Syringe.

Running the gel

1. Check that the buffers in the upper buffer chamber are full because leakage of the buffer may occur.

2. Place the lid on top of the lower buffer chamber. Make sure that the connection is correct, *i.e.* black-to-black and red-to-red.
3. Attach the electrical leads to a suitable power pack with the proper polarity (black to black and red to red). Run the gel at a constant current of 30 mA.
4. Stop the electrophoresis when the tracking dye reaches ~1 cm above the end of the glass plates.

Staining the gel

1. Remove the gel cassette from the electrophoresis chamber.
2. Loosen all four screws of the clamp assembly and remove the glass plate sandwich from it.
3. Push one of the spacers out to the side of the plates without removing it.
4. Gently twist the spacer so that the upper glass plate pulls away from the gel.
5. Cut the gel on one side to orientate the gel.
6. Remove the gel and place it in the container containing the fixing solution. Make sure that the gel is fully submerged in the staining solution. Agitate slowly for 2 hours on shaker.
7. Pour out the fixing solution. Flood the gel with Coomassie Blue stain solution and agitate slowly for 4 hours.
8. Pour out the staining solution. Rinse the gel with about 50 ml of fixing solution.
9. Pour out fixing solution. Cover the gel with destaining solution and agitate slowly for 2 hours.
10. Pour out destaining solution. Add fresh destaining solution and agitate till blue bands appear on clear background.
11. Store the gel in 7% acetic acid or water.

1.5.4 Troubleshooting guide

1. Acrylamide is a potent neurotoxin and can be absorbed through the skin. Wear gloves and a mask when weighing the reagents. Wear gloves when handling the solutions containing these reagents.
2. Work quickly after addition of TEMED to complete the gel casting, before the acrylamide polymerizes.
3. It is important to use the same batch of electrophoresis buffer in both of the reservoirs and in the gel. Small differences in ionic strength or pH produce buffer fronts that can greatly distort the migration of protein. Non-denaturing polyacrylamide gels are usually run at

voltages between 1 V/cm and 8 V/cm. If electrophoresis is carried out at a higher voltage, differential heating in the center of the gel may cause bowing of protein bands.

4. Acrylamide solution is light sensitive and should be stored in the dark at 4°C.

1.6 2-DIMENSIONAL GEL ELECTROPHORESIS

1.6.1 Introduction

Two-dimensional electrophoresis was first introduced by P. H. O'Farrell and J. Klose in 1975. In the original technique, the first-dimension separation was performed in carrier ampholyte-containing polyacrylamide gels cast in narrow tubes. Today, introduction of immobilized pH gradients and reagents have imparted superior resolution and reproducibility to first-dimension IEF (Isoelectric Focussing) of 2D electrophoresis.

Several forms of PAGE exist and can provide different types of information about the proteins. Non-denaturing PAGE, also called as native PAGE, separates proteins according to their mass-charge ratio. Denaturing and reducing SDS-PAGE, the most widely used electrophoresis technique, separates proteins primarily by mass. Two-dimensional (2D) PAGE separates proteins by isoelectric point in the first dimension and by mass in the second dimension.

A large and growing application of 2-D electrophoresis is proteome analysis. Two-dimensional electrophoresis is used in this technique due to its unparalleled ability to separate thousands of proteins simultaneously.

Two-dimensional electrophoresis has unique ability to detect post and co-translational modifications of proteins, which cannot be predicted from the genome sequence. Applications of 2-D electrophoresis include, proteome analysis, study of cell differentiation, detection of disease markers, monitoring therapies, drug discovery, cancer research, purity checks, and microscale protein purification.

Limitations of 2D electrophoresis

1. Some proteins such as membrane proteins are under represented in the 2D electrophoresis due to their low solubility.
2. Separation of proteins is limited by gel concentration and size.
3. Basic and membrane bound proteins are not well separated by 2D gel electrophoresis.
4. The sample capacity and detection sensitivity of the 2D electrophoresis is limited. Therefore low abundance proteins are poorly detected.

5. 2D electrophoresis is capable of revealing only a small portion of the whole proteome of an organism.
6. A single protein in 2DE can make multiple spots, so the actual number of proteins are less than spots.
7. Results are difficult to reproduce in 2D electrophoresis.

To overcome these limitations, strategies such as subcellular fractionation and narrow pH range gels are used. Another approach is pre-fractionation of complex protein mixtures by isoelectric focusing.

1.6.2 Principle and Working

In 2D electrophoresis, two electrophoresis modes are combined on a single gel. First separation is performed in one dimension followed by second separation perpendicular to the first one. After first separation, a single band may contain more than one protein species. They are further separated by second separation method based on a different parameter.

Using this technique, proteins are separated by two different properties. Initially proteins or polypeptides are separated on the basis of their net charges by isoelectric focusing. Gels are then turned 90° and separation continues based on their molecular weight (Fig. 1.2). Because it is unlikely that two molecules will be similar in both properties, molecules are more effectively separated in 2-D electrophoresis than in 1-D electrophoresis gels.

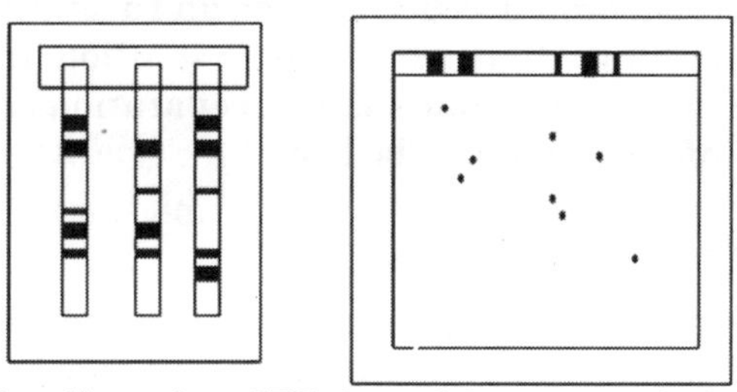

First dimension - IEF on strip gels   Second dimension - SDS-PAGE

Fig. 1.2: An overview of 2D gel electrophoresis

The 2D electrophoresis is governed by parameters like charge, mobility, size, and pI. All these parameters depend upon the mode of electrophoresis used.

In 2-D electrophoresis, proteins are initially separated by isoelectric focusing. This isoelectric focusing occurs across a pH gradient. The protein band will stop moving across the gel at its isoelectric point when the charge associated with the different amino groups is nullified by the pH. The gel is then turned 90° and the second electrophoresis occurs using SDS to separate the proteins by molecular weight. This second dimension of focusing gives a series of spots across the gel and each spot is a specific protein.

Using 2D electrophoresis, thousands of proteins or nucleic acids can be separated with high resolution. The first separation is performed in a single lane. In the second separation, proteins are separated by SDS-PAGE according to their molecular weight.

For analysis of complex proteins, they can be first digested with enzymes and the fragments can be analyzed. The whole proteomic analysis can also be done by taking whole protein samples from cell cultures. Typically 1000-2000 proteins can be resolved in a single run.

Appropriate sample preparation is absolutely essential for good 2-D results. Some proteins are naturally found in complexes with membranes, nucleic acids, or other proteins, some proteins form various non-specific aggregates, and some proteins precipitate when removed from their normal environment. Different treatments and conditions are required to solubilize different types of protein samples. In order to characterize specific proteins in a complex protein mixture, the proteins of interest must be completely soluble under electrophoresis conditions.

The effectiveness of solubilization of proteins depends on the choice of cell disruption method, protein concentration and dissolution method, choice of detergents, and composition of the sample solution. If any of these steps are not optimized for a particular sample, separations may be incomplete or distorted and information may be lost.

Cell disruption methods

In order to analyze all intracellular proteins, the cells must be effectively disrupted. Choice of disruption method depends on whether the sample is from cultured cells, solid tissue, or other biological material and whether the analysis is targeting all proteins or just a particular subcellular fraction. If only a fraction of the proteins in a tissue or cell type is of interest, pre-fractionation can be employed during sample preparation. If proteins from particular cellular organelles like nuclei, mitochondria, plasma membrane etc. are desired, the organelle of interest can be purified by differential centrifugation or by other methods prior to solubilization of proteins for 2-D electrophoresis.

The cells or tissue should be disrupted in such a way as to minimize proteolysis and other modes of protein degradation. Cell disruption should be done at low temperature and with a minimum of heat generation. Cell disruption should ideally be carried out directly into a strongly denaturing solution containing protease inhibitors.

Gentle lysis methods of cell disruption are,

- Freeze-thaw lysis method
- Detergent lysis method
- Enzymatic lysis method

Vigorous lysis methods of cell disruption are,

- Sonication
- Grinding in pestle and mortar
- Mechanical homogenization

After isolation and precipitation of proteins, the impurities in the protein samples such as detergents, salts, nucleic acids, polysaccharides and lipids need to be removed. Non-protein impurities in the sample can interfere with separation and subsequent visualization of the 2-D result, so sample preparation can include steps to free the sample of these substances.

In order to achieve a well-focused first-dimension separation, sample proteins must be completely disaggregated and fully solubilized. Complete denaturation ensures that each protein is present in only one configuration, and that aggregation and intermolecular interaction is avoided. Therefore, components such as urea and one or more detergents are always included in the protein samples. Urea solubilizes and unfolds most proteins to their fully random conformation, with all ionizable groups exposed to solution. A zwitterionic detergent is always included in the sample solution to ensure complete sample solubilization and to prevent aggregation through hydrophobic interactions. Therefore, either of two similar non-ionic detergents, NP-40 or Triton X-100, is used. A sample should remain in sample solution at room temperature for at least 30 min to achieve full denaturation and solubilization prior to centrifugation and subsequent sample application.

2D electrophoresis of nucleic acids

2D electrophoresis technique can also be used to separate nucleic acids like DNA and RNA. Long DNA fragments are usually analyzed after digestion with restriction enzymes. The first separation is carried out after treatment with one restriction enzyme. The bands that are separated are then treated with second restriction enzyme that cut at different sites than the first one. The DNA fragments obtained are then separated in second dimension. Transfer between two dimensions is the most challenging phase. The bands obtained in the first separation can be treated *in situ* with the second restriction enzyme. This can be done by soaking them in solution containing restriction enzyme or by incorporating the enzyme in the gel. 2D electrophoresis of DNA and RNA is usually done in agarose gels. 2D electrophoresis of RNA usually employs the native gel electrophoresis in the first separation where RNA molecules retain their folding. In the second dimension, the gel containing denaturing components is used. At this stage, RNA molecules undergo a change in configuration and can be separated based on molecular weight.

1.6.3 Procedure of 2-Dimensional Gel Electrophoresis

Materials required

- *Lysis buffer*

2M	Thiourea
7M	Urea
2% (w/v)	CHAPS 3-[(3-Cholamidopropyl) dimethylammonio] -1-propanesulfonate (CHAPS is a zwitterionic detergent)
0.8% (w/v)	Pharmalyte (pH 3-10) (for establishing pH gradient)
1% (w/v)	Dithiothreitol (DTT)
5 mM	Pefabloc (proteinase inhibitor)

  The proteinase inhibitor is added just before use. Lysis buffer should always be prepared fresh.
- Gradient mixer (2 × 15 ml)
- Glass plates with a 0.5 mm thick U-frame (200 × 260 mm)
- Plain glass plates (size 200 × 260 mm),
- Clamps
- GelBond PAGfilm (200 × 260 mm)
- Immobiline
- Repel silane
- Tetramethylethylenediamine (TEMED)
- *Acrylamide / Bisacrylamide solution (30% T, 3% C)*: The acrylamide solution contains 29.1% (w/v) acrylamide and 0.9% (w/v) N, N'-methylenebisacrylamide. To make 100 ml of the solution dissolve 29.1 g of acrylamide and 0.9 g of bisacrylamide in deionized water and make final volume up to 100 ml. This solution can be stored for one week at 4°C.
- *Ammonium persulfate solution (40% (w / v) in deionized water)*: To prepare 1 ml of the solution, dissolve 0.4 g of ammonium persulfate in 1 ml of deionized water. This solution should be prepared fresh just before use.

- *Stacking gel buffer*

Tris-HCl (pH 6.8)	0.5 M
SDS	0.4 % (w/v)

To make 100 ml, dissolve 6.05 g of Tris base and 0.4 g of SDS in about 80 ml of deionized water. Adjust to pH 6.8 with 4N HCl and fill up to 100 ml with deionized water. The buffer can be stored at 4°C for 2 weeks.

- *Resolving gel buffer*

Tris-HCl (pH 8.8)	1.5 M
SDS	0.4% (w/v)

To make 100 ml, dissolve 15.15 g of Tris base and 0.4 g of SDS in about 80 ml of deionized water. Adjust to pH 8.8 with 4N HCl and fill up to 100 ml with deionized water. The buffer can be stored at 4°C up to 2 weeks.

- *Rehydration solution*

Urea	8 M
CHAPS	1%
DTT	15 mM
Ampholyte	0.2%

To prepare 50 ml of the solution, dissolve 25.0 g of urea in deionized water and complete to 50 ml. To 48 ml of this solution add 0.5 g of CHAPS, 100 mg of DTT, and 0.25 ml IPG buffers. Rehydration solution should be prepared fresh.

- *Electrode buffer stock solution* (for horizontal SDS-PAGE)

  To make 1 litre of a 10x solution add 30.3 g of Trizma base, 144 g of glycine, 10.0 g of SDS and 100 mg of sodium azide. Dissolve in deionized water and fill up to 1000 ml and filter. Electrode buffer stock solution can be kept at room temperature for up to one week. Before use, mix 100 ml of the buffer with 900 ml of deionized water.

- *Fixative solution*

  Fixatives used to fix proteins to the gel are either 20% (w/v) trichloroacetic acid (TCA), or methanolic/ethanolic solutions of acetic acid.

- *Coomassie Blue Dye solution*

  Take 0.1% Coomassie Blue in 40% ethanol and 10% acetic acid.

Procedure

Sample preparation

1. In general, an adequate amount of deep-frozen cells or tissue like mouse liver, yeast, plant seeds etc. is disrupted by different techniques such as grinding in liquid nitrogen with mortar and pestle, sonication

with an ultrasonic probe in ice-bath, shearing-based methods, or mechanical homogenization.

2. After cell lysis, proteins are solubilized with sonication in lysis buffer.
3. Protein extracts are either used immediately or are stored at –80 °C.
4. For analytical runs, typically the sample solution is applied onto a single IPG (immobilized pH gradient) gel strip.
5. The amount of protein to be loaded onto a single IPG gel strip varies between 50–100 μg for analytical, and 0.5–10 mg for micro-preparative runs.

Casting of IPG gel

To prepare 15 ml each of acidic and basic solutions, mix the chemicals and reagent solutions. IPG slab gels with linear gradients such as pH 4–7, 4–9, 6–10, 6–12, 3–12 and 4–12 are cast. For better polymerization, the acidic and basic solutions are adjusted to pH 7 with sodium hydroxide and acetic acid, respectively. The acidic, dense solution is pipetted into the mixing chamber and the basic, light solution into the reservoir of the gradient mixer.

1. Assemble the polymerization cassette. Wet the plain glass plate (size 260 × 200 mm) with a few drops of water. Place the GelBond PAGfilm, hydrophilic side upwards, on the wetted surface of the plain glass plate. The GelBond PAGfilm should overlap the upper edge of the glass plate for 1-2 mm to facilitate filling of the cassette. Expel excess water with a roller. Place the glass plate, which bears the U-frame (0.5 mm thick) on top of the GelBond PAGfilm and clamp the cassette together. Put it in the refrigerator for 30 min.
2. To cast the IPG gel, pipette 12.5 ml of the acidic, dense solution into the mixing chamber of the gradient mixer. Outlet and valve connecting the mixing chamber and reservoir have to be closed.
3. Add 7.5 μl of TEMED and 12 μl of ammonium persulfate and mix. Open the connecting valve between the chambers for a second to release any air bubbles.
4. Pipette 12.5 ml of the basic, light solution into the reservoir of the gradient mixer. Add 7.5 μl of TEMED and 12 μl of ammonium persulfate and mix with a spatula.
5. Switch on the magnetic stirrer. Remove the polymerization cassette from the refrigerator and put it underneath the outlet of the gradient mixer. Apply the gradient mixture slowly, but steadily into the cassette from a height of about 5 cm just by gravity flow. Take care that the level in both chambers drops equally fast. Formation of the gradient will be complete in 2-3 min.
6. Keep the mould at room temperature for 15 min to allow adequate levelling of the density gradient. Then polymerize the gel for 1 hour at 50°C in a heating-cabinet.

7. After polymerization at 50°C, the cassette is allowed to cool down to room temperature. Then insert a spatula between the glass plates and the gel. Remove the glass plates and carefully remove the gel from the cassette.
8. Wash the IPG gel for 1 hour with water changes of 10 minutes interval with deionized water in a glass tray on a rocking platform.
9. Equilibrate the gel in 2% (w/v) glycerol for 30 min and dry it overnight at room temperature, or using a fan. Afterwards, protect the surface of the dry gel with a sheet of plastic film. The dried IPG gel can be stored in a sealed plastic bag at –20°C for at least several months without loss of quality.

Casting of horizontal SDS-PAGE gel

1. Assemble the polymerization cassette consisting of two glass plates, one covered with the PAG-film, the other bearing the U-frame (0.5 mm thick)
2. Immediately before casting the gel, add 5 µl of TEMED and 10 µl of ammonium persulfate to the gel solutions.
3. For casting an SDS gel (size 250 × 190 × 0.5 mm) with a stacking gel length of 50 mm, pipette 6.0 ml of stacking gel solution into the pre-cooled (4°C) mould.
4. Then cast the pore gradient on top of the stacking gel by mixing 9.0 ml of the dense (12% T, 25% glycerol) and 9.0 ml of the light (15% T, no glycerol) solution with the help of a gradient maker. The high glycerol concentration of the stacking gel solution allows overlaying of the pore gradient mixture without an intermediate polymerization step.
5. After pouring, leave the cassette for 15 min at room temperature to allow adequate levelling of the density gradient. Then place it in a heating cabinet at 50°C for 30 min for polymerization. The polymerized gel can be stored in a refrigerator overnight.

Performing IEF in IPG gel

1. Prior to IEF, the dried IPG gels or the ready-made Immobiline DryPlates- are cut into individual 3 mm wide strips with the help of a paper cutter.
2. Then the IPG strips are rehydrated to their original thickness of 5 mm, either in a rehydration cassette, or in a reswelling tray or in an IPG phor strip holder. In the former case, the thickness of the IPG gel is defined by the thickness of the spacer, whereas in the latter cases the correct IPG gel thickness is obtained by rehydrating with the proper amount of rehydration buffer.
3. IPG dry strips are rehydrated in a rehydration cassette, or a reswelling tray, followed by sample application by means of cup loading.

Rehydration of IPG strips in the IPG-phor strip holder

1. Solubilize proteins with sample solubilization (lysis) buffer and dilute the extract to 10 mg/ml with rehydration solution (dilution: 1:19 for analytical runs) to a final volume of 350-400 µl for 180 mm long IPG strips.
2. Apply the required number of strip holders onto the cooling plate-electrode contact area of the IPG-phor unit.
3. Pipette out 350 µl of sample-containing rehydration solution (for 180 mm long IPG strips) into the strip holder base.
4. Peel off the protective cover sheets from the IPG strip and slowly lower the IPG strip with gel side-down onto the rehydration solution. Avoid trapping air-bubbles. The IPG strip must be moveable and not stick to the tray.
5. Cover the IPG strips with 1-2 ml of silicone oil and apply the plastic cover. Pressure blocks placed on the underside of the cover assure that the IPG strip keeps in good contact with the electrodes as the gel swells. Apply low voltage of 30-50V during rehydration for improved entry of high molecular marker proteins.

Sample application by cup loading

For analytical purposes, typically 50-100 µg of protein are applied onto a single, 180 mm long IPG gel strip. Sample cup loading is accomplished with special cup-loading IPG-phor strip holders.

Performing IEF after sample application by cup loading

When protein separation is performed in basic pH ranges, much better separations are obtained by applying sample via cup loading on separately rehydrated IPG strips than by sample in-gel rehydration.

1. Rehydrate IPG dry strips with rehydration buffer.
2. After the IPG strips have been rehydrated, rinse them with deionized water for a few seconds and place them, gel side up, on a sheet of water saturated filter paper. Wet a second sheet of filter paper with deionized water and apply it onto the surface of the IPG gel strips. Blot them gently for a few seconds to remove excess rehydration solution in order to prevent urea crystallization on the surface of the gel during IEF.
3. Apply the required number of the cup-loading IPG-phor strip holders onto the cooling plate-electrode contact area of the IPG-phor electrophoresis instrument, and apply the rehydrated IPG gel strips into the cup-loading strip holders, gel side upwards and acidic ends facing towards the anode.
4. Moisten two filter paper electrode pads of size- 4 × 10 mm with deionized water. Remove excess liquid by blotting with a filter paper,

and apply the moistened filter paper pads on the surface IPG gel at the anodic and cathodic ends of the IPG strip.

5. Position the movable electrodes above the electrode filter paper pads and gently press the electrodes on the filter papers.
6. Position the movable sample cups near the anode or cathode, and gently press the sample cup onto the surface of the IPG gel strip. Pipette the sample in the sample cup and overlay the IPG strip with several ml of paraffin oil or cover fluid. Do not use silicone oil or kerosene.
7. Program the instrument according to the required voltage, time, voltage gradient, temperature etc. and run IEF according to the settings recommended.

Horizontal SDS-PAGE

1. Fill the buffer tanks of the electrophoresis unit with electrode buffer.
2. Soak 2 sheets of filter paper (size 250 × 200 mm) in electrode buffer and put them on the cooling block (15°C). Soak the electrode wicks (size 250 x 100 mm) in electrode buffer. Place them at the edges of the buffer-soaked filter papers and perform a pre-run at 30 mA for 3 hours to remove impurities from the electrode wicks.
3. Then remove the filter papers and discard them. Keep the purified electrode wicks in the electrode buffer tanks.
4. During the equilibration step of the IPG gel strips, open the polymerization cassette, pipette a few ml of kerosene on the cooling block (15°C) of the electrophoresis unit and put the SDS gel on it, keeping gel side up.
5. Apply the electrode wicks on the surface of the SDS gel so that they overlap the cathodic and anodic edges of the gel by about 10 mm.
6. Place the blotted IPG gel strips, with gel side down onto the SDS gel surface adjacent to the cathodic wick. No embedding of the IPG gel strip is necessary.
7. If it is desired to co-electrophorese molecular weight marker proteins, put a silicone rubber frame onto the SDS gel surface alongside the IPG gel strip and pipette in 5 µl of marker proteins dissolved in SDS-buffer.
8. Put the lid on the electrophoresis unit and start SDS-PAGE with 100 V for about 75 min with a limit of 20 mA. When the Bromophenol Blue tracking dye has completely moved out of the IPG gel strip, pause the run, remove the IPG gel strip and move the cathodic electrode wick forward for 4-5 mm so that it now overlaps the former sample application area.
9. Then continue the run at 600 V with a limit of 30 mA until the tracking dye has migrated into the anodic electrode wick. Total running time is approximately 6 hours.

Fixing of proteins to gel

After termination of the second dimension run of SDS-PAGE, fixing is necessary to immobilize the separated proteins in the gel and to remove any non-protein components, which might interfere with subsequent staining. Depending on gel thickness, the gel is submersed in the fixative for one hour at least, but usually overnight, with gentle shaking.

Coomassie Brilliant Blue (CBB) staining of gel

1. Prior to Coomassie Blue staining, proteins are fixed for more than 30 min in 20% TCA. After a 5 min washing with 40% ethanol and 10% acetic acid, the gel is saturated with the dye solution for 3 hours. Then the gel is destained twice for 20 min each with 40% ethanol and 10% acetic acid until the background is no longer stained dark blue.
2. Then the gel is immersed in 500 ml of distilled water containing a few ml of acetic acid for up to 48 hours to decrease background staining and to increase sensitivity.

Drying of gel

Gels must be dried before further analysis by autoradiography or for permanent storage. To dry Coomassie Blue or silver stained gels, they are first soaked for 0.5–2 hours in glycerol, sorbitol, or polyethyleneglycol. Horizontal, 0.5 mm thin SDS gels on plastic backing are usually impregnated in 30% glycerol for 30 min, air-dried for 2–3 hours and sealed in a plastic bag for storage.

1.6.4 Troubleshooting guide

1. A protein sample with high salt concentration can be diluted and analyzed, or desalted, then concentrated by lyophilization, or precipitated with TCA and ice-cold acetone and re-solubilized with rehydration solution.
2. Once purified, protein samples should be stored in aliquots at –80 °C. The samples should not be exposed to repeated thawing.
3. When the protein sample contains urea, it must not be heated over 37 °C. Elevated temperatures cause urea to hydrolyze to isocyanate, which modifies proteins by carbamylation.
4. Lysis buffer thawed once should not be refrozen again.
5. Problems with gel swelling or shrinking during fixing, staining and drying are usually encountered with gels not polymerized onto plastic backing
6. Make sure that the orientation of the IPG gel strips on the cooling block of the IEF chamber is correct (acidic end towards anode). Check the temperature of the cooling block (usually 20°C).

7. Sample solution should not be too concentrated (maximum of 10 mg protein/ml) to avoid protein precipitation at the sample application point.
8. Remove the IPG gel strip from the surface of the horizontal SDS PAGE gel as soon as the Bromophenol Blue dye front has migrated 4-5 mm off the IPG gel strip.
9. The glass plate that bears the U-shaped frame should be treated with Repel Silane to avoid sticking of the gel to the glass plate after polymerization.
10. If gel does not polymerize properly, the probable reason is that TEMED or ammonium persulfate is too old. Replace TEMED and persulfate.
11. IPG strips have to be rehydrated to their original thickness of 0.5 mm. IPG gel reswelling time depends on the composition of the rehydration buffer. If the rehydration solution contains high concentrations of urea > 8 M and detergents >1%, the rehydration should be performed overnight.
12. When sample is applied into cups, do not apply less than 20 µl of sample solution.
13. If IPG gel strips are found to burn near the electrode strips, then probable reason is that gel strips have dried out at the anodic or cathodic ends due to electroendosmotic flow. Add 10% glycerol to the reswelling buffer.

1.7 PULSED FIELD GEL ELECTROPHORESIS

1.7.1 Introduction

The original method of pulsed field electrophoresis was described by Schwartz and Cantor. This method could resolve DNA up to 2000 kb in length. However, current improvements in the technique allow resolution of up to 5000 kb. In 1982, Schwartz and coworkers introduced the concept that DNA molecules larger than 50 kb can be separated by using two alternating electric fields. Since that time, a number of instruments based on this principle have been developed, and the value of using pulsed fields has been demonstrated for separating DNA fragments from a few kilobasepairs to over 10 mega basepairs (Mb).

The technique of PFGE has extended the size range of molecules amenable to molecular analysis and has profoundly influenced the study of genes and genomes. For some organisms, intact chromosomes can be separated from each other by PFGE, allowing gene mapping by Southern hybridization and providing a source of purified chromosomal DNA. Application of PFGE can reveal chromosome length polymorphisms, thus facilitating evolutionary and population studies in a number of species. For chromosomal DNA that is too large to be separated as intact molecules

PFGE has surpassed traditional mapping techniques such as restriction and deletion mapping thus enabling the constructions of genetic maps.

The method of pulsed field gel electrophoresis was first utilized in 1982, and since then several apparatuses have been developed for separating large molecules of DNA, all using multiple electric fields. All systems separate DNA molecules within the same size range but differ in the speed of separation and the resolution.

PFGE is both preparative and analytical tool. It has played a fundamental role in the development of large-fragment cloning, detailed physical mapping of genomes of pathogens and as a powerful tool for monitoring bacterial epidemic spread.

PFGE has excellent ability to separate small, natural linear chromosomal DNAs ranging in size from 50-kb to multimillion-bp yeast chromosomes. However, intact human chromosomes range in size from 50 million to 250 million bp is still too large for direct PFGE separations.

Applications of pulsed field gel electrophoresis

- The advent of PFGE techniques for the resolution of large DNA molecules has provided a new analysis approach for bacterial genomes. The PFGE of DNA fragments obtained using different enzymes is a powerful technique for quick resolution of the bacterial genome into a small number of large fragments.
- The PFGE of DNA fragments obtained by using restriction enzymes produce a discrete pattern of bands useful for the fingerprinting and physical mapping of the chromosome.
- The PFGE technique is useful in establishing the genetic mapping and phylogenetic comparisons among different strains of the same species.
- Technique of PFGE has proved to be an efficient method for genome size estimation and the construction of chromosomal maps.
- PFGE has proven extremely powerful in the analysis of large DNA molecules from a variety of sources including intact chromosomal DNAs such as fungi, parasitic protozoa, viruses and specifically fragmented genomes of bacteria and mammals.
- This technique simplifies many previously laborious investigations and makes possible many new investigations. Yeast Artificial Chromosome (YAC) libraries have been constructed by using PFGE.
- The study of radiation-induced DNA damage and repair, size organization and variation in mammalian centromeres is also done using this technique.
- PFGE is used to determine the order of molecular markers more precisely and new markers can be quickly localized within that region.

- PFGE proves to be useful technique in genetic engineering of organisms. It allows easy isolation of the individual restriction fragments for further restriction mapping, gene insertion and functional gene mapping.

1.7.2 Principle and Working

The limit of resolution of PFGE depends on several factors, including the uniformity of the two electric fields, the duration of the electric pulses, ratio of the pulse times for each of the alternating electric fields, angles of the two electric fields to the gel, and the ratio of the strengths of the two electric fields.

In conventional electrophoresis involving a constant electric field, DNA migrates to a distance that is inversely proportional to the logarithm of its length. Thus, for smaller molecules, relatively small differences in length result in large differences in mobility. Due to the logarithmic relationship, sensitivity decreases with increase in size. Furthermore, all DNA molecules above a certain size limit that varies with the electrophoretic conditions, migrate together regardless of their length. This means that very large DNA molecules migrate abnormally fast. This makes almost impossible to determine the molecular weight of very large DNA molecules.

To overcome this problem, Pulsed-field gel electrophoresis (PFGE) was developed. In the method given by Schwartz and Cantor, pulsed, alternating and orthogonal electric fields are applied to a gel. Due to this large DNA molecules are trapped each time the electric field direction changes and again migrate after they have reoriented along the new field axis. The larger the DNA molecule, the longer is the time required for its reorientation. DNA molecules whose reorientation time is less than the duration of the electric pulse will therefore be fractionated according to their size. This forms the basis of PFGE.

In PFGE, DNA is subjected alternately to two electrical fields at different angles for a time called the pulse time. The molecules must presumably change direction prior to net translational motion. Each time the field is switched, larger molecules take longer to change direction and have less time to move during each pulse, and so they migrate slower than smaller molecules. Molecules so small that their reorientation time is short compared to the pulse time, will spend most of the pulse duration in conventional electrophoretic motion. As a result of this, resolution in PFGE is likely to be optimal for molecules with reorientation times that are comparable to the pulse time.

At applied field strengths of about 10 V/cm, 0.1 sec pulse time resolves DNA optimally in the 5 kb size range, while pulse times of 1,000 sec at 3 V/

cm are used to resolve large 3-7 Mb molecules. Pulse time is a crucial factor in PFGE. Pulse times are selected so that DNA molecules of a targeted size spend most of the duration of the pulse reorienting rather than moving through the gel, which accounts for the long periods of time, (usually days or weeks) that is needed to fractionate large DNA molecules. The *Saccharomyces cerevisiae* genome PFGE, in the 10 Mb range may take electrophoresis run of approximately one week.

Electrical field strength can be adjusted to tune the size range of effective PFGE resolution. The resolution of PFGE is affected by the number and configuration of the electrodes used, because these alter the shape of the applied electrical fields. The most critical parameter is the angle between the alternate electrical fields. The most effective electrode configurations yield angles of more than 110°. A continually increasing angle between the fields produces band sharpening that greatly enhances the resolution.

The angle between the alternate fields should be greater than 90° where good resolution is required. In cases of excellent resolution, field angles typically range from 120° to 150°. Angles of 90° or smaller are not effective, probably because the DNA molecules easily become oriented midway between the two applied fields.

While more complete studies on optimum angles are needed, it is found that angles in the range of 120°-150° provide very high resolution. Field strengths that decrease, or angles that increase, progressively along the direction of the net DNA motion produce band sharpening because the molecules at the front of each DNA zone always migrate more slowly than those at the rear. Most commercially available pulsed-field gel boxes use a fixed angle of 120° between the alternating fields.

As with switch time, the choice of the voltage used in PFGE must vary with the size of the DNA to be separated. In conventional gel electrophoresis, DNA molecules are run at room temperature but, in PFGE, DNA is run at a low temperature between 4ºC and 15°C to neutralize the heating caused by pulsating electric current. Temperature has a dramatic effect on DNA mobility in PFGE. Temperatures between 14°C and 22°C are generally regarded as the best compromise between speed and resolution.

The single most important determinant of mobility in PFGE is the interval at which the direction of the electric field is switched. If the switching interval is increased beyond the time required for a fragment to reorient, then the fragment will spend a large portion of the gel run migrating, as in conventional electrophoresis, with a resulting loss in resolution. The choice of an appropriate switching interval for PFGE must depend upon the size range of the fragments to be resolved.

Restriction enzymes are often used for production of fragments of DNA of different lengths. Any enzyme producing a large number of fragments smaller than 10 kb is not useful for PFGE and mapping.

There are different types of PFGE:

(a) Field-inversion gel electrophoresis (FIGE)

In 1986, Carle, Frank and Olson developed a simpler system, FIGE, in which the two fields were 180° apart. In this type of PFGE, electrode polarity is reversed at intervals, with a longer forward than reverse pulse time to generate a net forward sample migration (Fig. 1.3).

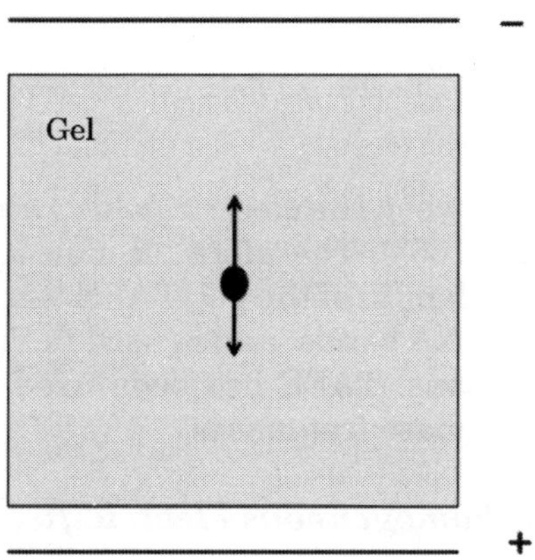

Fig. 1.3: Field-inversion gel electrophoresis (FIGE)

Net forward migration is achieved by increasing the ratio of forward to reverse pulse times to 3:1. To improve the resolution of the bands by FIGE, the duration of pulse time is increased progressively during a run. All that is needed in this technique are standard gel boxes and a pulse controller. Today, FIGE is very popular for smaller fragment separations.

(b) Transverse-alternating field gel electrophoresis (TAFE)

This form of PFGE allows separation of large DNA fragments in a simple, convenient format without the drawbacks of other pulsed-field techniques. In TAFE, the gel is placed vertically and a simple four-electrode array is placed in front and at the back of the gel. Thus, sample molecules are forced to zigzag through the thickness of the gel (Fig. 1.4). All lanes experience the same effects, so the bands remain straight. As the molecules move down the gel, they are subjected to continual variations in field strength and reorientation angle, but to all lanes equally.

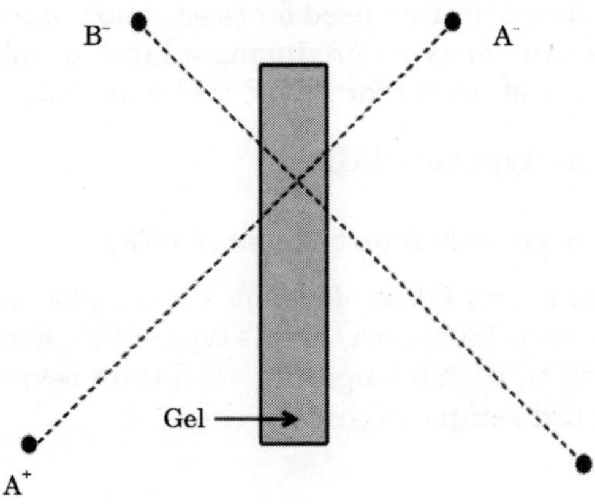

Fig. 1.4: Transverse-alternating field gel electrophoresis (TAFE)

However, the angle between the electric fields varies from the top of the gel (115°) to the bottom (165°). Therefore all molecules do not move at a constant velocity over the length of the gel. TAFE technology, gives regular and sharp separation of DNA bands, and is used in the study of genetics of many pathogenic protozoans. TAFE has been used for the separation of fragments up to 1,600 kilobase fragments.

(c) Contour-clamped homogeneous electric fields (CHEF)

CHEF is the most widely used apparatus in PFGE. The CHEF apparatus provides a more sophisticated solution to the distorting effects of both the edges of the chamber and the passive electrodes. CHEF has twenty-four point electrodes equally spaced around the hexagonal contour. All the electrodes are connected to the power supply via an external loop of resistors, all of which

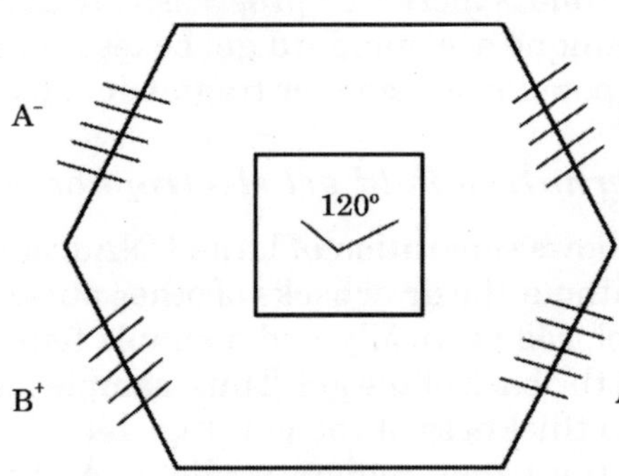

Fig. 1.5: Contour-clamped homogeneous electric fields (CHEF)

have the same resistance. This loop is responsible for setting the voltages of all the electrodes around the hexagonal contour to values appropriate to the generation of uniform fields in each of the alternate switching positions.

The CHEF system sets the voltages at these 24 points. This apparatus produces electric fields that are sufficiently uniform so that all lanes of a gel run straight. CHEF uses an angle of reorientation of 120° (Fig. 1.5). Molecules up to 7,000 kb can be separated by CHEF.

(d) Rotating gel electrophoresis (RGE)

In 1987, Southern described a novel PFGE system that rotates the gel between two set angles while the electrodes are off. This technique is called rotating gel electrophoresis. In RGE, the electric field is uniform and bands are straight because only one set of electrodes is used. This technique also enables users to study the effects of different angles. RGE uses a single homogeneous field and changes the orientation of the electric field in relation to the gel by discontinuously and periodically rotating the gel. Switch times are too long in RGE. The DNA molecules migrate in straight lanes, due to the homogeneous fields, and DNA molecules from 50 kb to 6,000 kb can be separated by adjusting the frequency of the gel rotation. In addition, the angle of reorientation can be easily altered simply by changing the angle of rotation.

(e) Programmable autonomously-controlled electrodes (PACE)

The PACE electrophoresis system offers precise control over all electric field parameters by the independent regulation of the voltages on 24 electrodes arranged in a closed contour. The flexibility of the PACE system derives from its ability to generate an unlimited number of electric fields of controlled homogeneity, voltage gradient, orientation and duration. The PACE system can perform all previous pulsed field switching regimens *i.e.* FIGE, OFAGE, unidirectional pulsing, as well as generate voltage clamped homogeneous static fields.

The PACE system separates DNA fragments from 100 bp to over 6 Mb. The ability to alter the reorientation angle between the alternating fields permits an increased speed of separation for large DNA molecules. It is an extremely useful tool for studying variables such as pulse time, temperature, agarose concentration, voltage and angles between fields affecting DNA migration in PFGE.

(f) Pulsed-homogeneous orthogonal field gel electrophoresis (PHOGE)

The major difference between this instrument and other gel boxes with homogeneous electric fields is that the field reorientation angle in PHOGE is 90. PHOGE uses 90° reorientation angle, but the DNA molecules undergo four reorientations per cycle instead of two. The DNA lanes in PHOGE do not run straight, a phenomenon, which has been described for gel runs involving multiple electric fields in this manner. This system separates DNA fragments of up to 1 Mb.

PFGE equipment

The basic components of a PFGE system consist of a gel box, temperature regulation system, switching unit for controlling the electric fields, a cooler and a power supply.

Gel box

The basic design of PFGE boxes consists of an immobilized gel within an array of electrodes and a means of circulating the electrophoresis buffer. Voltage gradients of 10 volts/cm are commonly used in PFGE. The temperature of the buffer is controlled by a heat-exchange mechanism. Generally, the buffer is recirculated throughout the gel box.

High voltage power supply

Precise control of the electric field gradient is necessary to obtain consistent PFGE separations. The output ratings of the power supply should therefore be high enough to meet, both the voltage and current requirements of the gel box. A typical PFGE gel box has electrodes that are 25 to 50 cm apart.

Switch unit

The ability to control the switch interval is critical for the separation. The relays are usually controlled by a computer. To overcome the drawbacks of electromechanical relays, electromechanical relays are used with commercial PFGE systems. These designs offer the advantages of improved reliability, the capability of high speed switching at milliseconds interval, ample voltage and current (0.5 A). These equipments have the ability to control the reorientation angles between electric fields.

Computer program/Automation

Careful control of the switch interval is crucial in controlling the resolution in PFGE. A versatile switching unit should have software with the same characteristics. The maximum run time should be about two weeks to allow for the separation of very large DNA molecules. This is controlled by a computer program.

Cooler

Buffer recirculation is an important factor, as it eliminates temperature variations within the gel so as to alleviate buffer breakdown due to electrolysis. DNA molecule migration is sensitive to temperature, and thus a uniform temperature across the gel is needed to ensure even migration in each of the lanes. The buffer is chilled in its reservoir tank by cold water (5°C) circulated through a glass tubing heat exchanger. Buffer temperature is thus maintained at 13-15°C throughout the run.

1.7.3 Procedure of Pulsed Field Gel Electrophoresis

Materials required

- GTBE buffer

TBE buffer	0.5x
Glycine	0.1 M

  Store at room temperature
- Lysis buffer

EDTA, pH 8.0	100 mM
Tris-Cl, pH 8.0	10 mM
Sarkosyl	1% (w/v)
Proteinase K	100 µg/ml

  Store at room temperature, without proteinase K. Add proteinase K just before use.
- Storage buffer

Tris-Cl, pH 8.0	10 mM
EDTA, pH 8.0	10 mM

  Store at room temperature
- A wide variety of molecular-weight markers available commercially are suitable for PFGE.
- Staining solution: 10 mg/ml ethidium bromide

Procedure

Preparation of DNA for PFGE

1. To avoid shearing during extraction of large DNA molecules, cells are lysed *in situ* in an agarose plug. Bacterial cells are resuspended in low melting temperature agarose and solidified in blocks. A variety of substances are infused into the plugs in order to lyse the cells and to remove proteins from the DNA. The DNA obtained by this procedure is both intact and susceptible to restriction in situ. After digestion with an appropriate restriction enzyme, the block can be melted and loaded on the gel.
2. Harvest a small amount of the bacterial culture with a sterile glass rod and resuspend in 10 ml of PBS.
3. Transfer 10 ml of the bacterial suspension to a 14 ml disposable

centrifugation tube and centrifuge at 1500 g at room temperature for 5 minutes. Decant the supernatant and invert the tubes on filter paper for at least 2 min to allow draining.

4. After use, clean blocks formers with 70% ethanol and rinse extensively with water. They are stored dry at room temperature. At this point, seal the bottom of the formers with autoclaving tape and label it.
5. Resuspend the cell pellet with 1 ml of 1% low-melting agarose stored in a water bath at 42°C with the aid of a pipette. Avoid the formation of air bubbles.
6. Pipette the agar suspension quickly into the block formers.
7. For each strain, prepare 5–6 blocks. Keep the block formers on ice for about 20 min.
8. Using a small spatula, poke the blocks out of the former into *E. coli* lysis buffer and incubate at 30°C overnight.
9. Replace the buffer with 2.5 ml ESP per tube. Incubate at 50°C overnight.
10. Wash the blocks at 37°C in 50 ml of TE buffer four times for 30 min each with shaking at 50–100 rpm. The blocks can then be stored at 4°C for at least 6 months.

DNA restriction

1. Cut 1/4th block on a sterile surface such as Parafilm with a sterile scalpel and place it in a 1.5 m Eppendorf tube.
2. Equilibrate the 1/4 block with 0.5 ml of digestion buffer at room temperature for at least 20 min. This step dilutes the EDTA inside the block.
3. Remove the entire equilibration buffer with a Pasteur pipette. Add 100 ml of the appropriate restriction mixture containing 10 ml of 10x restriction buffer, 1 ml of 100 mg/ml acetylated BSA, 5–10 Units of restriction enzyme and water to make 100 ml volume.
4. Incubate at 37°C overnight.
5. Stop the reaction by removing the restriction mixture and adding 0.5 ml of running buffer.

Preparation of the PFGE Gel

1. For a 15 cm gel, melt 100 ml of 1% agarose in a microwave oven or in a water bath. Allow the agarose to cool at room temperature while stirring.
2. Place the gel-supporting tray on a levelled table such that the slits are on the sides.
3. Clean the gel frame with ethanol. The conical holes at the bottom of the tray must be visible within the gel-casting frame. Test the assembly by putting a 0.5 mm comb in the position indicated by the

thick red line inside the square on the gel support tray. Adjust the plastic clips such that the teeth of the comb touch the bottom of the tray. With the plastic clips in position, remove the comb before pouring gelling mixture.

4. Pour the agarose mixture into the mould, making sure that the conical holes are filled. If the gel contains air bubbles, remove them with a Pasteur pipette. Place the comb in the gel and allow the gel to solidify for at least 30 min.
5. Remove the comb carefully.
6. Use the gel within 1–2 h or cover it with plastic wrap for later use. If you want to store the gel, flood it with buffer before covering it and store it at 4°C.

Sample and Marker loading

1. Remove all equilibration buffer from the tube containing the sample.
2. Incubate the sample for 15–20 min at 65°C till the agarose is melted.
3. Load 10 ml of melted agarose per well with a 10 ml micropipette. Loading should be very quick to avoid premature solidification
4. The markers are usually supplied in a GelSyringe dispenser. Using a razor blade cut a slice on parafilm as above that is slightly narrower than the width of the comb.
5. Transfer the slice to the gel with the help of a fine spatula and let it glide into the well with help of a second spatula. The slice should be placed in close contact with the front of the wells.

Running pulsed gel electrophoresis

1. Assemble the electrophoresis apparatus according to the manufacturer's instructions.
2. Start the cooling equipment at least 30 min before the beginning of gel electrophoresis to ensure a buffer temperature of 12°C.
3. Fill the gel chamber with 2.5 litres of running buffer and start the external pump that circulates buffer. Always make sure that there is buffer in the tank when the pump is running.
4. After the samples are loaded, place the gel-support tray into the tank. If the buffer is insufficient to cover the gel, add more.
5. Adjust the voltage to constant 165 V. Adjust the conditions required for the experiment.

Staining of gel

1. Turn off the power supply before removing the lid. To remove the gel, slide a scalpel under it to cut the agarose in the conical holes and release the gel from the bottom of the tray.

2. Then slide the gel into a plastic dish containing 0.5 mg/ml of EtBr in 0.5X TBE. Be carefully not to break the gel. Incubate at room temperature on shaker for 30 minutes.
3. Remove the gel, and wash it with distilled water. Visualize the gel with the help of UV transilluminator.

1.7.4 Troubleshooting guide

1. If problem of bands broadening or disappearing at top of gel occurs, decrease trapping by decreasing temperature or voltage.
2. In cases where bands are smeared, running samples into gel for longer time, at lower voltage, or both may help.
3. Problem of excessive smile. This occurs when more heat is being produced than that can be dissipated effectively. Either reduce the heating by decreasing voltage, gel thickness, or buffer depth, or increase buffer recirculation.
4. For alternating-angle electrophoresis, best resolution is obtained by using pulse times just long enough to resolve the largest molecules of interest
5. Pulse times are much more critical for field inversion electrophoresis, because a given reverse pulse time resolves a narrower range of sizes. Use a range of reverse times broad enough to separate all molecules of interest.
6. The ratio between forward and reverse times is generally in the range of 2.5:1 to 3.5:1. Lower ratios give better separation but make gels run more slowly. The most commonly used ratio is 3:1.
7. Low agarose concentration shifts the sizes of molecules resolved toward the larger range, and also speeds the migration of all molecules. Special pulsed-field grade agarose is available from several companies.
8. Because pulsed-field gels are usually run at higher voltages than standard gels, they have the potential to generate quite a bit of heat. For this reason, they are often run in 0.5x TBE buffer, which carries little current. DNA has higher mobility in TAE buffer, but TAE carries a lot of current, causing gels to heat hence should be avoided.
9. Chilling the agarose adequately is necessary for avoiding later problems with removing the agarose plugs from the formers.
10. Always use clean and sterile tools when handling the blocks to avoid bacterial contamination.
11. Correct loading of samples is critical for success. Small air bubbles present while pipetting, or damages to the wells, will result in irregular bands and smears.

12. Positioning the slice very close to the well front is important for obtaining sharp marker bands. Make sure that the slices are not wider than the well.
13. A frequent cause of poor results is inadequate temperature control because the pump breakdown or flow blocked by air bubbles. Proper buffer flow is indicated by fluid moving through the tubing connecting the gel chamber and the cryostat.
14. Check the temperature at the beginning and the end of each run.

1.8 ISOELECTRIC FOCUSING

1.8.1 Introduction

Isoelectric focusing or IEF offers a sensitive tool to study micro-heterogeneity of proteins, separation of isozymes as well as for the purification of proteins.

Separation of proteins in complex mixtures for analytical resolution can be achieved by isoelectric focusing (IEF), in which proteins are separated based on their net charge (isoelectric point, or pI) in the presence of a pH gradient.

Analytical focusing is carried out either in polyacrylamide gels prepared with immobilized pH gradients or in agarose gels prepared with mobile carrier ampholytes, which form a pH gradient when subjected to electrophoresis.

Compared to alternative methods, IEF is more efficient, expressive and economic since no sophisticated equipment required. It is also easy to comprehend, as it involves one-dimensional separation. This method is faster and more sensitive for separation of proteins over a narrow pH range. Separation of proteins larger than 2,000 kDa can be achieved using this method.

The technique of IEF is applied for immunofixation directly in the gel, and direct tissue isoelectric focusing.

1.8.2 Principle and Working

Isoelectric focusing (IEF) or electrofocusing is a method employed to separate proteins mixtures by electrophoresis in a pH gradient based on their isoelectric point (pl). The pH at the anode is lower than the cathode so creating a pH gradient; with the pH range chosen so that the proteins to be separated have their isoelectric points within this range. At a pH other than their isoelectric point, proteins will be charged. If they are positively charged they will move towards the more negative end of the gel and if they are negatively charged they will move towards the more positive end of the gel. At its isoelectric point, the protein will no longer carry a charge and will no longer migrate in the gel. Therefore proteins become focused into sharp stationary bands with each protein positioned at a point in the pH gradient corresponding to its pI. The technique is capable of extremely high resolution with proteins differing by a single charge being fractionated into separate bands. IEF is mainly used as a technique to assess the complexity or purity of protein samples and unlike SDS-PAGE will not yield information about the molecular weight of the separated proteins.

If the pH is less than the pI, the protein will have a net positive charge and move towards cathode. If the pH is greater than the pI, the protein will have a net negative charge and will move towards anode. If the pH equals the pI, the protein will have no net charge and hence it does not move in the electric field. If the molecule diffuses into any direction, it becomes charged again and moves back to its pI. Therefore, the sample can be introduced anywhere in the pH gradient and it will always move to exactly the same point.

IEF is an endpoint method. Once the band is focused it will not be affected by any band broadening factors over time. Bands as narrow as 0.01 pH units can be obtained. The resolution obtained in isoelectric focusing can be expressed as ΔpI, the minimum pH difference to resolve two compounds. If compounds have similar pIs, resolution can be improved by using pH gradients over narrow range and smaller increments. The ΔpI depends upon the pH gradient and voltage gradient at the positioning of the bands.

$$\Delta \text{pI} \propto \sqrt{\frac{\text{pH gradient}}{\text{voltage gradient}}}$$

Isoelectric focusing allows the separation of zwitterionic analytes such as proteins or peptides according to their isoelectric point (pI). The pI of a protein depends upon the sum of all charges as well as tertiary structure and post translational modifications such as glycosylation, phosphorylation etc.

In IEF, instead of using a buffer with a constant pH for the whole gel, a pH gradient is generated in which pH value increases from anode to cathode. When a zwitterionic compound like protein is placed in the pH gradient, it migrates until it reaches a point where its net charge equals zero. At this point, the pH equals protein's pI.

The proteins are driven to their isoelectric points by the electric field. The isoelectric points of the proteins can be estimated with a calibration curve using marker proteins. Marker proteins for various pH ranges are available. These proteins are chosen so that they can focus independently of the point of application. Standard marker proteins can not be used in urea containing gels, because their conformations are changed, and thus their isoelectric points (pI).

Native or denaturing conditions can be chosen by either omitting or adding high amounts of urea. The proteins become highly concentrated at their pIs. This results in a high sensitivity for detection. Small charge differences can be differentiated. For improvement of resolution, narrow gradients can be employed. The method is also suitable for preparative applications, when high amounts of proteins have to be purified.

As isoelectric focusing is in principle a non-denaturing method, the optimization of the running conditions is very important to prevent precipitation and aggregation of proteins, and to achieve good reproducibility.

In IEF, agarose and polyacrylamide are both used. Sufficiently large pores are required in the gel as this reduces the molecular sieving effect and hence shortens the separation time. With agarose gels, proteins larger than 800 kDa can be focused. Polyacrylamide gels are limited to lower molecular weight proteins.

Isoelectric focusing (IEF) is mostly run in horizontal gels because of several reasons, as IEF separates only according to the charge, the gel matrix must contain large pore sizes. Such a soft gel can slide down between vertical glass plates. The gel should preferably be cast on a film support. Isoelectric focusing requires efficient cooling and exact temperature control. Samples have to be applied on a defined pH location within the pH gradient in order to avoid aggregation and precipitation of some proteins. This is only possible on a horizontal gel with an open surface. Because sharply focused bands can only be obtained with high field strength, high voltages have to be applied. The power supply must provide enough voltage for obtaining sharply focused IEF zones and must preferably be programmable in order to run an IEF protocol automatically.

Stable and linear pH gradient is the key to the success of IEF. To create a stable pH gradient with constant conductivity, various components like carrier ampholytes or immobilized pH gradients and acrylamide buffer are important.

Carrier ampholytes or amphoteric electrolytes are mixtures of molecules containing multiple aliphatic amino and carboxylate groups. A mixture of hundreds of different carrier ampholytes, each with a different pI is used to form a pH gradient. The ampholytes must have a high buffering capacity at their pI. It is also important that each single compound in the ampholyte mixture has the same concentration and there are no gaps in the spectrum of pIs. These ampholytes are included directly in the IEF gel. Under the influence of electric currents, carrier ampholytes form smooth pH gradients increasing linearly from anode to cathode. The slope of the pH gradient is determined by the pH interval covered by the carrier ampholyte mixture and distance between the electrodes.

The choice of carrier ampholytes determines the pH range of the gradient. The pH gradient may be between 2.5 and 11. However, narrow range pH gradients give better resolutions hence are generally used.

The protein sample can be introduced at any place in the gradient. Sometimes, proteins are introduced in the gel together with ampholyte solution. The protein is then focused at the same time as the pH gradient develops. Due to the low diffusion coefficients of proteins, they form much sharper bands than the ampholytes. The ampholyte molecules have relatively lower molecular weights and have high diffusion coefficients. They form relatively broad bands that overlap.

In general, problems with carrier ampholytes can arise when long focusing times are necessary. For example, in the case of short intervals or in the presence of highly viscous additives such as urea or non-ionic detergents, the gradient slowly starts to drift in both directions but particularly towards the cathode. This leads to formation of plateau in the middle of the gradient with gaps in the conductivity. In this process, some of the proteins may be lost from the gel. It is important to limit the focusing time and run similar experiments for exactly the same time for reproducible results. Resolution depends on the various factors like pH gradient, thickness of the gel, time of electrophoresis, applied voltage and diffusion of the protein into the gel.

Polyacrylamide IEF gels are polymerized with initiator system that includes riboflavin for photopolymerization. Photochemical initiation of polymerization using components such as riboflavin, ammonium persulfate and TEMED, results in better polymerization of IEF gels.

The gel is immersed into a buffer with a medium pH, which contains 1-2% of carrier ampholyte mixture. The anode reservoir is filled with a low pH buffer called as 'anolyte' (Refer Table 1.4). It is important that the pH of the anolyte has to be more acidic than the lower pI in the ampholyte mixture. The cathode reservoir is filled with high pH buffer called as 'catholyte' (Refer Table 1.5). The pH of the catholyte should be more basic than the highest pI in the ampholyte mixture.

There are no fixed rules regarding application of protein samples in the gel. In general, samples should not be applied to the area where they are supposed to focus. To protect the proteins from exposure to extreme pH, the samples should be applied keeping at least 1 cm distance from each electrode.

Table 1.4: Examples of anolytes used in IEF system

Chemicals	*Concentrations*	*pH (25°C)*
Phosphoric acid	1 M	1.0
Sulfuric acid	0.2 M	1.6
Acetic acid	0.5 M	2.6
L-glutamic acid	40 mM	3.2
Indole acetic acid	3 mM	3.8
L-tyrosine	4 mm	4.5

Table 1.5: Examples of catholytes used in IEF system

Chemicals	*Concentrations*	*pH (25°C)*
Threonine	50 mM	5.8
Glycine	50 mM	6.5
Hepes	0.4 M	7.3
L-Histidine (free-base)	40 mM	7.35
Sodium hydroxide	1.0 M	13.0

1.8.3 Procedure of Isoelectric Focusing

Materials required

Gelling mixture

Acrylamide solution 30%

To make the solution, dissolve 29.1 g of acrylamide and 0.9 g of bisacrylamide in 70 ml of distilled water. Make final volume 100 ml.

Glycerol 50%

FMN (riboflavin-5'-phosphate) 0.1%

Carrier ampholytes 40%

Ammonium persulfate 10%

TEMED

Staining solutions for IEF gels

Stock solutions

TCA (Trichloroacetic acid): 100% TCA (w/v) 1 litre

Solution A: 0.2% (w/v) $CuSO_4$+ 20% acetic acid

Solution B: 60% (v/v) methanol

Solution C: Coomassie Blue stain in methanol.

Acetic acid	10%
Methanol	25%
Glycerol	5%

Procedure

Casting the gel

1. Assemble the gel casting apparatus according to manufacturer's instructions.
2. Combine the ingredients of the gelling mixture. Mix the components properly.
3. Transfer the gelling mixture to casting apparatus. Allow the gel to polymerize.
4. Place a fluorescent lamp 3-4 cm away from the gelling mixture and illuminate the solution for 1-2 hours.
5. The gel can be used immediately. For storage, the gel should be covered with plastic wrap and store at 4°C.

Assembling gel in electrophoresis apparatus

1. Set the refrigerated circulator bath to 10°C-15°C.
2. Spread 0.2 ml-0.3 ml of distilled water on the cooling plate of the IEF chamber.
3. Lower the gel with film-side down onto the wetted area. Avoid trapping air under the polyester film.
4. Wipe excess fluid from the edges of the film.
5. Blot the surface of the gel briefly with a sheet of fine-grained blotting paper.
6. If necessary, trim the edges of the gel parallel to the direction of focusing with a sharp razor blade to ensure the edges are even and free of cracks.
7. Cut two electrode wicks to the exact width of the gel.
8. Completely immerse one wick in catholyte solution.
9. Place the wick on blotting paper to remove excess fluid.
10. Place on the negative electrode contact of the gel.
11. Completely immerse a second wick in anolyte solution.
12. Place the wick on blotting paper to remove excess fluid.
13. Place on the positive electrode contact of the gel. The wicks must lie parallel to each other on the ends of the gel, evenly touching the surface.

14. Place a glass plate on top of the gel and wicks for approximately one minute. This ensures uniformity of contact between wicks and gel.

Application of samples

1. Place the sample applicator mask across the gel at least 1 cm from either wick for *e.g.* 3 cm from cathode.
2. Load sample and pI markers into the slots. The quantity of sample ranges between 2 µl-5 µl; with 2-10 mg/ml concentration of pI markers. In direct tissue isoelectric focusing, tissue samples may be placed directly onto the applicator slots.
3. Ensure that electrodes and electrical contacts are clean and there are no breaks in the wire.
4. Place the electrodes on the wicks and not the gel surface. Align them so that they lie in parallel upon the wicks.
5. Set the power supply at 1 W constant power. Apply power for 10 minutes.
6. Now turn power off.
7. Remove the sample applicator mask.
8. Gently remove any precipitated sample from the gel surface with blotting paper.

Staining of gel

Isoelectric focusing gels require special staining procedures: the proteins have to be efficiently fixed, whereas carrier ampholytes have to be washed out. A high detection sensitivity is obtained only with the procedure described below.

Coomassie Blue Staining of Agarose gels

1. Remove the gel carefully from gel cassette.
2. *Fixing*: Apply fixing solution enough to submerge the gel completely. Incubate for 30 min in 20% (w/v) TCA.
3. *Washing*: Wash the gel twice for 15 min in 200 ml of fresh solution of 10% acetic acid and 25% methanol.
4. *Drying*: Place 1 layer of moistened filter paper and 3 layers of dry filter paper on the gel, glass plate, and a 1 kg weight on top. Remove the filter papers and weight after 10 min and finish drying in the heating chamber.
5. *Staining*: Apply staining solution over gel to submerge the gel completely for 10 min in 0.5% (w/v) Coomassie blue in 10% acetic acid, 25% methanol.
6. *Destaining*: This is an important step. Submerge the gel in 10% acetic acid and 25% methanol till the background is clear.
7. *Drying*: Dry the gel in the heating cabinet.

Coomassie blue staining of polyacrylamide gels

1. *Fixing*: Apply fixing solution enough to submerge the gel completely. Incubate for10 min in 300 ml of 20% TCA.
2. *Washing*: Wash the gel for 2 min in 300 ml of solution A and B mixture (A and B, 1:1).
3. *Staining*: Apply staining solution over gel to submerge the gel completely for 15 min in 300 ml of staining solution A and C mixture (A and C, 1:1) at 50 °C with stirring.
4. *Destaining*: This is an important step. Submerge the gel for 15 to 20 min in the solution A and B mixture (A and B, 1:1)
5. *Impregnating*: for impregnation of stain, incubate the gel for 10 min in 200 ml of 5% glycerol and 10% acetic acid.
6. *Drying*: For drying, gel dryers can be used or it can be air-dried.

1.8.4 Troubleshooting guide

1. To prevent condensation on the gel and plate, do not circulate the coolant to the IEF chamber just prior to focusing.
2. If fuzzy bands appear, it could be due to incomplete run or focusing for a long time. Also insufficient fixation at the time of staining the gel can result in fuzzy bands.
3. Place the gel immediately in fixing solution after IEF run, to avoid fuzzy or smeared bands.
4. Sometimes streaks can appear that are perpendicular to the bands. This can happen due to various reasons such as particulate matter in the sample, very old samples, and precipitation of samples. To avoid this problem, carefully filter the sample before applying to the gel. Also try different application location.
5. Skewed bands may result from unclean electrode or uneven electrode contact. The skewed bands appear because of various other factors such as short electrode wick, incorrect electrolyte solution, overloading of samples etc.
6. The electrode wicks should be evenly wetted and placed parallel to each other on the gel.
7. The length of electrode wicks should be proper. They should extend to the edge of the gel.
8. The appearance of wavy band may be due to various factors like improper electrolyte used, excessive protein sample, presence of salt in sample, and unclean electrodes.
9. If the gel peels away from the support film, then perhaps wrong support film was used; or wrong side of the film was used.
10. If too low or no current, there may be poor or no contact between the

electrodes and electrode strips. Make sure that the electrode strips are correctly placed. If a small gel or part of a gel is used, place it in the middle.

11. If current rises during the IEF run, electrode strips or electrodes are probably mixed up. Always, acid solution wetted wick at the anode, basic solution at the cathode.
12. If gel swells around electrode strips, then there is too much electrode solution in the electrode strips. After soaking, blot the strips with filter paper.
13. If lines of condensation are found over the whole gel, it could be due to hot spots or conductivity gaps because of plateau phenomenon. This can also happen with too long focusing time, especially for narrow pH ranges.

1.9 REVIEW QUESTIONS

1. Describe the principle of electrophoresis.
2. What is the significance of electrophoretic techniques in molecular biology?
3. Compare agarose gel electrophoresis and PAGE.
4. State the difference between native PAGE and SDS-PAGE.
5. Describe principle and method of SDS-PAGE.
6. What is the principle of pulsed field gel elelctrophoresis?
7. Describe different variations of PFGE.
8. What is isoelelctric point? Describe in brief isoelectric focusing.
9. Give principle and method of 2D gel electrophoresis.

1.10 REFERENCES

Birren, B., Lai, E., Clark, S., Hood, L. and Simon, M. (1988). Optimized conditions for pulsed field electrophoretic separations of DNA. *Nucl. Acids Res.*, **16:** 7563–7581.

Bisht, D., Singhal, N., Sharma, p. and Venkatesan, K. (2006). Analysis of mycobacterial strains by two–dimensional gel electrophoresis. *J. Commun. Dis.,* **38(3):** 255–262.

Carle, G., Frank, M. and Olson, M. (1986). Electrophoretic separations of large DNA molecules by periodic inversion of the electric field. *Science,* **232:** 65–68.

Chu, G. (1989). Pulsed field electrophoresis in contour–clamped homogeneous electric fields for the resolution of DNA by size or topology. *Electrophoresis,* **10:** 290–295.

Chu, G., Vollrath, D. and Davis, R. (1986). Separation of large DNA molecules by contour clamped homogeneous electric fields. *Science,* **234:** 1582–1585.

Gardiner, K. (1991). Pulsed–field gel electrophoresis. *Analytical Chemistry,* **63:** 658–665.

Herschleb, J., Ananiev, G. and Schwartz, D. (2007). Pulsed field gel electrophoresis. *Nature Protocols,* **2:** 677–684.

Lai, E., Birren, B., Clark, S., Simon, M. and Hood, L. (1989). Pulsed–field gel electrophoresis. *Biotechniques,* **7:** 34–42.

Merril *et al.* (1979). Trace polypeptides in cellular extracts and human body fluids detected by two–dimensional electrophoresis and a highly sensitive silver stain. *Proc. Natl. Acad Sci. USA,* **76:** 4335–4339.

Schwartz, D. and Cantor, C. (1984). Separation of yeast chromosome–sized DNAs by pulsed field gradient electrophoresis. *Cell,* **37:** 67–75.

Smith, C., Coote, J. and Parton, R. (1986). R plasmid–mediated chromosome mobilization in *Bordetella pertussis*. *J. Gen. Microbiol.*, **1432:** 2685–2692.

Southern, E. and Elder, J. (1995). Pulsed–field gel electrophoresis. Edited by Monaco, A.P. Oxford, IRL Press. pp. 1–119.

Stellwagen, N. (2009). Electrophoresis of DNA in agarose gels, polyacrylamide gels and in free solution. *Electrophoresis,* **30(1):** 188–195.

Steward, G., Furst, A. and Avdalovic, N. (1988). Transverse alternating field electrophoresis (TAFE). *Biotechniques,* **6:** 68–73.

2

Techniques of Centrifugation

2.1 INTRODUCTION

Centrifuge was in use as early as 19th century Europe. It was first used to separate cream from milk, which made it a domestic apparatus. However, centrifuges used at this time were very much primitive in design and were hand operated. Centrifuge as a laboratory equipment was used much later. In 1869, Friedrich Miescher of Switzerland first used the centrifuge to isolate cell organelles, where he successfully isolated cell nuclei. At the beginning of 20th century, electrically operated centrifuges were invented. This gave a lot more power and speed to the rotor of centrifuge.

In 1924, Svedberg and Rinde developed electrically operated centrifuge, which was called a low-speed ultracentrifuge. The centrifuge developed by Svedberg was useful in determining molecular weight and subunit structures of highly complex proteins like haemoglobin. In 1926, Svedberg made the fundamental studies on measurement of molecular weights of haemoglobin and ovalbumin by sedimentation equilibrium. Svedberg was awarded Nobel Prize for his studies on colloidal systems based on ultracentrifugation technique. In 1950s, many important landmark achievements of molecular biology were made using the ultracentrifuge. Messelson and Stahl used density gradient ultracentrifugation to prove the semi-conservative method of DNA replication. Experimentations using ultracentrifugation by A. Tissier and J. Watson lead to the discovery of ribosomes in 1958. In 1971, differential sedimentation method was introduced that opened new avenues for studying differences between sedimentation coefficients. Density gradient centrifugation used for separating and isolating cells and subcellular components, was developed by Martin Behrens.

Today, centrifugation technique is inevitable part of any molecular biology experiment. Centrifugation forms integral part of the biotechnology and molecular biology experiments like protein isolation, nucleic acid extraction and separation, preparation of samples, proteomics studies, isolation of viruses, lipids and proteins, separation of subcellular organelles etc.

Undoubtedly centrifuge has remained a valuable instrument that separates particles of different sizes and shapes from solutions. The centrifuge is widely used in biochemistry, industry and clinical diagnostics laboratory for the separation and concentration of different substances. The most popular and widely used technique in biochemistry, cellular and molecular biology as well as medicine is of course, centrifugation. Centrifuge is used by researches in these areas of studies because during the process, it preserves the biological function and properties of cellular components. A centrifuge uses centrifugal force (g-force) to isolate suspended particles from their surrounding medium. Generally centrifugation is used when the material to be separated is too small so that separation by filtration is almost impossible.

2.2 THEORY AND PRINCIPLE OF CENTRIFUGATION

Centrifugation is used for preparative and analytical separation of solid molecules from liquid phase. It is therefore one of the most important separation techniques. The centrifugation technique is based upon the behaviour of particles under the applied centrifugal force and differences in their size, shape, density and molecular weight. If a solution of insoluble heavy particles is left undisturbed, it is observed that these particles sediment at the base of the container under the influence of gravitational force. For a given particle, the rate or velocity at which the particles sediment is proportional to the force applied. Therefore, when applied force is greater than the gravitational force, particles tend to sediment more rapidly.

Centrifugation produces a centrifugal force that is hundreds or thousand times of the gravitational force, which greatly speeds up the sedimentation process. The centrifugal force is usually measured in terms of revolutions per minute (rpm). Greater the number of revolutions per minute, greater is the gravitational force. The theoretical basis of the centrifugation technique is that two particles of different masses will sediment in a tube at different rates under the influence of gravity.

2.2.1 Principle of Sedimentation

Centrifugal force is consequence of rotation that implies acceleration. An object following circular path experiences an outward force that is directed outward from the centre of the circle. This force is called centrifugal force.

The force depends upon the mass of the object, speed of rotation and distance of object from the centre of the circle. Centrifugal force is calculated by the following equation,

$$F = mv^2/r$$

Where,

F= centrifugal force

m= mass of the object

v= velocity of rotation

r= radius or distance of the object from centre of the circle

When mass of the object is more, greater is the force of centrifugation. The greater the speed of rotation, the greater is the force; and the lesser the distance of object from the center, the greater the force. Therefore, while performing centrifugal separation of particles, rotor speed, radial dimensions and time of operation of rotor are important factors to be considered.

As biological and biochemical experiments are usually conducted with particles dissolved and/or suspended in solution, the rate of sedimentation of particles is not only dependent upon the applied centrifugal force but also upon the mass of the particles. Mass of a particle is expressed as a product of its volume and density. The rate of sedimentation is dependent upon the density and viscocity of the medium as well as the degree to which the particle deviates from perfectly spherical shape. When particle sediments, it displaces some of the solution in which it is suspended.

The denser components of the mixture migrate away from the axis of the centrifuge, while particles of the mixture with lower density migrate towards the axis. The particles that accumulate at the bottom of the tube are called pellet and remaining solution is called the supernatant liquid. The force exerted on the particles is called relative centrifugal force (RCF). Relative centrifugal force is defined as the ratio of centrifugal force to the force of gravity. RCF can be calculated from the following formula,

$$RCF = 1.119 \times 10^{-5} \times (rpm)^2 \times r$$

Where 'r' is expressed in centimeters, RCF is expressed as 'xg' (for *e.g.* 5000 g). The sedimentation of a particle depends upon the RCF and time for which centrifugal force is applied. If we assume that size and the dimensions of a centrifuge tube are the same, the product of the g-force and time taken for centrifugation can be used to determine the conditions required to sediment a particle of given size and density. For example, if it takes 60 minutes to completely sediment a particle at 1000 g, then it should take 10 minutes to sediment the same particle at 6000 g.

The mixture of heterogeneous almost spherical particles can be separated by centrifugal force based on their densities and/or sizes. The rate of sedimentation of a given particle is proportional to its size, difference between the densities of particle and medium, and the centrifugal force applied. The sedimentation rate is lower when the medium is more viscous. This can be overcome by increasing the centrifugal force because; sedimentation rate is higher if greater centrifugal force is applied. When densities of particle and the liquid medium are equal, then sedimentation rate is zero or no sedimentation occurs.

The rate of sedimentation (ν) can be calculated from the following formula,

$$\nu = 2r_p^{\,2} \times (\rho_p - \rho_m) \times \omega^2 r / 9\eta$$

Where,

r_p = radius of the spherical particle

ρ_p = density of the particle

ρ_m = density of the medium in which particle is suspended

ω = angular velocity of the rotor (radians/sec)

r = distance of particle from the centre of circle of rotation (cm)

η = viscosity of the medium

The differences between densities of particles and medium, differences between sizes of particles, alterations in rotor speed and time of rotation form the basis of separation of biological macromolecules and cell organelles from tissue sample homogenates. Svedberg and Pederson have given the sedimentation constant S of a particle as sedimentation velocity (ν) relative to the centrifugal acceleration.

$$S = \nu/\omega^2 r$$

The unit of Svedberg constant is 's' and for biological macromolecules, 1S is in the order of 10^{-3}s. Virus particles can have sedimentation coefficients of more than 1000S while small biomolecules like ribosomes have 70S or 80S sedimentation coefficients.

2.3 TYPES OF CENTRIFUGES

Every centrifuge has two basic components, electric motor with drive shaft to spin the samples and a rotor that holds the sample tubes. Selecting the proper centrifuge and rotor system for a specific application can sometimes be a challenge. A wide variety of centrifuges are available today like low speed centrifuges, laboratory centrifuges, ultracentrifuges etc. Understanding the various types of rotors available and the applications they are best suited for, requires a good understanding of centrifugation. There are mainly three types of centrifuge rotors,

1. Swinging bucket rotor
2. Fixed angle rotor
3. Vertical tube rotor

When selecting a rotor, it is important to know which type of separation you require, differential, isopycnic, or rate-zonal. It is also important to know the volume and RCF requirements for your experimentation for selecting the capacity of the rotor.

(a) Swinging bucket rotor

Swinging-bucket rotors are ideal for separating large-volume samples up to 10 litres at low speeds. A swinging-bucket rotor system consists of the rotor body that attaches to the centrifuge drive and has four or six arms to support the buckets, and the buckets are placed onto the arms of the rotor body. The buckets are attached to the rotor body by hinge pins or crossbars. These components form the basic swinging bucket rotor framework. The buckets swing to a horizontal position as the rotor accelerates and then the rotor body supports them (Fig. 2.1). Swinging bucket rotors are used for pelleting, isopycnic studies and rate zonal study based on sedimentation coefficient, harvesting of cells from growth medium, blood samples processing and large scale tissue culture processing. This type of centrifuge is ideal for large sample volumes.

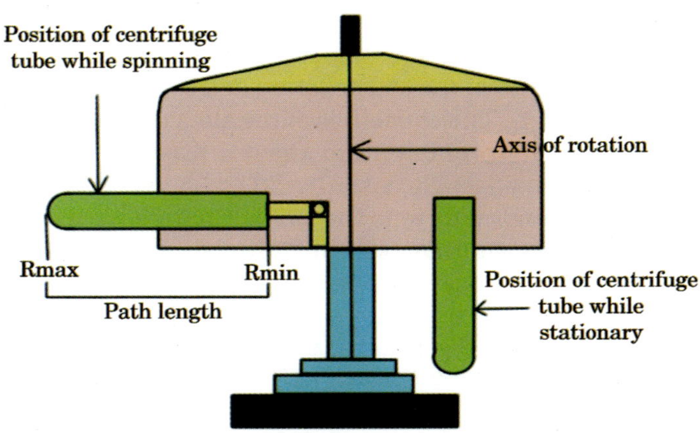

Fig. 2.1: Swinging bucket rotor

For ultracentrifuge applications a swinging-bucket rotor can support two types of separations, rate-zonal that is based on mass or size and isopycnic that is based on density. For rate-zonal separations, a swinging-bucket rotor is advantageous because the path length of the gradient is long enough for separation to occur.

Also, since the buckets are positioned at 90° during the run and returned to a vertical position at the end of the run, the sample retains its orientation, which minimizes any pellet or band disturbance. Swinging bucket rotors are best applied in the rate zonal studies in which maximum resolutions of the sample zones are needed. It is useful for pelleting, where it is required to form pellet at the exact centre of the bottom of the tube (Fig. 2.2). Swinging bucket rotor is most frequently used for density gradient separations.

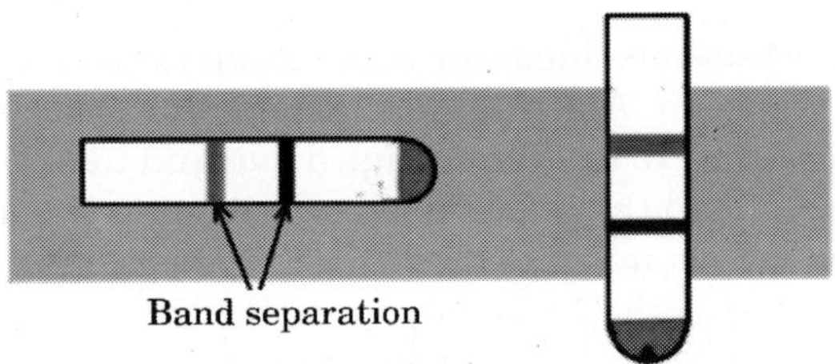

Fig. 2.2: Particle separation in swinging bucket rotor

(b) Fixed angle rotor

Fixed angle rotors are ubiquitously used in centrifugation experiments. Majority of them are used in the cell pelleting and separations by differential centrifugation (Fig. 2.4). The capacities of the slots for centrifugation tubes in this rotor can range from 0.2 ml to about 1 litre. In this type of rotor, tubes are held at a fixed angle, usually 20 to 35° to the axis of rotation (Fig. 2.3). This tube angle shortens the particle path length as compared to swinging bucket rotors, thereby reducing number of spins. One can choose a type of fixed angle rotor based on desired RCF and volume. An important specification while selecting this type of rotor is the K factor, which gives the pelleting efficiency of the rotor at top speed. A low K factor indicates higher pelleting efficiency.

Fixed angle rotors are general-purpose rotors that can be used for isopycnic centrifugation, separation of viruses and subcellular organelles.

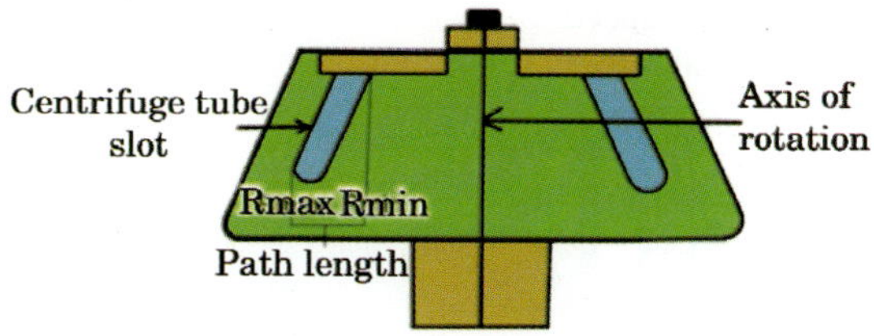

Fig. 2.3: Fixed angle rotor

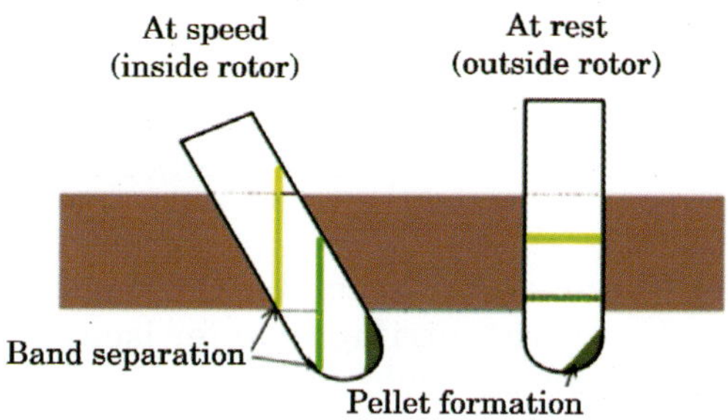

Fig. 2.4: Particle sedimentation in fixed angle rotor

Most fixed angle rotors come with a lid. These lids have O-rings that help maintain the atmospheric pressure while centrifugation.

(c) Vertical tube rotor

In vertical tube rotors, tubes are held parallel to the axis of the rotation (Fig. 2.5). Therefore, the bands separate across the diameter of the tube rather than length of the tube (Fig. 2.6). Vertical tube rotors are specially

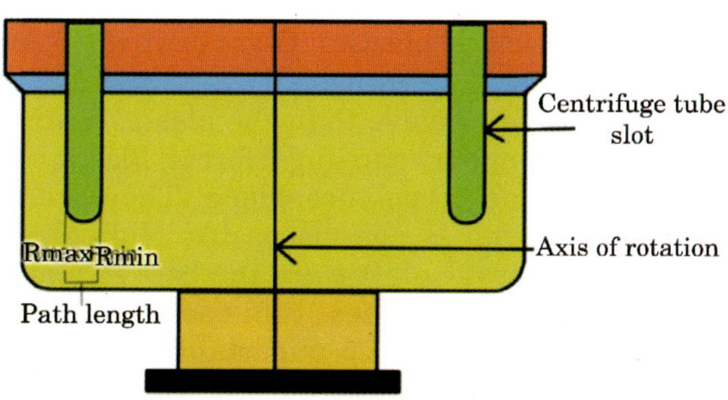

Fig. 2.5: Vertical tube rotor

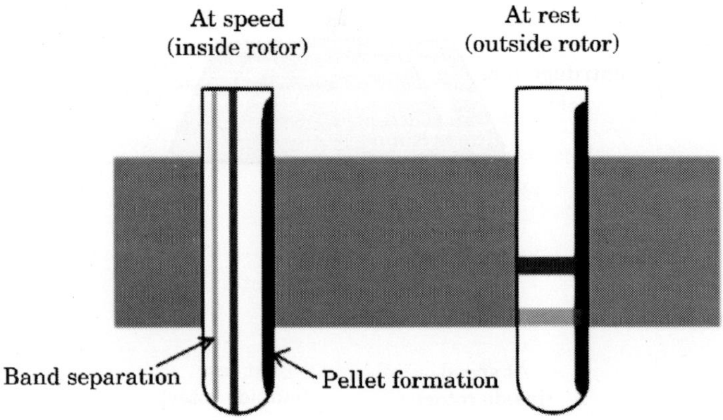

Fig. 2.6: Particle sedimentation in vertical tube rotor

used in isopycnic banding and rate zonal centrifugation. They are most commonly used during ultracentrifugation for isopycnic separation and banding of DNA by cesium chloride gradient. Vertical rotors have very low K factors; thereby it significantly reduces the run time.

The volume of the sample and speed of spin are deciding factors in this type of rotor. These rotors have plugs that are screwed in the rotor cavities over quick seal tubes. These plugs keep the tubes in position and provide support against the hydrostatic forces generated by centrifugation.

(d) Handling of centrifuges

Special type of tubes called centrifuge tubes are used for putting samples or specimen in the centrifuges. Centrifuge tubes should be chemically neutral to the solvents used, and they should not shed any colour or chemical in the solvents. They should be able to withstand the maximum centrifugal force and friction that can be caused inside the centrifuge. Centrifuge tubes are made from glass as well as plastics. Glass centrifuge tubes are made from high quality glasses that can withstand the friction and centrifugation. The glass tubes are more expensive than the plastic centrifuge tube and they don't perform well at higher centrifugal forces. Plastic centrifuge tubes are made from polystyrene or polypropylene. They are found in both variations, disposable as well as non-disposable. High quality centrifuge tubes with almost transparent plastic are the best as one can easily see pellet formation inside the tubes. These tubes are provided with covers to prevent the samples from spilling. It is important to tightly screw the caps of centrifuge tubes before placing them in the slots. This avoids spillage and contamination of other samples. Also spillage inside the centrifuge damages the rotor inside and reduces its life.

While loading samples in the centrifuge, care should be taken to ensure that it is properly balanced. Small differences in mass of the load can result in a large force imbalance when the rotor is at high speed. This force imbalance puts strain on the spindle and may result in damage to centrifuge or personal injury. When loading the samples, equally filled centrifuge tubes of the same type should be placed opposite to each other in the slots. If your sample size is very small, tubes filled with plain water can be used to balance the weight on the opposite side, provided the sample is almost of the same weight. Never put a single centrifuge tube to run the centrifuge, even on the low speed. Some centrifuges have automatic rotor imbalance detection. This control software immediately stops the run when imbalance appears.

Before starting the centrifuge, rotor lockage and lid lock should be checked. Centrifuge rotors should never be touched while spinning. There is a possibility of hand getting stuck with the rotor and causing serious injuries. Centrifuges usually have main lid locking system to avoid such accidents. Many centrifuges have software that detects main lid lock before starting spinning and gives error if the lid is not properly shut.

Centrifuges present dangers of high kinetic energy stored in the basket that is spinning at a very high speed. Mechanical failure of some part of the centrifuge can result in ejection of metal parts at extremely high velocity that can cause serious damage to the instruments or persons nearby. Also a defective centrifuge will give off sparks and heat due to friction after spinning. Although, every centrifuge is able to withstand a certain degree of vibration and imbalance, excessive and frequent imbalance will cause metal-metal contact, friction and local heating of parts. The friction greatly increases wear and tear of metal, bearings, rubber and plastic parts and adds up to higher maintenance.

2.3.1 Low Speed Centrifuge

Low speed centrifuge is used for routine sedimentation of heavy particles and cell pelleting. It has a maximum speed of 4000 to 5000 rpm and RCF values up to 3000 g. Usually this instrument comes without temperature control and therefore samples can be centrifuged at room temperature. As temperature is not controlled while centrifugation, samples are likely to get heated at the time of the spin.

Two types of rotors, fixed angle and swinging bucket may be used in this type of centrifuge. This type of centrifuge can come with tube capacity of 10 to 50 ml. Low speed centrifuges are used for rapid sedimentation of blood cells, bacterial cells or cells from plant/animal suspension culture.

2.3.2 Microcentrifuge

Microcentrifuge, like all other centrifuges, separates particles based on their size, shape and densities. This type of centrifuge is used to prepare very small sample volume. These are bench top centrifuges that are usually used in laboratories for molecular analysis and separation of biomolecules. Usually small capacity polypropylene tubes and eppendorf tubes are used in this type of centrifuge. This equipment is very common in microbiology, biochemistry and molecular biology labs. The tubes of small capacity up to 2 ml can be used in this centrifuge. It can generate speed of 15000 rpm and relative centrifugal force of up to 12,000 g. This equipment may come with or without temperature control. Usually low temperature (4°C to 10°C) centrifugation is followed using this equipment. This can also be used for preparative stages like cell pelleting of bacterial samples. Sedimentation usually occurs with pellet sticking to one side of the centrifuge tube wall.

The microcentrifuge tubes that are used in microcentrifuges are small plastic tubes with conical bottoms with typical press fit cap. They are popularly used in molecular biology and biochemistry experiments to store and centrifuge solutions. They are quite inexpensive, so they can be disposed after use, or can be autoclaved and reused again. The microcentrifuge tubes are generally used along with micropipettes because they can measure solutions in small quantities like in µl. Microcentrifuge tubes are made of polypropylene and can withstand very high and low temperatures. They withstand wet heat of autoclave and can be readily used to store samples at low temperatures such as –80°C and –196°C of liquid nitrogen. These tubes are neutral towards organic solvents such as phenol, ethanol and chloroform that are frequently used in nucleic acid isolation and other molecular biology experiments. The capacity of microcentrifuge tubes ranges from 250 µl to 2 ml, with most common being 1.5 ml. Eppendorf is a major manufacturer of these tubes, and therefore they are popularly called as Eppendorf tubes.

2.3.3 High-Speed Centrifuge

High-speed centrifuges have speed ranging from 10,000 to 25000 rpm. They may have maximum RCF values up to 50,000 g. High-speed centrifuges usually come in sophisticated form with temperature control. They are ideal for processing, sedimentation and separation of temperature sensitive biological samples. In many biological experiments, samples need to be processed at low temperature such as 4°C. In such cases, high-speed centrifuge is a preferred choice. Three types of rotors, fixed angle, swinging bucket and vertical tube rotors are available in this type of centrifuge. High-speed swinging bucket rotors are used for density gradient centrifugation. For differential pelleting of particles, high-speed centrifuges with fixed angle rotors are used. In the vertical tube rotor, sample tubes remain in the upright position, parallel to the axis of rotation. The high-speed centrifuge with vertical tube rotor is used for gradient centrifugation. High-speed centrifuges are required in the processes involving sedimentation of cell debris after cell homogenization, sucrose gradient centrifugation, and ammonium sulphate precipitation of proteins and differential separation of cell organelles.

2.3.4 Ultracentrifuge

Ultracentrifuges can be traced back in history to the work of Svedberg. In 1920s, Svedberg did an extensive study of biomolecules using the very technique of ultracentrifugation that is still in use today. It was Svedberg's vision of ultracentrifuge as a most important apparatus in revealing the structure of biomolecules. Today most sophisticated of the centrifuge equipments are ultracentrifuges. Ultracentrifuges can generate speed of about 75,000 rpm and RCF values of 6,00,000 g. Because of the high speed generated by this centrifuge, considerable heat is generated. Therefore, ultracentrifuges come with a spin chamber that is refrigerated and under high vaccum to reduce friction.

Ultracentrifuges come in two types, preparative and analytical ultracentrifuge. Preparative ultracentrifuges are used for the separation and purification of sample for further analysis. The ultracentrifuge consists of four major components, speed control, temperature control, rotor and vaccum pump. The rotor used in ultracentrifuge is invariably fixed angle rotor. At such high speed generated by the centrifuge, friction caused by air is quite significant. To eliminate this friction, rotor chamber is sealed and evacuated by vaccum pump that is integral part of the equipment. These centrifuges usually come with auto locking system that locks the lid once it is shut to avoid any hazards or accidents that could occur at such high speed. Although they may come with an emergency stop button, the lid can't be opened manually before the spin comes to a complete halt.

The analytical ultracentrifuges may come with an additional optical system (usually UV light) to record the distribution of concentration in the sample in the centrifuge cell. This system can be used to monitor the sedimentation of particles during the centrifugation process. For analysis, a sample of volume 0.1 to 1 ml consisting of nucleic acid or protein is placed in the analytical cell and rotated. Light is passed through the sample, parallel to the axis of rotation and measurements of absorption by sample molecules are made. The optical system is connected to computers that are capable of relating changes in absorbance readings to the rate of movement of particles in the sample. This optical system detects and measures the moving boundary of sedimenting molecules. These measurements lead to analysis of concentration distribution in the sample.

Ultracentrifugation is used for differential separation of subcellular portions like microsomes, ribosome, and membrane components. It is useful in molecular mass determination of macromolecules, exploring genomic properties of DNA in cesium chloride gradient, studying protein-DNA interactions and various studies of structural biology.

2.4 TYPES OF CENTRIFUGATION TECHNIQUES

There are different types of centrifugation techniques, and they can be classified in a number of ways. For research and laboratory purposes, centrifugation is generally classified in the two types, (i) differential centrifugation and (ii) density gradient centrifugation.

2.4.1 Differential Centrifugation

2.4.1.1 *Introduction*

The differential centrifugation was introduced for separation of subcellular components in the biology research in the 1940s. In 1947, Albert Claude, along with two other scientists, was awarded Nobel Prize in Physiology or Medicine for their discoveries concerning structural and functional organization of the cell. Albert Claude extensively used differential centrifugation technique to separate different cellular organelles.

2.4.1.2 *Principle and working*

In the differential centrifugation process, particles are separated based on their size. Differential centrifugation is usually applied using fixed angle rotor (Fig. 2.3). In the method of differential centrifugation, sample is uniformly distributed along the length of the tube and its concentration is same throughout before centrifugation. During centrifugation, larger particles sediment faster than the smaller ones and provide basis for obtaining crude organellar separations. By the technique of centrifugation, cell organelles are separated by increasing the speed after each run. Therefore it is seen that only heavy organelles like nuclei sediment first and can be found in the pellet while smaller molecules like ribosomes, sediment last. In this experiment, cells are first disrupted and the homogenate is subjected to low speed centrifugation to sediment heaviest particles. Then the supernatant of unsedimented lighter particles is taken in a new tube and centrifuged at higher speed. This process is repeated several times with increased centrifugation speed and time. This sequence is repeated till all particles had sedimented.

The cellular fractions obtained at each stage of centrifugation can be identified by microscope. Cell nuclei, mitochondria and chloroplasts can be seen under light microscope, however small molecules like ribosomes are best visualized with the help of electron microscope.

Some of the sedimented organelles obtained by this process may have partial purity and may be contaminated with other particles. This may occur because two types of cell organelles like mitochondria and lysosomes that have nearly same sedimentation coefficients. These partially purified organelles can further be purified by density gradient centrifugation. Different marker techniques like DNA markers for detection of nuclei and specific enzyme markers for detection of mitochondria can also be used to detect presence of organelles in cell fractions (Refer Table 2.1).

Table 2.1: Markers used to detect cell organelles from subcellular fractions of differential centrifugation

Cell organelle	*Marker*
Nucleus	DNA
Mitochondria	Succinate dehydrogenase
Lysosomes	Alkaline phosphatase
Chloroplasts	Chlorophyll
Peroxisomes	Catalase

2.4.1.3 *Applications of differential centrifugation*

There are numerous applications of differential centrifugation technique in biochemistry, clinical diagnostics as well as molecular biology and biotechnology experiments.

1. This type of centrifugation is commonly used for preparative stages in experimentation. It can be used for preparation of sample, cell pelleting, centrifugation of tissue homogenate etc.
2. This type of centrifugation is used for partially pure preparation of cell organelles and other macromolecules.
3. Differential centrifugation is widely used technique to separate cell organelles like cell nuclei, chloroplasts, mitochondria, lysosomes, peroxisomes, cytoskeleton, plasma membrane, endoplasmic reticulum.
4. It is even used to separate small biomolecules like ribosomes and viruses.
5. Differential centrifugation has been successfully used for selective isolation of microorganisms like Actinomycetes from soil and plant litter. It has proved to be an efficient enrichment technique, which enables rapid and selective isolation of diverse zoosporic Actinomycetes genera directly from soil and plant litter (Hayakawa *et al.*, 2000).
6. Yamamura *et al.* (2003) have successfully applied differential centrifugation for selective isolation of *Nocardia* spp. from soil. Here, differential centrifugation proved to be very efficient method for isolating *Nocardia* spp. populations that are suppressed by conventional dilution plating approaches. By this method, it is possible to differentiate *Nocardia* from other Actinomycetes taxa by the simple process of centrifugation.

7. Differential centrifugation is used for the isolation and purification of virus particles. Bacteriophages have been commonly isolated by differential centrifugation.

2.4.2 Density Gradient Centrifugation

2.4.2.1 *Introduction*

The technique of density gradient centrifugation was first introduced by N.G. Anderson and M.K. Brakke in 1940s. This method is used to purify subcellular organelles and other biomolecules. This technique has been commonly used for purifying proteins and nucleic acids.

2.4.2.2 *Principle and working*

In density gradient centrifugation, particles are separated based on their densities. In this method, sample is centrifuged in a liquid medium with increasing density from top to bottom of the tube. Therefore a density gradient is established in the centrifuge tube. Density gradient may be discontinuous or continuous type. Discontinuous density gradient is prepared by stepwise layering of solution of decreasing density in the centrifuge tube. Mixing dense and light solutions with various proportions and adding the solution in the centrifuge tube in continuous stream produces the continuous density gradient. The solutes that are used for making density gradient are chosen depending upon their solubility in water and their compatibility with the biomolecules that are to be separated. Various solutes like sucrose, cesium chloride, ficoll, percoll, metrizamide etc. are used to prepare density gradient.

2.4.2.3 *Types of density gradient centrifugation*

Depending upon the process of establishment of density gradient in the centrifuge tube, density gradient centrifugation is classified into two types, (i) rate zonal centrifugation and (ii) isopycnic centrifugation.

2.4.2.3.1 *Rate zonal centrifugation*

In the rate zonal centrifugation, density gradient is applied to the sample in the tube before centrifugation. Rate zonal centrifugation depends on both particle size and mass. Also path length of the gradient should be sufficient for separation to occur. In rate zonal centrifugation, usually swinging bucket rotor is used (Fig. 2.1).

In this experiment, density gradient is created in the centrifuge tube before centrifugation, by mixing high and low density solutions like sucrose solution and glycerol. Then the sample is applied as a thin layer on top of the density gradient. Under centrifugal force, the particles begin sedimenting through the gradient in separate zones according to their size, shape, and mass. If the particles have higher density than densest solution of density gradient, then they will reach bottom of the tube with sufficient centrifugation. If the time of centrifugation is adjusted carefully, then the particles will get distributed throughout density gradient according to their sedimentation rates. As different particles have different masses, they are sedimented at different rates. Rate zonal centrifugation establishes particles as bands of increasing densities from top to bottom of the tube. Therefore, if the gradient is collected as individual fractions, highly purified particles can be obtained in this way.

The speed at which particles sediment in the density gradient depends primarily on their sizes and mass instead of density. As the particles in the band move down through the density gradient, zones or bands containing particles of similar size form, because the faster sedimenting particles move ahead of the slower ones. As the density of the particles is always greater than the densest solution of the gradient, all the particles eventually form a pellet if centrifuged long enough. Therefore in order to obtain bands, the spin must be stopped before any particles reach bottom of the tube.

In rate zonal centrifugation, the problem of contamination of particles with different sedimentation rates is avoided by layering the sample in narrow zone on the top of the density gradient. Therefore, fast sedimenting particles are not contaminated by slow sedimenting particles, which occurs in differential centrifugation. However, this also limits the volume of the sample that can be applied to the density gradient.

2.4.2.3.2 *Isopycnic centrifugation*

The word 'isopycnic' means having same density. In isopycnic centrifugation, density gradient is formed during centrifugation process. In the isopycnic technique, the density gradient column encompasses the whole range of densities of the sample particles and the sample is uniformly mixed with the gradient material before centrifugation. Every particle sediments only to the region in the centrifuge tube at which the gradient density is equal to its own density. The sedimentation does not depend on the path length of the gradient. In isopycnic method, swinging bucket rotor or fixed angle rotors can be used (Figs. 2.1 and 2.3).

In this type of centrifugation, the particles in the sample have lower density than the densest region of the gradient. As a result, the particles never reach bottom of the tube even when centrifugation is done for a long period of time. The particles sink during centrifugation until a position is reached where the density of the surrounding medium is same as the density of the particle. At this stage, they are said to have reached equilibrium. After the particles reach equilibrium, they remain at the same position even on further centrifugation.

Isopycnic separation is also called buoyant or equilibrium density gradient separation. In this method, particles are separated solely on the basis of their density. Particle size only affects the rate at which particles move until their density is the same as the surrounding gradient medium. This method establishes bands of particles of increasing density from top to bottom of the tube. These bands can be collected separately that contain purified particles.

A common example of this method is separation of particles in cesium chloride density gradient. The centrifugation time needs to be sufficient for particles to form bands at their isopycnic point. Excessive centrifugation doesn't have any adverse effect on the formation of bands. However isopycnic separation can be an exhaustive process. For example, isopycnic separation of DNA can take about 36 to 48 hours of centrifugation with cesium-chloride gradient.

2.4.2.4 *Applications of density gradient centrifugation*

1. Rate zonal centrifugation is commonly applied for the separation of antibodies and proteins. As antibodies have similar densities but different masses, they can be readily separated by this method.
2. Rate zonal centrifugation is applied to determine the sedimentation coefficient (S) of the particles.
3. Density gradient centrifugation technique is commonly used for purifying proteins and nucleic acids.
4. Nucleic acids RNA and DNA are routinely separated and purified by isopycnic centrifugation.
5. Density gradient centrifugation is used to purify subcellular components like mitochondria, lysosomes, ribosomes, microsomes etc.
6. The classic experiment by Messelson and Stahl that showed semi-conservative replication of DNA employed the method of isopycnic centrifugation.
7. Density gradient centrifugation is used for determination of molecular weights of proteins (Siegel and Monty, 1966).
8. Isopycnic centrifugation is commonly used for isolation and purification of plant and animal viruses like tobacco mosaic virus, wheat streak virus, herpes simplex virus, poliovirus (Guskey and Wolff, 1972) etc.
9. Density gradient centrifugation is used to isolate purified cell organelles like mitochondria, lysosomes, flagella, chloroplasts and ribosomes. It is also used to separate all types of blood cells.
10. Density gradient centrifugation is used to isolate microorganisms like diatoms, blue green algae, mycoplasmas like organisms (MLO), dinoflagellates etc.

11. Mammalian metaphase chromosomes have been isolated by isopycnic centrifugation in nycodenz gradient.

2.5 REVIEW QUESTIONS

1. Describe principle of centrifugation in detail.
2. What are different types of rotors that are used in a centrifuge? Describe how swinging bucket rotor differs from fixed angle rotor.
3. Describe principle of differential centrifugation. In which experiments would you use differential centrifugation technique?
4. Describe principle of density gradient centrifugation.
5. What is isopycnic centrifugation? Where is it used?
6. Describe rate zonal centrifugation. What is its application in molecular biology?
7. Describe and compare different types of centrifuges.
8. Compare rate zonal and isopycnic centrifugation.
9. What measures would you undertake while handling centrifuges?

2.6 REFERENCES

Boyer, R. (2000). Modern experimental biochemistry. 3rd Edition. Pearson Education Inc. pp. 207–228.

Enerback, L. and Svensson, I. (1980). Isolation of rat peritoneal mast cells by centrifugation on density gradients of Percoll. *J. Immunol. Methods.,* **39:** 135–145.

Ghosal, S. and Srivastava, A. (2009). Fundamentals of bioanalytical techniques and instrumentation. PHI Private Learning Ltd. pp. 114–129.

Golin, S. and Wray, W. (1984). Isopycnic centrifugation of mammalian metaphase chromosomes in nycodenz. *Experimental Cell Research,* **152(1):** 204–211.

Guskey, L. and Wolff, D. (1972). Concentration and purification of poliovirus by ultrafiltration and isopycnic centrifugation. *Appl. Microbiol.,* **24(1):** 13–17.

Hayakawa, M., Otoguro, M., Takeuchi, T., Yamazaki, T. and Limura, Y. (2000). Application of a method incorporating differential centrifugation for selective isolation of motile actinomycetes in soil and plant litter. *Anton Van Leeuwenhoek,* **78(2):** 171–185.

Kumar, A. and Awasthi, A. (2009). Bioseparation engineering: A comprehensive DSP volumen. I. K. International Publishing House Pvt. Ltd. pp. 46–69

Serdyuk, I., Zaccai, N. and Zaccai, J. (2007). Methods in molecular biophysics: Structure, dynamics, function. Cambridge University Press. pp. 339–387.

Siegel, L and Monty, K. (1966). Determination of molecular weights and frictional ratios of proteins in impure systems by use of gel filtration and density gradient centrifugation. Application to crude preparations of sulphite and hydroxilamite reductase. *Biochimica et Biophysica Acta,* **112:** 346–362.

Yamamura, H., Hayakawa, M. and Limura, Y. (2003). Application of sucrose–gradient centrifugation for selective isolation of *Nocardia* spp. from soil. *J. Appl. Microbiol.,* **95(4):** 677–685.

3

Spectroscopic Techniques

3.1 INTRODUCTION

Light is in fact, electromagnetic radiation that has different energies and wavelengths. The light's wavelength is related to the energy of the light. The shortest wavelength and the most energetic electromagnetic waves are γ-rays. The longest wavelength and the least energetic electromagnetic waves are radio waves. The light spectrum can be divided into a number of regions, which include γ-rays, X-rays, UV, visible light, infrared (IR), microwaves and radio waves (Fig. 3.1).

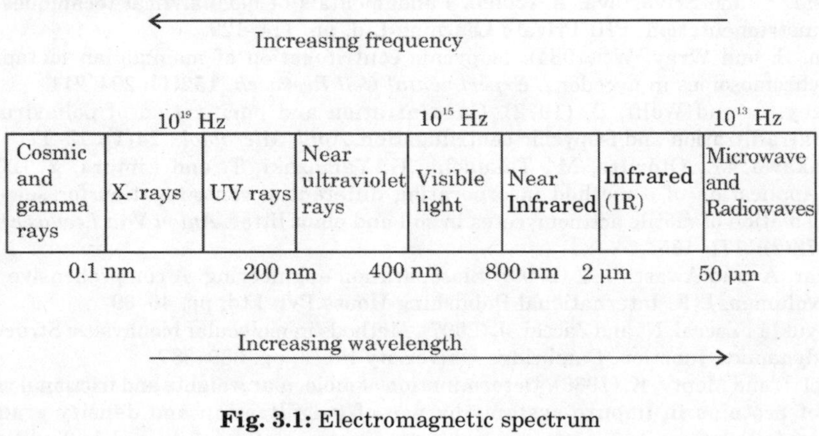

Fig. 3.1: Electromagnetic spectrum

Electromagnetic spectrum has many properties including wavelength, energy, frequency and trends. Trends related to electromagnetic radiation means the visible lights that produce different colours. Wavelength is the length to which a single wave travels and frequency is the amount of times a wave occurs. The shorter the wavelength, higher is the frequency. The electromagnetic spectrum uses energy because energy produces the electromagnetic radiation waves. Photon energy usually produces electromagnetic radiation waves.

When we see differently coloured objects, they are because of the visible light. Visible light is the only type of electromagnetic wave that is visible to the human eye. The coloured objects absorb particular wavelength and reflect others, making them appear as coloured object corresponding to the wavelength reflected (Refer Table 3.1). Microwaves are a type of electromagnetic waves.

Table 3.1: The spectrum of visible light

Wavelength range (nm)	***Colour of reflected light***	***Colour of absorbed light***
400–465	Violet	Yellow-green
465–482	Blue	Yellow
487–493	Blue-green	Red-orange
498–530	Green	Red-purple
559–571	Yellow-green	Purple
576–580	Yellow	Blue
587–597	Orange	Greenish-blue
617–780	Red	Blue-green

Microwaves can heat many molecules while passing through them. An example of this is the heating applied in a microwave oven. In this appliance, microwaves heat the water molecules present in the food items. Radar also uses microwaves to track echoes by the microwaves directed to the object and reflecting back to give the signal.

Infrared light is electromagnetic radiation and generally has a wavelength longer than visible light. Infrared light can be used for measuring the temperature of something, night vision and tracking devices that seek heat. Ultraviolet light contains electromagnetic radiation and has a shorter wavelength than visible light. The human eye cannot visibly see ultraviolet light. The ultraviolet light is known to produce DNA mutations and cause cancers in the individuals having prolonged exposure. The light from the sun is responsible for natural source of ultraviolet light. Gamma rays contain electromagnetic radiation and generally have short wavelength; therefore having high frequency. Like ultraviolet light, gamma rays can cause health problems in the human body. Gamma rays can be used sometimes to kill cancer cells.

X-rays are a type of electromagnetic radiation. Certain X-rays can travel through some solids depending on the wavelength. X-rays have longer wavelengths than gamma waves. X-rays have been used for many decades in medicine, diagnostics, visualizing fractures in bones, identifying cracks in metal sheets etc.

Out of all of the types of electro-magnetic radiations, the radiowaves have the lowest frequency and the longest wavelength. The atmosphere of the Earth is transparent to radio waves with wavelengths of a few millimetres to about twenty metres. The Radiowaves are used in radio telescopes for astronomic studies. The largest radio telescope in the world in located in Arecibo, Puerto Rico, which is 300 metres in diameter.

The spectroscopic techniques form the largest and most important group of techniques used in the field of chemistry and biochemistry. These techniques provide qualitative and quantitative information.

Spectroscopy simple means the study of spectra, especially experimental observation of optical spectra. It is the analysis of spectra, especially light or mass spectra, to determine properties of their source. In light or optical spectroscopy, the spectrum of light source is analyzed through a spectroscope to determine atomic composition of a substance. In mass spectroscopy, a spectrometer is used to determine the composition of ions or charged molecules in a sample. In other words, the study of the amount of light that is absorbed, given off, or reflected by an object is spectroscopy. Spectroscopy breaks down light and measures different wavelengths of visible and non-visible light. In medicine, different types of spectroscopy are being used to study tissues and to help make a diagnosis.

All spectroscopic techniques are more or less dependent on the emission or absorption of electromagnetic radiation that corresponds to certain energy changes in the atoms or molecules (Refer Table 3.2). The most important

Table 3.2: Various electromagnetic radiations and the respective spectroscopic techniques

Wavelength range (nm)	***Electromagnetic radiation***	***Atomic transitions***	***Type of spectroscopy***
Below 0.01	γ-rays	Transitions in nuclei	γ-ray spectroscopy
0.01–10	X-rays	Transitions of electrons in inner orbits	X-ray spectroscopy
10–400	Ultraviolet (UV)	Transitions of electron in outer orbits	UV spectroscopy
400–750	Visible light	Transitions of electron in outer orbits, excitations of electrons	UV-Visible absorbance spectroscopy
750–10^6	Infra-red	Molecular oscillations	Infra Red (IR) and Far Infra Red (FIR) spectroscopy
10^6–10^8	Microwaves	Molecular rotations and transitions of the magnetic moment of electrons	Electron Spin Resonance (ESR) spectroscopy
Above 10^8	Radiowaves	Transitions of magnetic moment in nuclei	Nuclear Magnetic Resonance (NMR) spectroscopy

use of spectroscopic techniques in analytical experiments is for the determination of atomic and molecular structures including functional groups and stereochemical arrangements. The instrument used in such studies is called spectrometer. Special devices are incorporated in spectrometer for automatic recording of spectra. The spectrum of a molecule is represented as a graph obtained by plotting either absorbance or transmittance of electromagnetic radiation as a function of wavelength or wave number.

3.2 UV SPECTROSCOPY

3.2.1 Introduction

Ultraviolet spectroscopy or UV spectroscopy refers to the absorbance spectroscopy or reflectance spectroscopy falling in the ultraviolet-visible spectrum region of light. Ultraviolet and visible spectrometres have been in general use for the last many years and over this period have become the most important analytical instrument in the modern day laboratory. In many applications other techniques could be employed but UV spectrometry is superior to all due to its simplicity, versatility, speed, accuracy and cost effectiveness.

In a historical context, the introduction of UV spectroscopic analysis during 1930s was as important milestone in the science of analytical chemistry. The vast saving of time and ability to work with very small amount of samples demonstrated the power of spectroscopic analysis to biochemists, who wanted to use the technique for various purposes without being concerned about its physical background. UV spectroscopy is the first method used for the elucidation of structure of organic compounds. The limitation of this method is that it does not give any information on the molecule as a whole but on its chromophoric moieties.

Visible light absorption is known to all of us, because this is what causes objects to be coloured. For example, a blue dye appears blue because the light at the red end of the spectrum is absorbed by the dye, leaving the blue light to be transmitted. The absorption or reflectance in the visible range directly affects the perceived colour of the material involved. In this region of electromagnetic spectrum, molecules can undergo electronic transitions. The energy of visible light depends on its frequency, and is approximately equivalent to 170 kJ/mol (mole of photons) for red light and 300 kJ/mol for blue light. The promotion of electrons to different energy levels is not restricted to electromagnetic radiation in the visible part of the spectrum but it can also occur in the ultraviolet region. UV spectroscopy is commonly used in analytical chemistry, biochemistry and biotechnology in quantitative analysis of organic compounds, biological macromolecules including proteins, enzymes, DNA, RNA, lipids and carbohydrates.

3.2.2 Theory and Principle

Ultraviolet-visible spectroscopy involves the absorption of ultraviolet light by a molecule causing the promotion of an electron from a ground electronic state to an excited electronic state. The structural features of the molecules responsible for absorption of UV light are referred to as chromophores. The wavelength of the maximum of the broad absorption is called λ_{max}.

Wavelengths in the UV-visible spectral region range from 200 to 800 nm. The relation of nanometres with other units of measurements is as follows,

$$\begin{aligned} 1 \text{ nm} &= 10^{-9} \text{ m} \\ &= 10^{-7} \text{ cm} \\ &= 10 \text{ A}^{\circ} \\ &= 1 \text{m}\mu \end{aligned}$$

The substance to be analyzed is generally dissolved in a solvent and pure solvent is kept as control. A good solvent should not absorb ultraviolet radiation in the same region as the substance, whose spectrum is being determined. Care must be taken when choosing a solvent, because many solvents absorb in the ultraviolet region. Usually solvents, which do not contain conjugated systems, are most suitable for this purpose, although they vary as to the shortest wavelength at which they remain transparent to ultraviolet radiation. The solvents most commonly used are water, 95% ethanol, and n-hexane. Polar solvents may not form hydrogen bonds as readily with excited states as with the ground states of polar solvents.

On the other hand, in some cases the excited states may form stronger hydrogen bonds than the corresponding ground state. In such cases, a polar solvent would shift in absorption to longer wavelength, since the energy of the electronic transition would be decreased. Transitions of the type are shifted to longer wavelengths by polar solvents. Ethanol is almost transparent to UV light above 200 nm (Table 3.3). Therefore ethanol is excellent as solvent for UV spectroscopy.

The different wavelengths at which different solvents are used can be described as follows,

Table 3.3: Absorbance of various solvents at their respective wavelengths

Solvent	***Minimum absorbance wavelength (nm)***
Water	191
Cyclohexane	195
Hexane	201
Ethanol	204
Dichloromethane	220
Trichloromethane	237

In UV-visible absorption spectroscopy, light from a suitable source is passed through a prism or monochromator and then passed through the sample. This beam of light then reaches detector that is placed on the other side of the sample slot. This arrangement minimizes exposure of sample to the light of wavelengths that might cause photodecomposition.

In the simplest type of UV spectrophotometer, the light source, sample and the detector are in the single beam sequence. Such devices are the cheapest, robust and very easy to maintain and use. Double beam spectrophotometres facilitate the measurement of absorption spectra as the light from the source is split onto two equal beams after passing thorough the monochromator. These two beams pass through the test sample and reference sample and the spectrum obtained is thus compensated for changes in some of the system variables.

The crucial optical components of the UV spectrometer are the light source, monochromator and detector. In standard UV spectroscopy, analysis is carried out with a dilute solution of the sample in solvent and is referred to the same pure solvent for comparison. It must be remembered that ordinary glass is opaque to wavelengths below 300 nm. Therefore UV analysis has to be carried out in special quartz cuvettes, which are transparent to lower wavelengths till 170 nm.

The fundamental relationship between UV light absorption and the concentration of the substance being analyzed can be explained by the Beer-Lambert's Law. The Beer-Lambert Law states that absorbance (A) of a solution is directly proportional to the path length (l) of the cell containing the solution (in cm), and the concentration of the absorbing molecules (c) in moles per litre according to the equation,

$$A = \varepsilon cl$$

Where ε is the molar absorptivity (cm^2/mol) of the absorbing molecule or molar extinction coefficient, which is characteristic to the molecule. This equation assumes that all molecules of the sample contribute to the absorption of incident beam and that no absorbing molecule is in the shadow of the other. For this assumption to hold true, the sample are usually very much diluted before reading in the spectrometer.

This value of molar extinction coefficient is constant for a given molecule at a given wavelength. The molar extinction coefficient is the measure of the ease of transition caused by the absorption of the radiation. If a particular electronic transition can take place readily, then the light will be absorbed strongly and A values will be high. On the other hand, if the transition does not occur readily, then the A values will be low. Substituents that increase the extinction coefficient of the basic chromophore are said to have a hyperchromic effect and those that decrease the extinction coefficient have hypochromic effect.

The absorption of light will occur only if the energy of the photon corresponds exactly to the energy of the transition. A higher energy and shorter wavelength of light is necessary to excite σ-bonds than π-bonds as there is larger energy difference between the bonding and antibonding σ-orbitals as compared to bonding and antibonding π-orbitals. The electronic transitions are often distinguished as 'allowed' and 'forbidden'. For example, $\sigma \rightarrow \sigma^*$ and $\pi \rightarrow \pi^*$ transitions are allowed while $n \rightarrow \pi^*$ transitions are forbidden. The * corresponds to the antibonding orbital or excited state and n denotes non-bonding orbital. The allowed processes have ε values greater than 10^3 while the forbidden processes have ε values less than 100.

3.2.3 Applications of UV Spectroscopy

Majority of the drug molecules possess sufficiently intense UV spectra and quantitative determination by UV spectrometer is routinely performed. It is of great use in pharmaceutical sciences. Samples are taken from various sites of the body and their solutions are analyzed to determine the amount of drug reaching those parts of the body. A useful feature of this type of analysis is the ability to calculate very small concentrations in the order of 0.0001 mol/dm^3, with extreme accuracy. It is important that the absorbance of the solution remains below two for quantitative measurements because of limitations of the instrument and solute-solute interactions that can cause deviations from the Beer-Lambert law.

UV spectrophotometres are widely used as detectors in HPLC and other separation techniques. UV spectroscopy is an effective tool in identification of key chromophores of unknown molecules.

When isolating and purifying DNA, RNA or oligonucleotides it is often necessary to obtain information about the purity of the respective DNA preparation (Figs. 3.2 and 3.3). In practice the absorbance ratio, A_{260}/A_{280} is

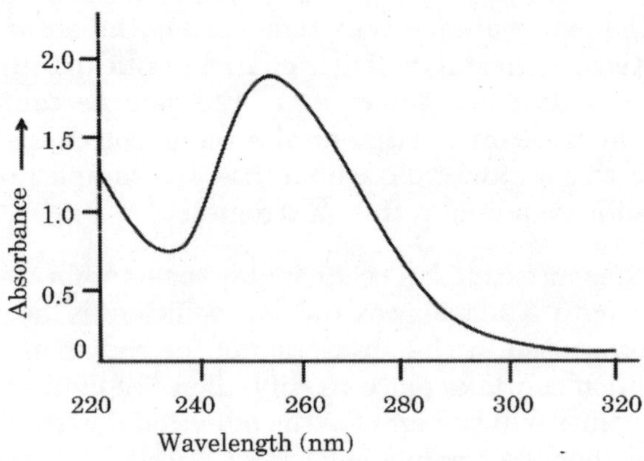

Fig. 3.2: Ultraviolet light absorption spectrum of DNA: DNA absorbs UV light in the range of 240 to 280 nm, highest being at 260 nm.

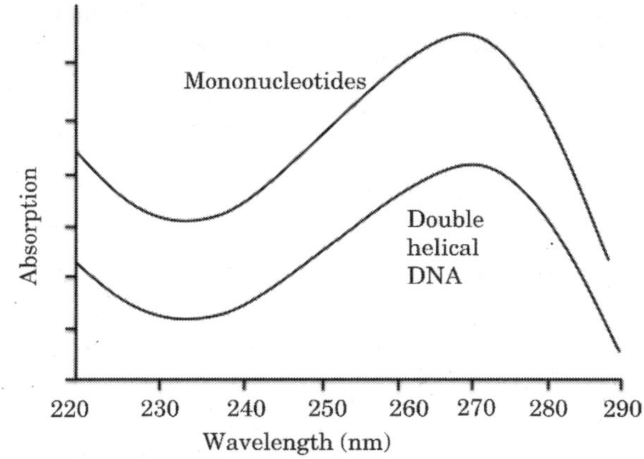

Fig. 3.3: Graph representing light absorption by nucleic acids in UV spectrometry

used as a qualitative measure of DNA purity. If the ratio is between 1.7 and 2.0 the DNA preparation is considered pure. A higher ratio indicates RNA contamination and lower ratio indicates protein or phenol impurities. The concentration of RNA in solution is determined by measuring the absorbance at 260 nm that is A_{260}.

When absorbance at A_{260} is 1, the concentration of RNA is about 40 µg/ml. However the actual value depends on the base composition, since the bases differ in extinction coefficient. The molar concentrations of oligonucleotides can be estimated in a UV spectrometer by measuring the number of absorbance units at 260 nm and by relating the value obtained to the sum of molar extinction coefficient of the individual nucleotides. Extinction coefficient is a parameter, which shows how strongly a substance absorbs light at a given wavelength per molar concentration.

The concentration and purity of DNA in solution can be measured by spectrophotometer. The concentration of DNA in solution is determined by measuring the absorbance at 260 nm that is A_{260}. When absorbance at A_{260} is 1, the concentration of DNA is about 50 µg/ml. In case of single stranded DNA, concentration can be calculated by measuring the absorbance at 260 nm. When the absorbance is 1, concentration of ssDNA is 33 µg/ml.

Quantitative analysis using UV spectroscopy is routinely used in biochemical studies to measure the rates of enzymatic reactions. It is also used in the determination of kinetic parametres Km (Michaelis Menton constant) and the turnover rate of an enzyme for a number of enzymatic reactions.

The molar extinction coefficient ε, of a known molecule is constant under identical conditions of solvent, concentration and path length. This principle can be used to identify amount of a pharmaceutical in a tablet. Such procedures form the basis of many quality assurance procedures in the pharmaceutical industry.

UV spectroscopy is exploited for the detection of analytes in a number of separation techniques like thin layer chromatography (TLC) and high performance liquid chromatography (HPLC). In organic and medicinal chemistry research, compounds that contain and aromatic ring or UV absorbing system, can be visualized on TLC plate under UV lamp.

UV spectroscopy is routinely used for the quantitative determination of analytes solutions in transition metal ions and highly conjugated organic compounds. Solutions of transition metal ions can be identified because the electrons within the metal atoms can be excited from one electronic state to another. UV spectroscopy is also used in environmental chemistry to determine the concentration of various metal ions, sometimes involving absorption spectra for organic complexes with the metal ions. The trace metal content of an alloy, such as manganese in steel, can be determined by firstly reacting the sample to get the metal into the solution in the form of ions. The ions are then complexed or made to react so that they are in a form that can be measured; for example, manganese as the magnate ion. When the spectrum is recorded, the most useful information obtained is the absorbance; because if the absorption coefficient of the chromophore is known the concentration of the solution can be calculated, and hence the mass of the metal in the sample can be calculated.

In practice, ultraviolet spectrometry is mostly limited to conjugated systems. A more widespread use of UV spectroscopy, however, is for determining the concentration of an unknown sample. Using calibration curve of λmax versus concentration of standards, concentration of an unknown sample could be determined.

In research, ultraviolet-visible spectroscopy is used more extensively in assaying than in identification. UV-Vis spectroscopy is a widely used analytical technique with many potential applications in biochemistry and biotechnology. UV spectroscopy is a non-destructive method for small sample volumes whose measurement is very easily and quickly done. Purified nucleic acids and most other biological macromolecules are colorless. Therefore, these molecules do not exhibit any absorption in the visible range of the spectrum. However, DNA exhibits a characteristic UV absorption band with a maximum around 260 nm and a minimum at 230 nm. Proteins, which have a band with absorption maximum at about 280 nm, are a common impurity in DNA preparations. The important wavelengths for the analysis of nucleic acids and proteins are given in Table 3.4.

Table 3.4: The important wavelengths for analysis of various biomolecules

Biomolecules	***Wavelengths (nm)***
Minimum absorbance wavelength for nucleic acids-DNA and RNA, peptide bonds of proteins	230
Nucleic acids have highest absorbance	260
Aromatic compounds, aromatic amino acids of proteins have highest absorbance	280
Neither nucleic acids, nor proteins show absorbance	320

All the absorption wavelengths mentioned in the table play an important role for quantitation and purity check of the respective biopolymers. Nucleic acids absorb maximum of UV light around 260 nm. Purine based nucleotides exhibit maximum absorption below 260 nm, whereas pyrimidine based nucleotides show maximum absorption above 260 nm.

Purine based dNTPs have significantly higher molar extinction coefficients as compared to pyrimidine based dNTPs. Double-stranded DNA gives a lower absorbance than single-stranded DNA. Single-stranded DNA again gives a lower absorbance than the completely hydrolyzed DNA. This decrease in absorbance accompanying the formation of more ordered structure is termed as hypochromic effect. This hypochromism results from interaction of the π-electrons of the bases, which are in close proximity due to the helical structure of double-stranded DNA.

UV spectroscopy is also used in the melting point determination of DNA. During the unwinding process of the double helix, a change in the molar absorption coefficient or extinction coefficient (ε) is observed. When a solution of double stranded DNA is heated, the absorption at 260 nm (A_{260}) increases drastically over a very narrow temperature range. This increase of absorption upon denaturation is termed as hyperchromic effect. The temperature corresponding to the point of inflection of the absorbance curve is decided as the melting point of DNA. This melting point determination is important for hybridization experiments and polymerase chain reactions.

3.3 NUCLEAR MAGNETIC RESONANCE (NMR)

3.3.1 Introduction

NMR spectroscopy technique relies on the magnetic properties of atomic nucleus. Nuclear magnetic resonance spectroscopy is the most powerful and versatile analytical technique today. NMR spectroscopy is the principle technique used to obtain physical, chemical, electronic data and three-dimensional structural information about biomolecules like proteins, carbohydrates and nucleic acids. In 1945, the research groups of Bloch and

Purcell independently obtained first nuclear resonance signal. They were awarded Nobel Prize in 1952 for this discovery.

The high resolution NMR spectroscopy can provide information about a protein's secondary and tertiary structure. Gonzalez-Vela and coworkers did first application of ^{13}C-NMR spectroscopy. They used ^{13}C-solution state NMR to study humic acid. Later, Fourier transformation technique developed by Ernst and Anderson provided breakthrough in NMR technology. This advancement permitted recording of NMR spectra of samples that were of low concentrations.

NMR and Mass Spectrometry (MS) are complementary techniques. MS is destructive, whereas NMR is not. However, a much smaller amount of material is needed for MS techniques. The MS can provide information about the molecular weight of a molecule; NMR can differentiate between structural isomers, and provide information about connectivity between atoms within a molecule. NMR technique correlates individual atoms within the molecule, rather than groups of atoms as in case of UV and IR spectroscopy.

Therefore nuclear magnetic resonance spectroscopy is considered as the most powerful experimental method available for the elucidation of atomic and molecular structure.

3.3.2 Theory and Principle

The phenomenon of nuclear magnetic resonance occurs because the nuclei of certain atoms possess spin. This spin is characterized by nuclear spin quantum number 'I'. Nuclei with zero spin having I=0, are not amenable to NMR observation. Most of the elements have at least one isotope that has nuclear spin. The most commonly studied elements in NMR are hydrogen ^{1}H and ^{13}C.

Many atoms such as ^{1}H, ^{13}C, ^{15}N, ^{17}O, ^{19}F, ^{31}P etc. behave as if a positively charged nucleus is spinning around an axis. This spinning charge, like an electric current, creates magnetic field. When a spinning nucleus having $I \neq 0$ is placed in a static magnetic field B_0, the magnetic moment m aligns with the magnetic field. When such atoms are exposed to strong external magnetic field, the nuclei try to align with it. This is similar to a simple bar magnet aligning itself with an external magnetic field. This motion is known as Larmor procession and it occurs at the Larmor frequency ν, which is directly proportional to the strength of the magnetic field applied.

$$\nu_0 = \frac{\gamma B_0}{2\pi}$$

Where ν_0 is the procession rate or Larmor frequency in Hertz, γ is the strength of nuclear magnet or the magnetogyric ratio, and B_0 is the strength of external magnetic field. The nuclear magnetic resonance follows following equation.

$$\omega_0 = -\gamma B_0 \Delta m_1$$

Where Δm_1 is the energy difference between ground state and excited state. Quantum mechanical rules state that only those transitions are allowed in which $\Delta m_1 = 1$. The principle equation of nuclear magnetic resonance states that for a spin transition between energy levels in a magnetic field B_0 to occur, an energy quantum of ω_0 is necessary. The transition from lower energy to higher energy level is detected as absorption of energy, while transition from higher energy to lower energy level is detected as emission of energy. The net amount of all transitions occurring in a sample is detected and identified as NMR spectrum as resonance signals occurring at specific resonance frequency of nucleus, ω_0. The magnetic field strength and resonance frequencies are directly proportional. Therefore, greater the field strength B_0, higher is the resonance frequency ω_0.

The signals detected in pulse Fourier Transform-NMR are considered to be emission signals. After the pulse, nuclei begin to relax and emit energy at specific Larmor frequency, which is detected as emission signal, and Fourier transformed to yield the spectrum.

The chemical shift

When a nucleus is placed in magnetic field, it processes at a rate determined by the magnetogyric ratio of the nucleus and strength of the magnetic field. The resonant frequency varies slightly depending upon the position of the atom within a molecule, which is termed as chemical environment. This occurs because the bonding electrons create their own magnetic field that modifies external magnetic field in the vicinity of the nucleus. This subtle variation is called the 'chemical shift'. The chemical shift provides valuable information about the structure of molecules.

NMR spectrometer

NMR spectroscopy takes advantage of the magnetic properties of the nuclei that are influenced by the chemical and physical interactions with their environment. These interactions can be measured by specific NMR constants such as coupling constants, relaxation times and changes in chemical shifts.

NMR spectrometer contains a superconducting magnet, a probe, a radio transmitter, radio receiver, analog-to-digital converter (ADC) and computer (Fig. 3.4).

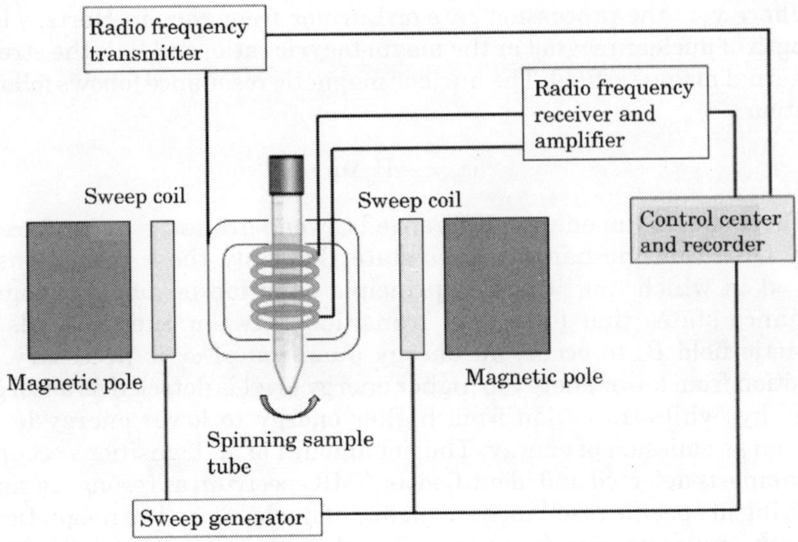

Fig. 3.4: Nuclear magnetic resonance (NMR) system

This technique relies on the ability of atomic nuclei to behave like a small magnet and ability to align themselves in an external magnetic field. To get the nuclei of a molecule to align in the same direction, a very strong magnetic field is generated using a superconducting electromagnet. This superconducting electromagnet requires extremely low temperatures to function. The coils of the magnet are surrounded by liquid helium (–269°C), which is prevented from boiling off too quickly by a surrounding layer of liquid nitrogen (–196°C).

These coolants are all contained in double-layer steel with a vacuum between the layers to provide insulation just like a thermos. In the NMR apparatus there is a narrow hole through the middle of the magnet, and the sample tube and radio frequency coils or more popularly called as probe, are located there. The probe is practically a coil of wire positioned around the sample that alternately transmits and receives radio frequency signals.

The computer directs the transmitter to send high-power short duration pulse of radio frequency to the probe coil. Immediately after the pulse, a weak signal is received. This weak signal received by the probe coil is amplified, converted into audio frequency signal and sampled at regular frequency by ADC. The computer processes the signals provided by ADC, performs Fourier transformation and displays the spectrum on the computer screen.

How to read the spectrum results

An NMR spectrum appears as a series of vertical peaks/signals distributed along the x-axis of the spectrum. Each of these signals corresponds to an atom within the molecule being observed. The position of each signal in the spectrum gives information about the local structural environment of the atom producing the signal.

NMR spectroscopy of carbohydrates

For NMR study of carbohydrates and glycoconjugates ^{1}H and ^{13}C have proved to be extremely valuable to determine primary structures. For the study of the interaction of carbohydrates with complementary compounds NMR spectroscopy can be a valuable tool. In fact the characterization of structures of glycoprotein-derived glycans has greatly facilitated the unraveling of biosynthetic pathways and studying the functional roles of these glycans in complex biological systems. Another important aspect concerns the confirmation of the identity of glycan structures that are supposed to be identical to known compounds.

Owing to the inherent flexibility of carbohydrate chains, the characterization of the three-dimensional structure in solution is rarely feasible into the same detail as for proteins and nucleic acids. The labelling of carbohydrates and glycan chains with isotopes is a bit more cumbersome for carbohydrates than for other bio-macromolecules.

In addition to ^{1}H and ^{13}C spectra, the nuclei of elements such as ^{15}N, ^{17}O, ^{19}F, ^{31}P and several other metal ions in carbohydrates have been recorded. It is important to note that, resolution and sensitivity are different for the various nuclei. Monosaccharides are in general stabilized by cyclization to form a hemiacetal or hemiketal, but all six possible tautomers of the D-aldohexoses in $2H_2O$ solutions have been detected by ^{13}C NMR at 30°C. The ^{13}C chemical shifts of mono- and oligosaccharides have been shown to be exquisitely sensitive to the nature of the isotope of hydrogen in the hydroxyl group. Recently, ^{13}C- and ^{1}H-labelled monosaccharides have been used to establish the concentrations of minor tautomers and provide additional structural restraints.

The characterization of the spatial structure in solution of glycans, free or in a complex with complementary molecules, or covalently bound to other molecules like proteins or lipids, is relevant for gaining insight into the functioning of these molecules.

3.3.3 Applications of NMR

NMR can be applied to determine primary chemical structure of proteins, and for the confirmation of their secondary and tertiary structures. It also provides useful information about interaction between molecules, rate processes and molecular motion.

The high-field NMR spectroscopy can be used for the assessment of impurity profile of the mixtures of biomolecules without any preliminary separations and in metabolomic studies. In addition to this, NMR spectroscopy is used for the quantitative pharmacological analysis.

NMR spectroscopy has provided most of the experimental data that enables the complex equilibria of interconverting forms of reducing sugars to be unraveled. As the sensitivity of NMR methods has improved recently in conjunction with the availability of ^{13}C-labeled compounds, even minor components of those equilibria have become observable, and the conformational space occupied by carbohydrates has become better defined.

Modern high-field NMR is a suitable tool for investigating a large number of problems in drug research and analysis, in stereochemistry and conformational characteristics of small molecules. NMR is an effective technique in the identification of drug receptor binding sites of proteins, determination of liquid-state dynamics and three-dimensional structure of ligand-receptor interaction.

Besides biochemists, NMR spectroscopy technique is utilized by physicists, chemists, biotechnologists, pharmaceutical chemists and medical scientists. The NMR spectroscopy technique is utilized in the research involving structural biology. With the help of NMR, three-dimensional structure of proteins can be obtained with high resolution, which is similarly achieved in X-ray crystallography. The NMR spectroscopy of nucleic acids such as DNA and RNA is called nucleic acid NMR. The structures of polymers of nucleic acids can be elucidated using this technique.

3.4 X-RAY SPECTROSCOPY

3.4.1 Introduction

Vast information has been accumulated since Röntgen first discovered X-rays. X-rays were first discovered by Wilhelm Röntgen in 1895. The earlier researches on characteristic x-radiation were carried out by using the fluorescence method using ordinary X-ray tube as primary source. At this time the X-ray tubes, with cold cathode, mostly had an anti-cathode of platinum or platinized plates of copper, nickel, iron or silver.

The discovery of the diffraction of X-rays solved the controversy regarding the nature of these rays and proved that they were electromagnetic waves. This meant that the X-ray spectra are of the same character as the optical spectra except that their wavelengths are far shorter than even the ultraviolet region of the optical spectroscopic field. We now know that the X-ray spectra have wavelengths ranging from hundred Angstrom to tenths of an Angstrom (Å). X-rays of up to 10 to 0.10 nm wavelength are called as soft' X-rays, and of 0.10 to 0.01 nm wavelength as 'hard X-rays, according to their penetrating abilities.

Hard X-rays can penetrate most solids and liquids, and all uncompressed gases, and their most common use is to image of the inside of objects in diagnostic radiography and X-ray crystallography. The images produced by X-rays are due to the different absorption rates of different tissues. Calcium in bones absorbs X-rays the most, so bones look white on a X-ray film recording of the X-ray image, called a radiograph. Fat and other soft tissues absorb less, and look gray. Air absorbs the least, so lungs look black on a radiograph.

In the first stages of development of X-ray spectroscopy by means of diffractions in crystals, the radiation from ordinary technical X-ray tubes was used. The subsequent construction of technical X-ray tubes of high power, and the development of new, much more sensitive detectors for X-rays, like Geiger, proportional, and scintillation counters, made it possible to use the fluorescence method of excitation for the analysis.

3.4.2 Theory and Principle

Atomic X-rays are emitted during electronic transitions to the inner shell states in atoms of modest atomic number. These X-rays have characteristic energies related to the atomic number, and each element therefore has a characteristic X-ray spectrum.

X-rays arise from displacement of inner extra-nuclear electrons. The electrons with principle quantum numbers 1, 2 and 3 in an atom are supposed to occupy shells- K, L and M respectively. When a electron from an external source having sufficient energy to displace K-shell electron in the target atom is bombarded from the source, then the vacancy created by the displacement of K-shell electron is filled by L-shell electron and an X-ray of appropriate wavelength is emitted.

The energy transition from L to K is governed by quantum rules. These rules follow the equation $E=h\nu$ and hence the frequency and wavelength of the emitted X-ray can be determined. X-rays can be absorbed by matter and this gives rise to X-ray absorption spectra. The rules applying to the relationship between an incident beam of monochromatic x-radiation and the transmitted portion are similar to the Beer-Lambert Law.

X-rays can be generated by X-ray tube, which is a vaccum tube that uses high voltage to accelerate the electrons, released by hot cathode with very high velocity. Medical X-rays are produced by letting a stream of fast electrons come to a sudden stop at a metal plate. The high velocity electrons collide with a metal target, the anode creating the X-rays. In medical sciences, X-ray tube target is usually tungsten or a more crack-resistant alloy of rhenium and tungsten; but sometimes molybdenum is also used for more specialized applications, such as production of soft X-rays. In X-ray crystallography, a copper target is most common, while cobalt is often used when fluorescence from iron containing sample poses a problem.

In practice, a variety of X-ray techniques and methods are used such as X-ray absorption, fluorescence and diffraction. X-ray absorption technique is similar to other absorption methods used in other regions of electromagnetic spectrum. In this method, a beam of X-ray is allowed to pass through the sample. The attenuation or fraction of X-rays absorbed is considered as measure of the concentration of absorbing substance. This technique is used to find imperfections in the internal structures of a substance.

The X-ray diffraction method is based on the scattering of X-rays by crystals. By this method, one can identify crystal structures of various solids. This method is extremely important as compared to other methods such as X-ray absorption and X-ray fluorescence. In X-ray fluorescence methods, X-rays are generated within the sample. If X-rays have wavelength that is shorter than the K-absorption edge of an atom, then it is possible for the incident radiation to dislodge K-electrons. This results in the emission of X-rays of a frequency different from that of the incident beam. This phenomenon is called X-ray fluorescence. By measuring the intensity and wavelength of the generated X-rays, one can perform qualitative and quantitative analysis. X-ray fluorescence is a non-destructive method requiring very small amount of sample.

X-ray radiation may result from the interaction of high-energy particles or photons with matter. Bremsstrahlung is the name given to the radiation emitted as a result of the retardation of high-energy particles by matter. This is radiation given off by the electrons as they are scattered by the strong electric field near the high-z (atomic number) nuclei. These X-rays produce a continuous spectrum. Synchrotron radiation results from the acceleration of charged particles in circular orbits by strong electric and magnetic fields. Bremsstrahlung and synchrotron radiation are used in X-ray spectroscopy as sources of continuous X-ray radiation.

X-ray absorption spectroscopy

X-ray absorption spectroscopy is a powerful tool for elucidation of both electronic and physical structure. This technique originated in early 1970s and in recent years it has become an important analytical technique in various fields. X-ray absorption spectrum arises from core-level excitation from absorption of X-rays and is associated with absorption edge.

X-ray absorption spectroscopy is element specific and can be used to investigate solids, liquids, gases and their mixtures. It probes all of an element within a sample with moderate sensitivity and is applicable to a wide range of elements. The experiment is usually performed with synchrotron radiation sources, which provides intense and tunable X-ray beams. Samples in the gas-phase, solution, or condensed matter can be analyzed by this technique. By examining the of the X-ray absorption edge where the onset of absorption occurs one can determine the oxidation state of the metal ion of interest.

Energy dispersive X-ray spectroscopy

Energy-dispersive X-ray spectroscopy is an analytical technique used for the elemental analysis or chemical structure characterization of a sample. It relies on the investigation of the interaction of X-ray excitation source and the sample. Its characterization capabilities are based on the fundamental principle that each element has a unique atomic structure that allows a unique set of peaks on its X-ray spectrum.

To stimulate the emission of X-rays from a specimen, a high-energy beam of charged particles such as electrons or protons, or a beam of X-rays, is focused on the sample being studied. At rest, an atom within the sample contains electrons at ground state having discrete energy levels. The incident beam may excite an electron in an inner shell, ejecting it from the shell while creating an electron hole where the electron was. An electron from an outer higher-energy shell then fills the hole, and the difference in energy between the higher-energy shell and the lower-energy shell may be released in the form of an X-ray.

The number and energy of the X-rays emitted from a specimen can be measured by an energy-dispersive X-ray spectrometer. As the energy of the X-rays emitted from an atom is characteristic of the difference in energy between the two electron shells, and of the atomic structure of the element from which they were emitted. This allows elucidation of the elemental composition of the specimen.

X-ray fluorescence spectroscopy

The field of X-ray fluorescence (XRF) spectroscopy is the most upcoming one today. XRF was originally used to analyze geological samples. As the technique developed with the advancements in computers and technology, XRF is now used extensively in analytical laboratories. With advantages such as easy sample preparation, non-destructive and rapid multi-element analysis, and the ability to screen unknowns in a wide array of samples has proved to be advantageous. The samples in different forms such as liquids, solids, slurries, powders, pastes, thin films, air filters, and many others can be used for the analysis. XRF offers a perfect compliment to other types of analytical equipment found in the analytical laboratory. Because of these advantages the technique has a broad appeal to research, industrial, and quality assurance analysts.

X-ray spectrometer

A typical X-ray spectrometer consists of a suitable X-ray source that can be focused into the specimen chamber and specimen chamber contains the substance being analyzed. A monochromator is required to disperse the emitted or fluorescent radiation and a detector and data processing facility is used.

3.4.3 Applications of X-Ray Spectroscopy

X-ray absorption studies are helpful in elemental studies and measurement of thickness of materials. This technique is used in different research and scientific fields including molecular and condensed matter physics, chemistry, materials engineering, earth science and biology.

The X-ray spectroscopy has wide applications in forensic sciences and environmental pollution studies as it enables many elements to be detected and concentrations measured. It can also be used for the detection and measurement of trace elements in the fertilizers, as these elements most often find their way in the food chain and potential toxic components can interfere with the normal metabolism of animals. Example of such an application is the study of the uptake of lead in plants from the polluted soil.

XRF is applied in the analysis of various day-to-day items. Determination of iron content in milk powder and essential minerals found in vitamin tablets are examples of how XRF plays a role in our health. The ability to screen unknown samples in their raw form, aids in protecting us from hazardous chemicals, as well as identifying the elemental composition of evidence from a crime scene.

The clinical application of X-ray spectroscopy for performing bone densitometry measurements involves either single photon or dual energy X-ray absorption spectrometry. This study is useful in monitoring hormone replacement therapy in female patients.

X-ray fluorescence spectroscopy (XRF) is applied in various industries such as cosmetics, pharmaceutical, agricultural and food industry. Since many additives in cosmetics are mineral or metallic in nature, or are organics with a heavier element attached they can be measured by XRF. For example, titanium dioxide is used in most sunscreen lotions to help prevent UV radiation burns. Their concentration is high and easily measured by XRF.

XRF has been used to measure total iron, and other major components. Cosmetic products are usually analyzed for toxic components to ensure that they are safe. The sensitive XRF analyzers can be used for measuring toxic metals such as lead, cadmium, mercury, and arsenic in cosmetic products. Some of the brightly colored cosmetics like eye shadows and fingernail polishes contain metallic dyes that can be measured by XRF.

There are numerous regulations regarding a limit to the amount of toxic compounds and metals that can be released into the atmosphere. The technique of X-ray spectroscopy is useful in the toxic materials monitoring and monitoring of environmental pollution. XRF is commonly used as a screening tool to identify contaminated soils that must be remediated. XRF can also be used for analyzing metals in wastewater. Apart from the environmental data, tissue samples are monitored for toxic metal contamination as a way to monitor the safety of the environment. Fish are commonly analyzed for metals like mercury in order to determine water quality and ascertain if the fish are edible.

In the field of agriculture and food industry, X-ray spectroscopy again has a promising advantage. Grains and processed flour have long been fortified with iron, and calcium is added to a number of dietary supplements. In those cases, XRF instruments are used to monitor the products. Routine monitoring of the product with XRF can control the nutritional value of pet foods and animal feed. Metal concentrations in vitamin tables can be measured and monitored with many higher resolution XRF instruments.

X-ray absorption spectrometry is extensively used in medical field for medical diagnostic imaging. Soft X-rays are used for this application. X-ray spectroscopy is also applied for determination of iron content (haemoglobin) in blood.

In the field of forensic sciences, X-ray spectroscopy is frequently used. Because XRF is a non-destructive method and can identify a vast number of elements simultaneously, it is an excellent tool for fingerprinting all kinds of materials analyzed in forensic analysis. Calcium and/or magnesium are found in high concentrations in most antacid products and can be measured with many XRF instruments. Selenium is a common additive in shampoos to prevent dandruff. XRF has long been used in this application.

3.5 MASS SPECTROMETRY

3.5.1 Introduction

Mass spectrometry is an analytical laboratory technique that is used to separate the components of a sample by their masses. By the invention of mass spectrometry, scientists have practically developed a method of weighing molecules. Wilhelm Wien, who received the Nobel Prize for Physics in 1911, laid the foundation of mass spectrometry with his discoveries concerning the laws of thermal radiation.

In 1893 Wien who worked on the ionized gases, published a theoretical paper in which he presented his displacement law. J. J. Thompson constructed the first mass spectrometer in 1912. The early mass spectrometres were primarily used by physicists to study the atomic weight of the elements and the natural relative abundance of elemental isotopes (Willard *et al.*, 1988). However, these early instruments could not be used to study biomolecules. In 1959, MS was first used to sequence peptides and oligonucleotides, and in 1962 it was used to study the structure of nucleotides. Ionization of larger molecules such as proteins became possible after 1981, when the fast atom bombardment ionization method was introduced.

The ability to ionize larger molecules was further improved with the advent of electrospray ionization (ESI) by Fenn and coworkers in 1988. Time-of-flight (TOF) mass analysis had been developed and commercialized by 1956, but relatively poor mass resolution was a problem until improvements were made in the early 1970s. An ionization technique for biological molecules that could be used with TOF analysis was introduced in 1991. The matrix-assisted laser desorption and ionization (MALDI), was the result of work in Germany by Hillenkamp, Karas, and coworkers and in Japan by Tanaka and coworkers.

In mass spectrometry, individual molecules are measured in contrast to the other forms of spectroscopy like infrared, NMR and UV, where the bulk of the sample is analyzed. Since the 1990s there has been rapid growth in the field of biological mass spectrometry. Ionization methods like electrospray ionization (ESI) and matrix-assisted laser desorption have revolutionized the creation of gas-phase ions from high-molecular weight compounds.

3.5.2 Theory and Principle

A device called mass spectrometer is used to analyze the samples be the technique of mass spectrometry. The mass spectrometer determines the molecular weight of ions in vaccum. To avoid collision between ionic particles, a high vaccum of about 10^{-5} Pa is required. The sample is vaporized into gas and then ionized in ion source. These gaseous ions are then accelerated through a potential difference and focused into a beam.

The ion beam passes through a magnetic field, which bends the charged stream. Lighter components or components with more ionic charge will deflect in the field more than heavier or less charged components. Therefore the particles are separated on the basis of mass and charge. A detector counts the number of ions at different deflections and passes the data to computer for analysis and recording of the spectrum. The data can be plotted as a spectrum of different masses.

The important components of a mass spectrometric system are ion source, mass analyzer and detector.

The purpose of the ion source is to ionize and in some cases, to vaporize the sample. As the ions are very reactive and short-lived molecules, their manipulation can be conducted only in vaccum. The ionization of the molecules can be done by several techniques such as electron-impact ionization (EI), chemical ionization (CI), fast atom bombardment (FAB), electro spray ionization (ESI) and matrix-assisted laser desorption ionization (MALDI).

Electron impact ionization (EI)

When individual molecules of the sample are bombarded with high-energy electrons, they cause ejection of the one or more electrons from those molecules on impact. This technique is known as electron impact ionization. This emission of electrons from the molecules of the sample produces positively charged cations. Electron impact ionization is the most common ionizations technique used in mass spectrometry. It works well with many gas phase molecules and the mass spectra are reproducible.

The electrons used for ionization are produced by passing a current through a wire filament. The number of electrons emitted depends upon the amount of current applied. The electric field accelerates these electrons to produce a beam of high-energy electrons. When analyte molecule passes through this electron beam a valence shell electron is removed producing ions of analyte molecules.

However, this technique has some limitations, as it produces fragmented ions. Molecular ions are intact ions of the analyte while fragmented ions are formed from further fragmentation of these molecular ions.

Chemical ionization (CI)

In chemical ionization, molecules of the sample are collided with a reactive gas. Chemical ionization produces ions with little excess energy. As a result, very low number of fragmented ions are produced. As the reagent gas ions have low excess energy and are relatively stable, therefore fragmentation is minimized in CI.

In CI, reagent gas is ionized with electron beams to produce a cloud of ions. The reagent gas ions in the cloud react and produce ions like methane ions that are excellent proton donors. When analyte molecules are introduced in this cloud of ions, the reagent gas ions donate a proton to the analyte molecule and thus produce a molecular cation.

The energetics of proton transfer can be controlled by using different reagent gases. The most commonly used reagent gases in chemical ionization method are methane, hydrogen, ammonia and isobutane. Out of these, isobutane and ammonia cause weaker ionization while methane being the strongest proton donor, causes strong ionization.

CI is an excellent method for the determination of molecular mass of an unknown analyte. This method can be applied to those analytes that can not withstand the rigorous impact of the electrons.

Fast atom bombardment (FAB)

Fast atom bombardment method uses high-energy atoms to sputter and ionize the sample molecules. In this technique, a beam of rare gas neutrals such as argon gas is focused on the liquid or solid sample. The impact of this high-energy beam causes analyte molecules to sputter into gas and get ionized at the same time. In this system, no heating is required to transform solid or liquid substance into gaseous ones; therefore this method is very useful in case of temperature sensitive samples.

In FAB, the analyte is dissolved in a liquid matrix. A drop of analyte and matrix material is mixed together and fast atom beam is focused on this droplet to produce analyte ions. In the liquid matrix low vapour pressure liquids like glycerol are usually used.

FAB is considered as a hard ionization technique since it leads to the breakdown of sample molecules into smaller fragments. However these fragments can serve as a fingerprint of the sample and may help in detection of unknown components.

Electro-spray ionization (ESI)

Electrospray ionization (ESI) is a process that introduces solution-bound molecules into a mass spectrometer as gas-phase ions. In the electro-spray ionization, dissolved sample is sprayed into the electric field. The electrospray produces multiply charged ions and this causes high molecular weight molecules to show lower mass to charge (m/z) ratios. This enables analysis of high molecular weight analytes with less expensive mass spectrometer. Certain biomolecules like proteins can be decomposed by heat or ionization be other methods.

This difficulty is overcome by electrospray technique. In this, the molecules of the sample are directly transferred to the ionized gaseous phase by spraying the liquid sample through a very fine needle held near the electric potential of about 4 kV. This produces a mist of fine droplets that get instantly ionized and produce an ionized aerosol. By this technique, proteins up to the size of 150 kDa can be analyzed. ESI is combined with quadrupole analyzer for analysis of biomolecules like proteins. Volatile organic solvents such as methanol, ethanol and acetonitrile are typically contained in the sample solution for electro-spraying.

Excellent sensitivity with minute quantities of material is achieved when ESI is performed at low flow rates. One of the biggest benefits of ESI is the creation of a robust interface between liquid separation techniques and the mass spectrometer.

Matrix-assisted laser desorption ionization (MALDI)

This technique is used to analyze extremely large molecules with molecular weight over 2,00,000 Da. This technique directly vaporizes and ionizes the analyte. This technique is used for the analysis of natural and synthetic polymers and proteins. In 1987, Koichi Tanaka introduced the technique of soft laser desorption ionization (SLD), for which he received Nobel Prize in 2002. An improved version of SLD, which is now most commonly referred as technique of MALDI was introduced afterwards by Michael Karas and Franz Hillenkamp.

In MALDI, both desorption and ionization is introduced by a single laser pulse. The analyte is mixed with the matrix component that can absorb the laser wavelength. This prepared sample is placed on the probe tip and dried. A laser is focused on this dried sample and the energy from this laser is absorbed by the matrix components of the sample mixture.

This energy ejects the analyte ions from the mixture. The photon energy deposited into the matrix converts to thermal energy, causing rapid evaporation of matrix and analyte. Photoionization processes also occur and may contribute to the ionization of analyte. Ions are produced primarily as singly protonated molecules.

The advantage of MALDI method is that it does not produce fragmented ions. Unlike other types of ionizations, MALDI tolerates moderate concentrations of buffers and salts in the analyte mixture.

Mass analyzer

After ions are formed in the ion-source region, they are accelerated into mass analyzer by electric field. The mass analyzer separates ionized species according to the mass to charge (m/z) ratio. The analyzers are either continuous or pulsed type. The continuous analyzers include quadrupole filters and magnetic sectors. These analyzers are similar to the monochromator used in the optical spectrometry.

They transmit a single selected m/z ratio to the detector and the mass spectrum is obtained by scanning the analyzer. Since single m/z ratio is taken, ions at other ratios are lost. On the other hand, pulsed mass analyzers collect an entire mass spectrum in one single pulse of ions. The pulsed mass analyzers include quadrupole ion trap, ion cyclotron resonance and time of flight (TOF) analyzers. MALDI is usually combined with TOF analyzer for analysis of biomolecules.

In the time-of-flight analyzer, ions are separated based on time as they travel down the flight tube. This type of spectrometer works on the principle that all singly charged ions subjected to the same potential difference will achieve the same translational energy that is measured in volts. However, it implies that lighter particles will be accelerated faster than the heavier ones and will pass the flight path in shorter duration. In other words, they will have shorter time-of-flight than the heavier ones.

The performance of a mass analyzer is described by three characteristics: the upper mass limit, the transmission, and the resolution. The upper mass limit is the largest m/z that can be measured. The transmission is the fraction of ions produced at the source that reach the detector. The resolution is a measure of the ability to differentiate two m/z that are close together.

Detector

Detection of the ions is usually based on their charge. For large signals, Faraday Cup is used to collect ions and measure the current. Some detectors amplify the ion signal by using electron multiplier or multichannel plates.

Interpretation of results

The results obtained from mass spectrometric analysis are in the form of graph. One must learn to how to interpret these results and draw substantial conclusions about the analyte of interest. Since the mass spectrum is the fingerprint of the molecular structure, comparison of the obtained results o the computer database will help to identify the unknown compound. This is usually done using probability based matching (PBM). Advances in data processing have enabled the high-throughput analysis of mass spectra.

However, it is important to remember that a computer only compares the unknown spectrum to the library spectra and shortlists compounds with similar spectra. As already described, the information given by mass spectrometry is the molecular weight of the substance as ratio of m/z.

However, deciding which peak actually corresponds to the molecular ion is not necessarily straightforward and depends upon the ionization technique. In case of EI, fragmented molecular cation $M^{\bullet+}$ can be recorded. If softer ionization techniques such as CI, FAB or electrospray are used, one can measure quasi-molecular ions such as MH^+, MNH_4^+, or MNa^+ depending upon the technique used.

3.5.3 Applications of Mass Spectrometry

Mass spectrometry is an important tool in protein and DNA analysis. It is used in proteomics and polymorphism studies of nucleic acids.

By utilizing mass spectrometric technique, one can determine molecular weights of biomolecules as large as 5,00,000 Da with great accuracy. This technique is also used for structural determination of amino acid sequences in proteins or sugar sequences in oligosaccharides.

Mass spectrometry is combined with various other biochemical techniques such as liquid chromatography, gas chromatography and high performance liquid chromatography for analysis of biomolecules. These methods effect a temporal separation based on the chemical properties of the molecules and a physical separation of the m/z ratios is obtained in the mass spectrometer. Two types of mass spectrometry experiments are in common use for biological studies.

The first experiment measures molecular masses of intact molecules such as peptides and proteins. Such molecular measurements are often performed using a TOF mass spectrometer. Structural information can be obtained directly in mass spectrometry experiments by using a second type of instrument called tandem mass spectrometer. This instrument couples together ion isolation and ion activation with mass analysis to provide detailed structural information.

Mass spectrometry measures individual atoms and therefore this technique enables isotopic constitutions of the sample to be measured. Isotope-ratio MS (IRMS) is a technique that measures the relative stable isotopic abundance of elements. The elemental isotope ratio can be analyzed for a complex system, for a specific compound within a mixture, or for a specific position within a compound.

For example, the $^{13}C/^{12}C$ ratio can be measured for a tissue sample, for a specific fatty acid within the tissue, or for a specific carbon position within the fatty acid. IRMS can reveal information regarding the origin or state of a complex system. For instance, the ratio of $^{13}C/^{12}C$ in apple juice can indicate if the juice has been adulterated with corn syrup. The ratios of isotopes such as $^{13}C/^{12}C$, $^{15}N/^{14}N$, and $^{87}Sr/^{86}Sr$ can pinpoint the geographic origin of ivory, which can help prevent future poaching.

Mass spectrometres can be used to measure relative amounts of molecules present in experiments. By using stable isotope-labelled molecules, the relative quantities of proteins between two different states can be determined. In this process, proteins from the two different states are combined and the mass spectrometer is used to measure the relative heights of the same peptide or protein ion in the mass spectrometer. By measuring changes in the relative abundance for proteins in different cellular states, the dynamics of processes can be better understood.

For the analysis of biomolecules like DNA and proteins, soft ionization techniques like ESI and MALDI are used. MALDI technique is mainly used to determine molecular weights, to obtain structural information and to investigate posttranslational processes of proteins and peptide mixtures. Molecular weights of the proteins can be determined accurately with only small amount of sample in few milligrams.

Protein structure can be revealed by digesting proteins partially with enzymes and analyzing the peptide obtained. With the help of MALDI mass spectrometry, post translational modifications like formation of sulfide bonds or glycosylation can be identified in proteins.

ESI technique is suitable for analysis of most types of biomolecules such as peptides, carbohydrates, DNA and lipids. Molecular weight determination, as well as sequencing of peptides and DNA fragments can be done by ESI mass spectrometry.

Because of high sensitivity of the techniques, mass spectroscopy has become the powerful analytical tool for monitoring environmental pollutants, pesticides residues in food, water and soil up to parts per billion or parts per trillion levels. Mass spectrometry has unique advantages such as high accuracy and high sensitivity.

Applications of mass spectrometry include identifying and quantitating pesticides contamination of water samples, identifying steroids in blood samples of athletes, determining metal contamination at extremely low amounts as in ppq (parts per quadrillion) levels in water samples, carbon-dating of the rare and most delicate of relics and artifacts like in case of Shroud of Turin, and studying the effect of molecular collision angle on reaction mechanisms.

Mass spectrometry has tremendous applications in pharmacology and toxicology. This technique can be used to analyze complex samples such as urine and blood plasma. Identification and quantitation of drugs, hormones, anesthetics and other trace biomolecules can be done effectively by using this technique.

Therefore, mass spectrometry combined with chromatography techniques is used in various fields of medicine, diagnostics, forensic sciences, biotechnology, environmental sciences, proteomics and genomics. In the field of biotechnology, mass spectrometry is used for the analysis of proteins, peptides, oligonucleotides and DNA. In pharmaceutical sciences, MS is applied for drug discovery, combinatorial chemistry, in pharmacokinetics and drug metabolism.

In clinical diagnostics, MS is used for neonatal screening, haemoglobin analysis, drug testing etc. in the field of biochemistry and biotechnology, accurate molecular weight measurements of biomolecules, (sample confirmation, determination of purity of a sample, verification of amino acid substitutions, and detection of post-translational modifications, all these studies are possible because of MS analytical technique. Moreover, MS is also used in reaction monitoring, monitoring of enzyme reactions, chemical modification, protein digestion, amino acid sequencing, sequence confirmation, *de novo* characterization of peptides, identification of proteins by database comparison and oligonucleotide sequencing.

Mass spectrometry is helpful in revealing protein structure, (protein folding monitoring, protein-ligand complex formation under physiological conditions and macromolecular structure determination. The ability to identify proteins rapidly and automatically, directly from mass spectrometry data is a powerful capability. To accomplish this, the proteins are first separated by SDS-polyacrylamide gel electrophoresis. Then a separated band on the gel containing protein is excised from the polyacrylamide matrix and digested with a protease to create peptides. Typically, protease such as trypsin is used because it creates peptides in the molecular mass range of 800 to 4000 Da by cleaving after arginine or lysine residues. Peptides are then analyzed by MALDI-TOF to generate a molecular mass map.

3.6 REVIEW QUESTIONS

1. Give an account of electromagnetic radiations. Describe different types of spectroscopic techniques related to them.
2. What is the range of ultraviolet and visible light? Give principle of UV spectroscopy.
3. What are the applications of UV spectroscopy?
4. What is NMR spectroscopy? Give theory and principle of NMR.
5. How are X-rays used in spectroscopic techniques?
6. What are applications of X-ray spectroscopy?
7. What is mass spectrometry? Where is it used?
8. Describe various techniques of ionization of molecules for mass spectrometry.

3.7 REFERENCES

Bakhtiar, R. and Tse, F. (2000). Biological mass spectrometry: A primer. *Mutagenesis* **15:** 415–430.

Barber, M., Bordoli, R., Sedgwick, R. and Tyler, A. (1981). Fast atom bombardment of solids as an ion source in mass spectroscopy. *Nature,* **293:** 270–275.

Brenna, J., Corso, T., Tobias, H. and Caimi, R. (1997). High-precision continuous-flow isotope ratio mass spectrometry. *Mass Spectrom. Rev.,* **16:** 227–258.

Clark, B., Frost, T. and Russell, M. (1993). UV spectroscopy: Techniques, instrumentation, data handling. Chapman and Hall, London. pp. 1–15.

Fenn, J., Mann, M., Meng, C., Wong, S. and Whitehouse, C. (1989). Electrospray ionization for mass spectrometry of large biomolecules. *Science,* **246:** 64–71.

Ferentz, A. and Wagner, G. (2000). NMR spectroscopy: A multifaceted approach to macromolecular structure. *Quarterly Reviews of Biophysics,* **33:** 29–65.

Hammes, G. (2005). Spectroscopy for biological sciences. *John Wiley and Sons, Inc.* p. 172.

Humeny, A., Beck, C., Becker, C. and Jeltsch, A. (2003). Detection and analysis of enzymatic DNA methylation of oligonucleotide substrates by matrix-assisted laser desorption ionization time-of-flight mass spectrometry. *Anal. Biochem.,* **313:** 160–166.

Jardetzky, O. and Jardetzky, C. (1957). An interpretation of the proton magnetic resonance spectrum of ribonuclease. *J. Am. Chem. Soc.,* **79:** 5322–5323.

Little, D., Thannhauser, T. and McLafferty, F. (1995). Verification of 50- to 100-mer DNA and RNA sequences with high-resolution mass spectrometry. *Proc. Natl. Acad. Sci. USA.* **92:** 2318–2322.

Medzihradszky, K., Campbell, J., Baldwin, M., Falick, A., Juhasz, P., Vestal, M. and Burlingame, A. (2000). The characteristics of peptide collision-induced dissociation using a high-performance MALDI-TOF/TOF tandem mass spectrometer. *Anal. Chem.,* **72:** 552–558.

Munson, M. and Field, F. (1966). Chemical ionization mass spectrometry I: General introduction. *J. Am. Chem. Soc.,* **88:** 2621–2630.

Natelson, S. and Sheid, B. (1961). X-ray spectroscopy in the clinical laboratory IV phosphorus: total blood iron as a measure of hemoglobin content. *Clinical Chemistry,* **7(2):** 115–129.

Pappin, D., Bleasby, A., Sutton, C. and Cottrell, J. (1993). Rapid identification of proteins by database matching of proteolytic peptide masses. *Protein Science,* **2(supplement 1):** 90.

Pellecchia, M., Sem, D. and Wüthrich, K. (2002). NMR in drug discovery. *Nature Rev. Drug Discovery,* **1:** 211–219.

Savukov, I., Lee, S. and Romalis, M. (2006). Optical detection of liquid-state NMR. *Nature,* **442:** 1021–1024.

Schweppe, R., Haydon, C., Lewis, T., Resing, K. and Ahn, N. (2003). The characterization of protein post-translational modifications by mass spectrometry. *Acc. Chem. Res.,* **36:** 453–461.

Shevchenko, A., Jensen, O. and Podtelejnikov, A *et al.* (1996). Linking genome and proteome by mass spectrometry: large-scale identification of yeast proteins from two dimensional gels. *Proceedings of the National Academy of Sciences of the USA,* **93:** 14440–14445.

Sickmann, A., Mreyen, M. and Meyer, H. (2002). Identification of modified proteins by mass spectrometry. *IUBMB Life,* **54:** 51–57.

Tanaka, K., Waki, H., Ido, Y., Akita, S., Yoshida, Y. and Yoshida, T. (1988). Protein and polymer analysis up to m/z 100,000 by laser ionization time-of-flight mass spectrometry. *Rapid Commun. Mass Spectrom.,* **2:** 151–153.

Thomas, M. (2008). Ultraviolet and visible spectroscopy. 2nd Edition. John Wiley and Sons, LTD. p. 322.

Stephenson, G., Robert, A. and Grübel, G. (2009). X-ray spectroscopy: Revealing the atomic dance. *Nature Materials,* **8:** 702–703.

Wende, H. (2004). Recent advances in X-ray absorption spectroscopy. *Rep. Prog. Phys.,* **67(12):** 2105.

Willard, H., Merritt, L., Dean, J.Jr. and Settle, F.Jr. (1988). Instrumental methods of analysis, Wadsworth Publishing Co., Belmont, CA. pp. 465–507.

Wilson, K. and Walker, J. (2005). Principles and techniques of biochemistry and biochemical biology. 6th Edition. Cambridge University Press. pp. 551–620.

4

Deoxyribonucleic Acid and Extraction of DNA

4.1 WHAT IS DNA? ITS STRUCTURE AND PROPERTIES

DNA or deoxyribonucleic acid is the genetic material of all living organisms. Nucleic acids are essential for life just as proteins. They form the genetic material of all living beings like plants, animals, fishes, insects, and microbes and even of the simplest viruses. The term nucleic acid comes from the fact that DNA is found mainly in the nucleus of eukaryotes. Stains for DNA show nuclei very clearly under light microscope.

Discovery of the structure of DNA marked the turning point in biology because it revealed how living organisms store the information needed to control their activities and passed this information to next generations.

The DNA is a polymer of nucleotides linked together by phosphodiester bonds. DNA consists of four bases namely adenine, guanine, cytosine and thymine. The sugar found in the nucleotide is pentose sugar, deoxyribose. DNA molecules are usually double stranded. Adenine and guanine are purines while cytosine and thymine are pyrimidines. When purine or pyrimidine bases are joined to deoxyribose sugar, they are called nucleosides. When nucleoside is attached to phosphoric acid, it is termed as nucleotide. DNA is a polymer of four nucleotides A, C, G and T, which are joined by a backbone of alternating phosphate and deoxyribose sugar residues. These nitrogen-containing bases occur in complementary pairs by forming hydrogen bonds between them. Adenine always pairs with thymine by 2 hydrogen bonds and guanine always pairs with cytosine by three hydrogen bonds. The spans of A:T and G:C hydrogen bonded pairs are nearly identical, allowing them to bridge the sugar-phosphate chains uniformly.

Structure of DNA

The widely accepted three-dimensional structure of DNA is deduced from Watson and Crick model of DNA. In 1953, James D. Watson and Francis H. Crick proposed a three-dimensional structure of DNA. Their structure of DNA was based on X-ray crystallographic studies of DNA done by Rosalind Franklin in 1951 and observations of Erwin Chargaff. In 1948, Erwin Chargaff observed that in naturally occurring DNA, amount of T equals to amount of A and amount of C equals to amount of G. Watson and Crick who shared a Nobel Prize in 1962 for their discovery of DNA, observed that two strands of polynucleotides coil around each other and form a double helix. These two complementary strands of double helical DNA run in opposite direction as determined by the orientation of the 5'-3' phosphodiester bonds. The sugar-phosphate chains run along the outside of the helix and the bases lie on the inside, where they are linked to complementary bases on the other strand through hydrogen bonds.

In the double helix of DNA, nucleotides are stacked on top of each other. The distance between two base pairs is 0.34 nm. There are 10 base pairs in the single turn of DNA. Therefore each turn of DNA is 3.4 nm in length. Each base pair rotates about 36° along the vertical axis with respect to adjacent pair. The two strands of double helix twist around one another forming a wide major groove and narrow minor groove. The double helix is about 2 nm wide. The two DNA strands are antiparallel, which means that they run in opposite directions with respect to the orientation of sugars. One end of each strand has an exposed 5' hydroxyl group with phosphate attached while the other end has free 3' hydroxyl group. At one end, 5' end of one strand and 3' end of other is visible.

Properties of DNA

Physical and biochemical properties of DNA have been a subject of study for last many years. Any person handling or analyzing DNA should be aware of the following properties of DNA.

Denaturation of DNA

Denaturation of DNA refers to the separation of two strands of double helix. This involves breaking of hydrogen bonds between the bases of the double helix of DNA. In order to denature DNA, it is required to overcome the energy of hydrogen bonds between base pairs. In general, stacking energies are less in A:T rich regions since they have only two hydrogen bonds as compared to G:C, which have three. Therefore the sequence TATATA would be expected to melt quickly. Denaturation of DNA can occur due to following three factors.

1. Heat
2. Chemicals
3. Enzymes

Heat

When solution of DNA is heated, it reaches a temperature where DNA melts or two strands of DNA separate. The base pairs are separated as the H-bonds between them are broken and strands unwind, resulting in single stranded DNA. This separation of base pairs can be monitored spectrophotometrically by observing change or increase in absorbance at 260 nm. Unstacked bases, which are randomly oriented, absorb more light than neatly stacked base pairs. This results in melting curve.

The absorbance increases until all of the DNA is melted and then remains constant on further heating. This is called the hyperchromic effect. The absorbance of single stranded DNA (ssDNA) is usually around 50% greater than that of the corresponding duplex DNA. This process of DNA is reversible and if DNA is cooled slowly, complementary strands will pair up again to form double stranded DNA (dsDNA). This process of re-annealing of strands of DNA is called renaturation. The renaturation time of DNA can be used to estimate its base composition as well as the presence of repetitive fractions in the sequence.

Melting Temperature of DNA (Tm)

The melting temperature of DNA is the temperature at which 50% of DNA is melted or single stranded. Melting temperature (Tm) depends upon the length of the DNA molecule and its GC content. When the increase in absorbance at 260 nm is plotted on graph against increase in temperature, we get melting curve that is S-shaped (Refer Fig. 4.1). Melting temperature (Tm) is the midpoint of the melting curve.

Melting temperature is affected by many factors such as,

1. *Base composition*: Tm is greatly affected by the base composition of the DNA. Higher the G-C content, higher is the Tm.
2. *Ionic strength*: As the ionic strength increases, Tm increases. Double helical DNA is stabilized by cations such as divalent magnesium ions.
3. *Organic solvents*: Organic solvent formamide lowers melting temperature by weakening hydrophobic interactions. Formamide is also known to stabilize ssDNA.

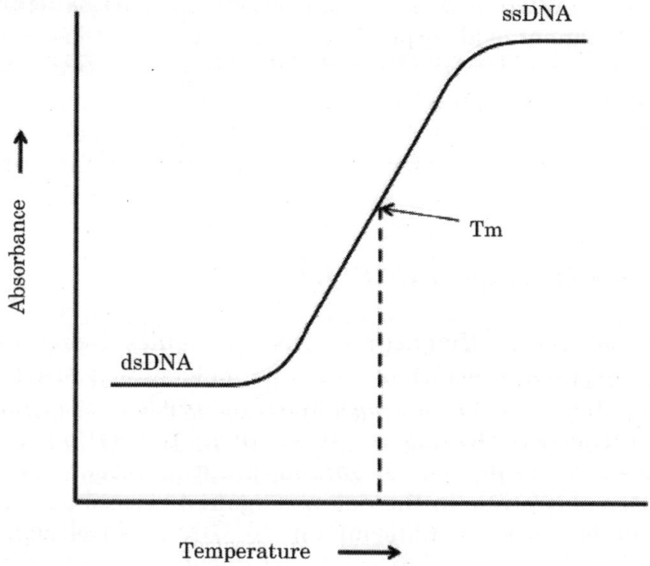

Fig. 4.1: Graph representing change in absorbance upon melting of double stranded DNA

Chemicals

The double helical structure of DNA is remarkably stable. This stability is derived from two chemical forces, hydrogen bonding and base stacking interactions. The two strands of DNA come apart readily on incubation at pH>12 or pH<2, due to ionization of the bases. Acid treatment of DNA leads to depyrimidation and depurination, and loss of bases by cleavage of the glycosidic bonds. The phosphodiester bond is more susceptible to hydrolysis. Thus strong acids will degrade DNA. Since single strands of DNA are relatively stable in alkali, alkaline denaturation is usually followed in DNA analysis process.

Enzymes

The two strands of double helical DNA are separated at the time of DNA replication and transcription by enzymatic action in the cell. At the time of replication, helicase enzyme helps unwind and separate DNA strands while at the time of transcription, RNA polymerase unwinds the DNA for initiation of RNA synthesis.

UV absorption

DNA readily absorbs UV light in double stranded conformation. The aromatic ring structures of bases in DNA are responsible for absorption of

UV light. However, when DNA is single stranded, the unstacking of bases leads to enhancement of absorption, a phenomenon called as hyperchromic effect. Therefore the extent to which DNA is single stranded or double stranded can be determined by monitoring UV absorption. DNA absorbs UV light, which can be said to be highest at 260 nm. The absorption profile of DNA can be used as a measure of the concentration and purity of DNA sample.

4.2 ENZYMES THAT DEGRADE DNA

DNases are the enzymes that degrade DNA molecules. Deoxyribonucleases can break phosphodiester linkages of DNA molecules to bring about their degradation. One of the major concerns of the nucleic acid isolation is the ubiquity of nucleases. DNases are also actively present on the surface of the skin. DNases are present in animal, plant and microbial cells alike. These DNases play a role in the defense against infectious agents, which have DNA as their genetic material, for *e.g.*, DNA viruses or bacteria.

DNases have been classified into two types:

1. DNase I enzymes
2. DNase II enzymes

DNase I enzymes have an optimum activity at approximately 7.0 pH and require magnesium ions (Mg^{+2}) for their activity. DNase I was isolated and crystallized from bovine pancreas by Kunitz in 1948. Recently this enzyme has been subjected to extensive investigation. According to biochemical and immunological evidences, actin is widely occurring inhibitor of DNase I. Deoxyribonuclease I is an endonuclease that catalyzes the hydrolytic cleavage of double stranded DNA in presence of divalent cations like Ca^{+2} and Mg^{+2}. Although DNases are part of every cell, some microorganisms, cells of mammalian organs like pancreas, stomach and some plant tissues, are known to produce and secrete DNase I extracellularly where they are thought to have a digestive function.

DNase II enzymes have optimum activity at 5.0 pH and can be active without presence of any divalent cations like Mg^{+2}. Deoxyribonuclease II that was discovered more than 50 years ago is a mammalian endonuclease. DNase II is acidic endonuclease, active in lysosome. Its lysozomal localization and ubiquitous tissue distribution suggests that this enzyme may play a role in the degradation of exogenous DNA encountered by phagocytosis. It acts as a barrier to the transfection of viral DNA or any other vector entering the cell. Investigation of DNase II activity inside cells have also suggested that DNase II is important for DNA fragmentation and degradation during apoptosis and cell death.

Various inhibitors of DNases are 2-mercaptoethanol, actin, EDTA, SDS etc. EDTA binds to ions such as magnesium and calcium ions that are required for activity of DNase I.

To avoid degradation of DNA by DNases while extracting DNA from cells, the samples must be lysed fast and completely, and lysis buffers must be able to inactivate DNases. Most of these lysis buffers or DNA extraction buffers contain protein denaturing and enzyme inhibiting components. DNases are said to be easier to inactivate as compared to RNases. Care should be taken to not to reintroduce DNases, once the DNA is purified. All glassware and plasticware should be autoclaved before use.

DNase contamination is the possible cause if you get a smear instead of DNA bands in gel. Failure to amplify DNA in PCR is also indicative of nuclease contamination.

DNase contamination can be detected by following procedure,

1. Incubate small aliquot of DNA sample at 37°C for a few hours or overnight.
2. After incubation, run the samples in agarose mini-gels. Failure to gel DNA band in the gel indicates DNase contamination.

4.3 PRESENCE OF DNA IN PROKARYOTIC AND EUKARYOTIC CELL

4.3.1 Linear Chromosomal DNA

Eukaryotic DNA is packaged into a set of linear chromosomes that are present inside the nucleus. Each chromosome consists of enormously long double helical DNA molecule that is wound on histone proteins. Winding of DNA on histone proteins forms extremely compact structure. This complex of DNA and proteins is called chromatin.

The entire human genome contains about 6 billion base pairs of DNA packaged into 23 pairs of chromosomes in each cell. Moreover, it is estimated that human body contains about 50 trillion cells. So, if that entire DNA was unwound and stretched linearly, it would measure approximately 100 trillion meters of DNA per human.

So how does this gigantic amount of DNA fit such a small subcellular space? Histone proteins help in compact arrangement of DNA into chromosomes to fit such a minute space of eukaryotic nucleus. When observed under electron microscope, chromatin appears like beads on string. The protein component of chromatin is of two types,

1. Histone proteins
2. Non-histone proteins

Non-histone proteins are those that are involved in the gene regulation. Histone proteins are H1, H2A, H2B, H3 and H4. Two of each H2A, H2B, H3 and H4 together form an octamer on which DNA is wound. DNA and histone complex is called nucleosome. Histone octamer wraps about 1.7 turns or 146 base pairs of DNA. The H1 histone protein wraps another 20 base pairs. Therefore DNA of about 166 base pairs is present in nucleosome. Each chromosome contains hundreds and thousands of nucleosomes. A strand of DNA that is about 20 base pairs long joins these nucleosomes. This DNA is called linker DNA. The chromatin is further coiled into shorter and thicker fibers of approximately 30 nm in diameter.

When chromosomes are stained with dyes, they appear to have alternatively light and dark coloured bands. The lightly stained regions contain transcriptionally active DNA and are called euchromatin. The darkly stained regions are usually repetitive sequences that are inactive. These regions are called heterochromatin.

4.3.2 Circular DNA of Bacteria

Prokaryotes such as bacteria essentially lack membrane bound cell organelles such as nucleus. Genetic material of bacteria is present in the form of circular DNA. Usually entire genome of bacteria is presented as a single circular DNA with some exceptions. In bacteria, DNA binding proteins help in forming loop and coiling of DNA for compaction.

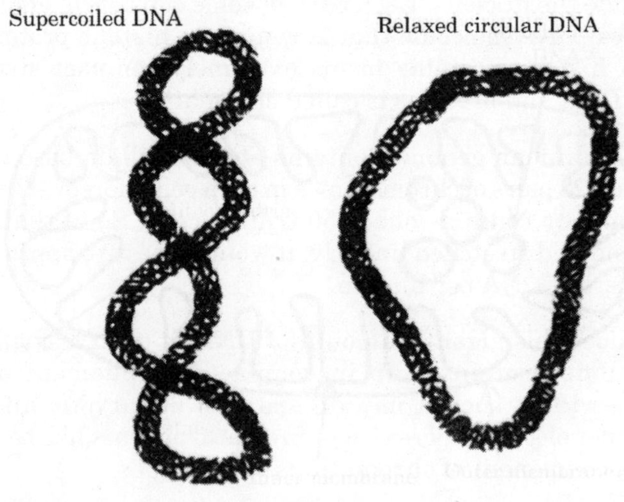

Fig. 4.1: The circular chromosomal DNA of bacteria. Supercoiling induces twists in the bacterial chromosome

The commonly known circular chromosome structure of bacteria was derived from the studies of *E. coli*. Prokaryotic chromosomes differ significantly from eukaryotic chromosomes on many aspects. Prokaryotic chromosome is closed circle, lacks histone proteins, nucleosome and chromatin structure.

The mechanism of DNA packaging in prokaryotes is supercoiling (Fig. 4.1). Underwound DNA forms negative supercoils while overwound DNA forms positive supercoils. The physical chemistry of DNA suggests that underwinding of the strands of double helical DNA favours reactions such as transcription, transposition, replication and recombination. Negative supercoils are put in DNA by ATP-dependent type-II topoisomerase enzyme called DNA gyrase. Negative supercoils are removed from DNA by type-I topoisomerase enzyme. A homeostatic balance between the countervailing activities of topoisomerase-I and DNA gyrase is thought to maintain negative supercoiling in bacterial cells.

4.3.3 Plasmid DNA

Plasmid DNA is the extrachromosomal genetic material that occurs in many bacterial strains. Plasmids are small circular DNA molecules that are capable of autonomous replication (Fig. 4.2). They are not essential for

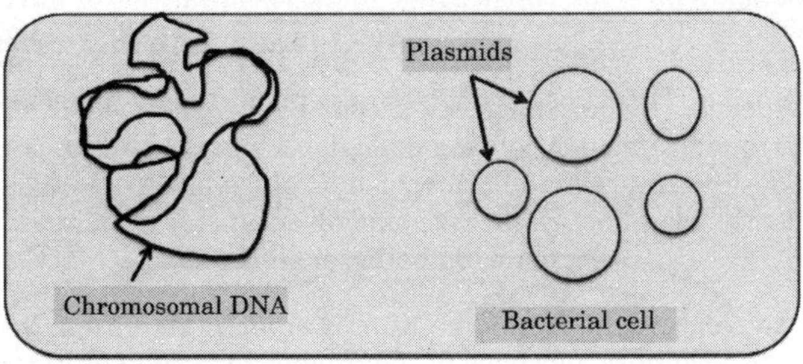

Fig. 4.2: Internal structure of bacterial cell. Plasmids exist as small circular DNA molecules inside bacterial cell

Table 4.1: Examples of plasmid vectors (Casali and Preston, 2003)

Host	***Plasmid***	***Size (kbp)***
Bacillus subtilis	pUB110	2.3
E. coli	ColE1	6.6
E. coli	pMB1	8.5
S. aureus	pSK41	46.4
Staphylococcus aureus	pT181	4.4
Pseudomonas sp.	pADP-1	108.8

bacteria but may confer a selective advantage for example, by imparting antibiotic resistance. Plasmid DNA can exist, as a single molecule in some bacteria while in others, there are numerous plasmids per bacterial cell. Plasmids are extensively used in genetic engineering experiments as vectors of gene transfer (Refer Table 4.1). Plasmids serve as an important tool in genetics and biotechnology experiments where they are commonly used to multiply and express genes of interest.

4.3.4 Mitochondrial DNA (mtDNA)

Mitochondria are membrane bound cell organelles present in most of the cells of eukaryotes. Mitochondria contain numerous enzymes that take part in the oxidative steps of Krebs cycle in the respiratory process. The high energy phosphate compounds such as adenosine diphosphate (ADP) and adenosine triphosphate (ATP) are synthesized in mitochondria. Mitochondria are popularly called as 'powerhouse of the cell' because these phosphate compounds liberate tremendous amount of energy upon breakdown.

Mitochondria have their own DNA that is different from nuclear DNA. Nuclear DNA is packed in the form of chromosomes. Mitochondrial DNA (mtDNA) is not part of the nuclear DNA. It consists of small circular molecule of DNA that exists inside mitochondria (Fig. 4.3). In mammalian cells, mitochondrial DNA constitutes about less than 1% of nuclear DNA.

Mitochondria produce energy by the process of oxidative phosphorylation. This process utilizes oxygen and carbohydrates to create

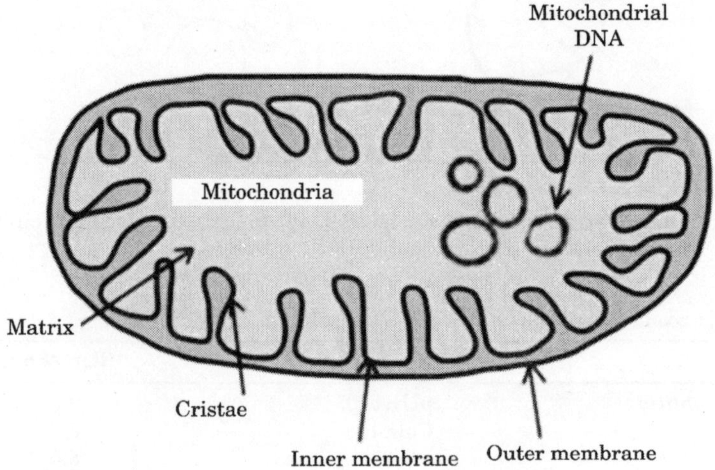

Fig. 4.3: Internal structure of mitochondria showing circular mitochondrial DNA (mtDNA)

ATP. Mitochondrial DNA is present in the matrix of mitochondria. Mitochondrial DNA is very similar to DNA of prokaryotes in many aspects. It is circular in shape and replicates independent of the nuclear DNA. Therefore it is hypothesized that mitochondria once existed as free-living bacteria, which were taken up by primitive ancestors of eukaryotic cells, by endosymbiosis. The primitive host cell with a nucleus must have provided ready source of energy rich nutrients to mitochondria and they in turn provided energy, using oxygen. This attribute must have become a key to survival as the primitive atmosphere changed from reducing to oxidizing.

Like prokaryotic genome, mitochondrial DNA has very few or no non-coding regions. Contrary to eukaryotic genomes, mitochondrial genomes lack introns and non-coding intergeneric regions.

In 1961, at the fifth Annual Meeting of the American Society of Cell Biology in Chicago, Hans Ris showed electron micrographs of mitochondria with inclusions resembling the DNA containing nucleoids of bacteria and made a proposal that mitochondria have their own DNA.

Mitochondrial DNA can be isolated from subcellular fractions prepared by cell disruption and differential centrifugation. Mitochondrial DNA can also be separated from nuclear DNA and from chloroplast DNA (when it is present) by density centrifugation in a gradient of cesium chloride. This separation is possible because base composition of mtDNA is different from the bulk of nuclear DNA.

The mitochondrial genome contains 37 genes, which are involved in production of energy and storage of ATP. Mitochondria have their own protein synthesizing machinery, own tRNAs, small 55S ribosomes, amino acids and enzymes. Mammalian mtDNA also codes for different amino acids than nuclear genes , for example, UGA codes for typtophan, AUA codes for methionine and AGA is stop codon. In mitochondrial genomes, 22 genes code for tRNAs, one for each 22 types of tRNAs. There are two rRNA coding genes that code for 16S rRNA and 12S rRNA, which form large and small subunit of mitochondrial ribosomes.

Mitochondrial genome sequences are becoming increasingly important for comprehensive evolutionary and population studies. For the past few decades, mtDNA has been widely regarded as molecular marker of choice for the phylogenetic analysis in metazoans due to its abundance in animal tissues, small genome size, faster rate of evolution and evolutionarily conserved gene products.

4.3.5 Chloroplast (DNA)

Chloroplast is a cell organelle responsible for photosynthesis in green plants, algae and some protists. The average size of chloroplasts of higher plants is

4-6 μ in diameter and thickness of 1-3 μ. The cells in most algae have only a single chloroplast while cells of higher plants may have 20-40 chloroplasts.

Cytological, electron microscopic, autoradiographic and biochemical studies have established the presence of chloroplast DNA or cpDNA inside chloroplasts. The cpDNA appears as a circular molecule in the stroma of chloroplast. Chloroplasts have their own ribosomes and RNA. The plastid ribosomes appear frequently as free 70 S particles. The chloroplasts seem to possess the essential attributes of prokaryotic cells. This can be said from the fact that they possess their own DNA, which replicates itself independent of the nuclear DNA. They have their own RNA, ribosomes, enzymes and protein synthesizing machinery. It is hypothesized that chloroplasts have evolved from prokaryotic blue green algae or cyanobacteria that developed a symbiotic relationship and have permanently settled in plant cells.

The genome of chloroplast is similar to that of mitochondria with respect to the fact that they consist of circular DNA molecules present in multiple copies per organelle. The chloroplast genome is however, larger and more complex than mitochondrial genome. Chloroplast genome contains about 150 different genes that aid in synthesis of rRNA, tRNA, various enzymes required for photosynthesis. Chloroplast genome contains genes that code for large subunit of enzyme RuBiSCO and components of photosystem II and I.

Over the past years, chloroplast genomes of several plants have been completely sequenced, leading to identification of many genes.

4.3.6 DNA of Viruses

DNA viruses are those types of viruses that have DNA as their genetic material. Unlike DNA of all other organisms, viral DNA can be single stranded or double stranded. This single stranded DNA of viruses, usually becomes double stranded, once it enters inside host cell. The double stranded DNA or dsDNA of viruses can be circular or linear.

Replication or multiplication of viruses always occurs inside the host cells. DNA viruses replicate with the help of DNA polymerase. Examples of DNA viruses are adenovirus, herpes virus, baculovirus, poxvirus etc. It is a well known fact that all viruses lack any machinery to replicate their own genetic material. The viral DNA once inside the host cell, can code for DNA polymerase that aids in replicating its own DNA. Examples of ssDNA viruses are circovirus and parvovirus, while dsDNA viruses are poxvirus, papilloma virus, herpes virus and adenovirus etc.

Viruses are a matter of great interest of study since they cause a number of human, animal and plant diseases. Today recombinant DNA techniques use biological vectors like plasmids and viruses to carry foreign genes into cells. Viruses can carry foreign genes or genes of interest into recipient host cell in the process of infecting that cell. Various methods for the engineering of site-specific insertions or deletions based on the use of selectable markers have been developed. These methods are applied for the elucidation of the function of specific DNA sequences, identification of genes non-essential for virus growth in cell culture and expression of foreign genes. With these methods it is possible to create viral vectors capable of delivering genes specifying antigens for the prevention of infectious diseases in humans and animals.

4.4 COMPOSITION OF CELL WALL AND CELL MEMBRANES

4.4.1 Composition of Bacterial Cell Wall

The cell envelop of bacteria contains three major components- cell membrane, cell wall and capsule. Some bacterial cells are surrounded by a viscous substance, which covers the entire cell. This layer is called capsule. This outermost layer or capsule is not present in all bacteria. Most bacterial capsules are composed of polysaccharides. Based on their composition, they can be classified as homopolysaccharides or heteropolysaccharides. A few types of capsules can also be made of polypeptides. In *Bacillus anthracis*, capsule is composed of a polymer of glutamic acid.

The cell wall of a bacterium is composed mainly of peptidoglycan, which is a protein-sugar molecule. Peptidoglycan is an insoluble, cross-linking polymer that gives rigidity and shape to the bacterial cell. Peptidoglycan is found only in prokaryotes and is absent in higher organisms. The composition of cell wall in bacteria differs from species to species. Gram-positive bacteria have greater amount of peptidoglycan than gram-negative bacteria. In addition to peptidoglycan, polysaccharides also occur in the cell walls of bacteria.

Peptidoglycan is a vast polymer of interlocking chains of peptidoglycan monomers. Peptidoglycan monomer consists of two amino sugars, namely N-acetylglucosamine (NAG) and N-acetylmuramic acid (NAM). The peptidoglycan monomers are linked by glycosidic bonds, to create a growing chain of peptidoglycan. These chains are further joined to one another by peptide cross-links between the peptides coming off of NAM. By this linking and cross-linking, the peptidoglycan provides tremendous strength to the bacterium.

The walls of gram-negative bacteria like *E. coli* are more complex than that of gram-positive bacteria (Fig. 4.4). In gram-negative bacteria, an outer membrane is present that surrounds the underlaying thin peptidoglycan layer. This membrane is rich in lipids and lipopolysaccharides. This

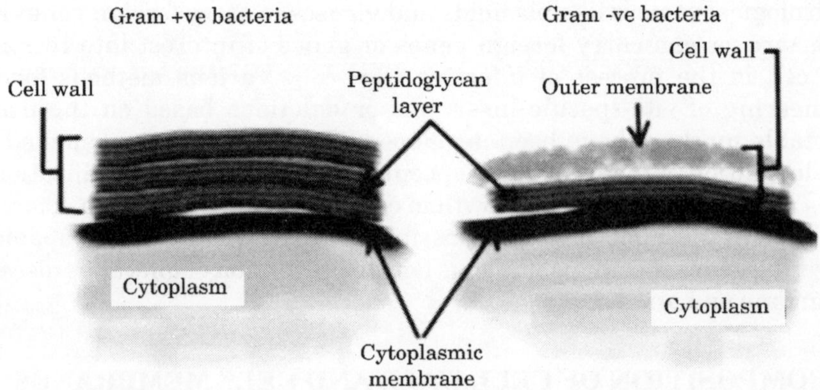

Fig. 4.4: Comparison between cell wall components of gram positive and gram negative bateria

membrane is bilayer, consisting of phospholipids, proteins and lipopolysaccharides. This outer membrane of gram-negative bacteria protects from harmful chemicals and enzymes that may damage the bacterial cell wall. Gram-positive bacteria lack this outer membrane. Hence, their cell walls are easily damaged by treatment with Lysozyme.

The cytoplasmic membrane of bacteria, which is the innermost layer, is mainly composed of phospholipids and proteins. The phospholipids form a bilayer in which integral proteins are embedded. This layer can be destroyed by treatment with detergents.

4.4.2 Composition of Animal Cell Membrane

All animal cells lack cell wall. A layer of cell membrane encloses the cytoplasm of animal cell. Animal cell membrane, also called as plasma membrane, has primarily two functions. It provides a boundary and holds all cytoplasmic components inside. Secondly, being semipermeable, it allows nutrients and other essential elements to enter the cell. It selectively allows entry and exit of various molecules. Most commonly, cell membranes are bilayers of phospholipids with embedded proteins, and carbohydrates attached to the periphery.

This plasma membrane is however flexible and not rigid, as a result of which animal cells do not have a rigid shape. Most animal cells are somewhat spherical, but various shapes can be seen in various tissues like nerve cells, muscle cells etc. Lipids constitute about 50% of the most cell membranes, however this percentage depends on the cell type.

It is hypothesized that eukaryotic animal cells have probably descended from prokaryotic cells that lost their cell walls. With only flexible membrane left as outer covering, they were able to expand in size and complexity. This might be the reason that eukaryotic cells are greatly larger in size than prokaryotic cells and more complex.

Analysis of mouse fibroblast (L cells) showed that neutral lipids make up 40% of the total membrane lipids, with cholesterol and triglycerides accounting for 20% and 13% respectively. The plasma membrane contained 4.7% protein and 13.8% of the lipid of the whole cell (Weinstein *et al.*, 1969).

Jonathan Singer and Garth Nicolson proposed the model of plasma membrane structure in 1972. This model is popularly known as fluid mosaic model that shows a lipid bilayer with embedded integral proteins. Most integral membrane proteins are transmembrane proteins, which span the lipid bilayer with their portions exposed on both sides of the membrane.

4.4.3 Composition of Plant Cell Wall

Plant cell wall is a unique structure, which is different from all other prokaryotic and eukaryotic cell types. Plant cell wall is greatly tougher and more rigid than bacterial cell walls. Thick cell wall of plant cells that provides immense mechanical strength, has helped evolve the gigantic plant species. The plant cell wall is located outside of cell membrane, provides shape and structural support in addition to providing protection.

Plant cell wall consists of three layers,

1. Middle lamella
2. Primary cell wall
3. Secondary cell wall

Middle lamella is the first layer formed during cell division. It makes up outermost covering of cell wall and is shared by the adjacent cells. Middle lamella is composed of pectins and proteins. Primary cell wall is formed after the middle lamella. It consists of rigid skeleton of cellulose-microfibrils embedded in matrix of pectic compounds, hemicellulose and glycoproteins. Secondary cell wall is formed after cell enlargement is complete. Secondary cell wall is hard and rigid and provides strength to the cell. It is composed of cellulose, hemicellulose and lignin. Secondary cell wall formation takes place inside the primary cell wall.

The cell walls are important features of plant cells that perform a number of essential functions including providing shape to form tissues and organs of plants. Forming the interface between adjacent plant cells, plant cell walls play an important role in intercellular communication. The components of plant cell walls are cellulose, hemicellulose, pectins, lignins, proteins, wax, cutin etc.

4.5 ENZYMES THAT DEGRADE CELL WALL

4.5.1 Cellulase

Cellulases are enzymes that are produced by fungi and bacteria. Cellulases bring about degradation of polysaccharide cellulose. Cellulase catalyses hydrolytic cleavage of β-1,4-glycosidic bonds that are present in cellulose. Cellulase enzyme is produced by many saprophytic and rot causing fungi and bacteria, which use cellulose of plant origin as an energy source. However, plants are not known to produce cellulase. Enzymes that degrade hemicellulose are called hemicellulase and are usually classified under cellulases in general.

Plant cell wall is essentially made of cellulose and hemicellulose. Therefore many pathogenic fungi produce cellulases that destroy the important barrier of cell wall that protects the cytoplasmic contents.

4.5.2 Pectinase

Pectinase is an enzyme that degrades pectic compounds. Pectinases such as pectolyases, pectozyme and polygalacturonase, are produced by various fungi and bacteria. Pectinase enzyme is however also produced in plants. Pectinase helps in softening of fruits at the time of ripening. At the time of ripening, Pectinase dissolves pectins, which is a major component of middle lamella, a glue that holds adjacent cells together. As a result, the fruit becomes softer when it is ripe. Due to enzymatic activity of pectinase, it can definitely weaken plant cell wall. Therefore many saprophytic bacteria and fungi like *Aspergillus*, are known to produce pectinase.

4.5.3 Lysozyme

Lysozyme is an enzyme that digests and breaks down cell wall of bacteria. Lysozyme is an antimicrobial enzyme that can be found in a wide variety of organisms including birds, mammals, plants, insects etc. Laschtschenko extracted lysozyme from chicken egg for the first time in 1909. Lysozyme is also found in body fluids like saliva, mucus and tears of animals.

Lysozyme is a glycosidase. Lysozyme cleaves β-glycosidic bonds between N-acetylglucosamine and N-acetylmuramic acid that are the components of peptidoglycan. As peptidoglycan is an important component of bacterial cell wall, lysozyme has lethal effect on most of the bacteria. Therefore lysozyme is very effective against gram-positive bacteria and can readily destroy their cell walls. In case of gram-negative bacteria, if their outer lipid membrane is damaged, then the lysozyme can reach underlaying thin peptidoglycan layer and destroy their cell wall.

4.6 METHODS FOR LYSING CELL MEMBRANES AND DEGRADING CELL WALLS

4.6.1 Enzymatic Method

Enzymes that attack cell walls and cell membrane components are often added to lysis buffer. Lysozyme attacks and degrades peptidoglycan, which is an important component of bacterial cell wall. Many methods based on lysozyme extraction for bacterial DNA isolation have been reported. Enzymes zymolase and murienase are used to dissolve cell wall of yeast. Proteinase K cleaves glycoproteins and hence is used in lysis buffers.

The use of lytic enzymes provides biological specificity to the process of cell lysis and product release. Lytic enzyme systems are usually specific either for yeast or for different types of bacteria.

Disadvantages of enzymatic method of cell lysis:

1. Cell lysis by enzymatic method may pose problems of enzyme stability, as all enzymes are active within a narrow range of temperature and pH.
2. Enzymatic method of cell lysis is not applicable to large-scale preparations. Large-scale preparations by enzymatic method tend to be very costly and non-reproducible.
3. After the cell wall lysis, enzymes have to be removed or inactivated to permit the isolation of the desired material.

4.6.2 Sonication

Sonication method of cell lysis applies ultrasound waves to sonicate the biological tissue samples. These high frequency waves are generated electronically and mechanical energy is transmitted to the sample via a metal probe that oscillates with high frequency. This probe is placed into the sample and high frequency sound waves are generated. These waves cause localized high pressure region, resulting in breaking open of cells. The ultrasound waves produce gaseous cavitations in the liquid. The term 'cavitation' describes formation of small bubbles of dissolved gas or vapours due to alteration in pressure. These bubbles are capable of undergoing resonance vibration and produce vigorous eddying or microstreaming, which is sufficient to break the cells. Sonication is a popular method for small-scale cell disruption. This method is useful for microbial cells and suspended animal cells.

Disadvantages of sonication method:

1. Heat is generated by ultrasound waves, which may become problematic in processing of samples. Therefore effective cooling is required if sonication is to be followed.
2. Sonication method gives variable yields; therefore this method is not reproducible.
3. Sonication may damage some intracellular enzymes, which may result in production of free radicals, oxygen, or hydrogen peroxide. These free radicals and oxidizing agents pose a potential threat to the intracellular components that may get damaged due to oxidation.
4. Due to sonication, there is a possibility that fragmentation of DNA may take place as a result of mechanical stress caused by the bubbles.

4.6.3 Grinding Tissue in Liquid Nitrogen

Grinding in liquid nitrogen for cell and tissue disruption is an effective method for preparation of plant tissues. Nitrogen gas is liquid at very low temperature. The temperature of liquid nitrogen is –196°C. Liquid nitrogen is colourless, odourless and nontoxic liquid. Liquid nitrogen is a cryogenic liquid which is used to snap-freeze living tissues. Cryogenic grinding is a very effective technique for homogenization of hard tissues like plant stems, roots, animal bones, organs etc. By grinding in liquid nitrogen, these tissues can be ground to a fine powder. The tough polysaccharides of plant tissues become very fragile and brittle at –196°C and easily shatter. The tissues are ground in a pre-chilled pestle and mortar. Liquid nitrogen immediately forms vapours when it comes in contact with a warmer object. Therefore grinding in liquid nitrogen is a rapid process.

However, when liquid nitrogen comes in contact with skin, it can cause severe cold burns. Therefore following precautions should be taken while handling liquid nitrogen.

1. Always wear gloves while handling liquid nitrogen.
2. Make sure the cryocan or the can in which liquid nitrogen is contained, is tightly sealed.
3. Cryogenic cans are equipped with a pressure relief device to control the internal pressure. Under normal conditions, these containers will periodically vent the gas. Do not plug, remove or tamper the pressure relief device.
4. Do not touch any uncovered part of your body directly to any vial containing poured liquid nitrogen. This vial is probably at several degrees below freezing temperature (0°C).
5. If you accidentally touch the vial of liquid nitrogen with bare hands and if hands get stuck due to frozen tissue, simply pour normal tap

water or lukewarm water to thaw. Do not pull your hand away from vial when it is stuck as it may tear the skin and cause more damage.

6. Never put your hand inside the cryocan to draw out liquid nitrogen or anything that has fallen accidentally inside the can. Always use long handled vessels (that usually come along with the cryocan) to draw more liquid nitrogen.
7. Nitrogen vapours should never be inhaled. Nitrogen is a non-toxic gas; however large amounts of nitrogen vapours may displace oxygen around you and cause suffocation. Nitrogen is considered as asphyxiant.
8. Never open cryocan of liquid nitrogen in a closely confined area. Always make sure that the area is ventilated before using liquid nitrogen.

4.7 PROTOCOLS FOR EXTRACTION AND PURIFICATION OF DNA

4.7.1 DNA Extraction Protocol- I: Extraction of DNA from Animal Tissue

4.7.1.1 *Introduction*

The fundamental principle of DNA isolation is simply, lysing of the cells and separating DNA from all other contaminants. The relative amounts of these contaminants vary from species to species and in different tissues of the same individual. For example, components of fat cells differ significantly from muscle or nerve cells.

DNA used for genetic diagnosis and forensic studies is derived from various sources like fingerprints, blood samples, hair roots, buccal swabs etc. Traditionally PCR based analysis techniques have utilized white blood cells fractionated from whole blood to isolate DNA. However, this method has inconvenience of drawing the blood, risk of exposure to pathogens, storage of blood samples etc.

Different methods of DNA extraction are used depending upon the source and size of the sample. Although various methods are used, the basic principle behind those methods is the same. All these methods aim at lysis of cells by disruptions of cell membranes and separation of DNA from proteins, RNA and other contaminants. Isolation of animal DNA is needed for genetic analysis, which is used for scientific, medical or forensic purposes. In scientific expeditions, DNA isolation and purification is a crucial step for genetic, molecular and evolutionary studies. The best example where human DNA isolation and purification was frequently required can be of Human Genome Project. Also in studies and production of cloned animals or transgenic animals, DNA extraction is an important stepping stone. In medicine, DNA extraction is often used for diagnosis of genetic diseases and disorders. On the other hand, forensic science needs to recover DNA for identification of individuals, production of DNA fingerprints, paternity determination etc.

Presence of proteins, lipids, polysaccharides and some other contaminants in DNA preparations can interfere with DNA analysis and downstream processing, especially in the techniques involving PCR.

4.7.1.2 *Principle and working*

Isolation of DNA begins with breakdown of tissues. While isolating DNA from hard tissues or bones, it is important to macerate and break down that tissue preferably in liquid nitrogen. DNA can be isolated from fresh or frozen samples. Stored samples can come from frozen blood or tissue samples, bones or even ancient human or animal remains.

Cells are often lysed with lysis buffer containing detergents. Detergents are used to solubilize cell membranes. Chemicals such as SDS and CTAB are commonly used for this purpose. Denaturants such as guanidinium, urea etc. are applied to lysis buffer as they lyse cell membranes and inactivate cellular enzymes. PEG (polyethylene glycol) selectively precipitates high molecular weight DNA but is also more difficult to dry and can interfere with downstream processing (Hillen *et al.*, 1981). Bones of animals are rich in calcium and minerals. Therefore mineral ions have to be removed from the samples so that they will not interfere with later downstream processes.

The lysed samples are then mixed with phenol-chloroform-isoamyl alcohol, for separation of DNA and proteins. These organic solvents denature proteins. When the sample is centrifuged, DNA remains in the upper aqueous layer and phenol forms the bottom layer, while proteins, lipids and cell debris form the interphase.

The major difficulties encountered during animal DNA preparations are high viscocity of genomic DNA, which makes it difficult to redissolve after alcohol precipitation and its high molecular weight makes it sensitive to shear forces (Birnboim, 1992).

4.7.1.3 *Protocol for animal DNA extraction and purification*

Materials required

- Animal tissue samples
- Extraction buffer

Tris-HCl (pH=8.0)	10 mM
EDTA	100 mM
RNase A	20 μg/ml
SDS	0.5 %

Store at 4°C

- Phenol: chloroform: isoamyl alcohol (25:24:1) (v/v)
- Proteinase K 20 mg/ml
- TE buffer (pH=8.0)
  Tris-HCl 10 mM
  EDTA 1 mM
- Liquid nitrogen
- NaCl 5 M

Procedure

1. Take pre-chilled pestle and mortar. Take a piece of animal tissue of about 200 mg and immediately grind it in liquid nitrogen to a fine powder. Add more liquid nitrogen if necessary.
2. Take this tissue powder in a sterile beaker. Add extraction buffer slowly. Add about 3 volumes of extraction buffer mix the contents thoroughly.
3. Transfer the contents to polypropylene centrifuge tubes and incubate for 1 hour at 37°C in waterbath.
4. Now add proteinase K (100 µg/ml) and incubate at 50°C for three hours in waterbath. Mix the contents thoroughly.
5. Cool the solution to room temperature. Add equal volumes of Phenol: chloroform: isoamyl alcohol (25:24:1) (v/v). Gently mix the contents by rotating the tube slowly.
6. Centrifuge the contents at 10000 g for 15 minutes.
7. Carefully transfer the viscous aqueous phase to a clean centrifuge tube.
8. Repeat Phenol: Chloroform: Isoamyl alcohol extraction twice or until the interphase is not visible.
9. Add 1/12th volume of 5M NaCl, mix slowly by turning the tube. Add 2.5 volumes of ice cold 100% ethanol. Mix the contents gently.
10. Incubate the contents for 30 minutes at 4°C
11. Centrifuge at 10000g for 10 minutes at 4°C
12. Decant the supernatant. Add 70% ethanol and mix the contents.
13. Repeat step11
14. Decant the supernatant. Air dry the DNA pellet, by storing the pellet in open tube until all traces of ethanol have evaporated
15. Dissolve the DNA pellet in TE buffer (pH=8.0). Store DNA aliquots at -20°C.

4.7.1.4 *Troubleshooting guide*

1. All the chemicals and materials required for DNA extraction should be prepared in advance to avoid wasting of time and errors.

2. Gloves and safety goggles should be worn to avoid skin/eye irritation while handling chemicals.
3. While extracting DNA containing aqueous layer, do not immerse micropipette too deep, to avoid sucking of non-aqueous particles.
4. For high yield of DNA with minimal contamination, it is advisable to use fresh animal tissue samples.
5. Shearing of DNA: Large DNA molecules like genomic DNA may get sheared during DNA purification process. Vortexing, repeated pipetting or some other mechanical stress might cause this shearing of DNA. When DNA sample is run in electrophoresis, sheared DNA gives multiple bands, while whole genomic DNA gives a single band in a single lane of agarose gel.
6. Chemical contaminants: Many times properly isolated DNA samples don't give the desired results, due to numerous chemical contaminants from the DNA isolation and purification process. The reagents of DNA purification process usually contribute towards the chemical contaminants in the isolated DNA samples.
7. A mixture of chloroform and phenol is often used to precipitate DNA during DNA isolation process. Residual chloroform can be problematic and may interfere with the downstream application of DNA.
8. Presence of trace amounts of phenol in purified DNA solution can become problematic as oxidation products of phenol can damage nucleic acids.
9. Storage of DNA: The quality of purified DNA in a solution can be hampered by factors such as DNases, extreme pH such as below 6.0 and above 9.0, heavy metals, oxidation by free radicals, high temperature and UV light.
10. EDTA protects nucleic acids against nucleases and heavy metals, therefore it is usually added in the form of TE buffer. Storage of DNA dissolved in TE buffer helps maintain the pH and will not generate free radicals.
11. Low temperature is very important for long-term storage of DNA. Storage of DNA aliquots at 4°C is only recommended for short-term storage such as a few days. Medium duration storage of a few months can be done at –20°C while for long-term storage, –80°C temperature is recommended.

4.7.2 DNA Extraction Protocol- II: Extraction and Purification of DNA From Plant Tissues

4.7.2.1 *Introduction*

With the increasing use of recombinant DNA techniques in plant research, the preparation of long, pure DNA has become a major concern. Extraction of DNA is the starting point for genotype analysis. Since the same forces that are required to break cell walls can also shear DNA, considerable care must be taken and one must often reach a compromise between DNA yields and length. Polysaccharides and tannins present in plant cells pose a major problem because they are difficult to separate from DNA. The most commonly followed method for the isolation of total genomic DNA is CTAB method (Doyle and Doyle, 1987).

Applications of plant DNA extraction

1. DNA from plants can be used for further molecular analysis in various techniques such as southern blot, PCR, molecular marker studies etc.
2. Analysis of ancient DNA from fossil records which may be valuable in studies of evolution.
3. Diagnosis of plant diseases often requires preparation of plant samples for the analysis of DNA from pathogens.
4. Analysis of pollen DNA can be of great value in population genetics and gene mapping.
5. Genotyping of plant may help in establishing the accuracy of labeling in food processing.
6. DNA extraction and analysis is commonly done to confirm the incorporation of gene of interest in genetic transfer studies.
7. DNA extraction is pre-requisite in case of gene mapping and gene tagging which is used in identification and tagging of various genes of agronomic importance like, genes for disease resistance, disease tolerance, salinity tolerant, insect resistance etc.
8. DNA extraction and purification is a basic requirement in genetic studies such as genetic linkage mapping, marker-assisted selection, population and conservation genetic studies.

4.7.2.2 *Principle and working*

Polysaccharides have viscous, glue like texture, which makes the DNA unmanageable for pipetting. These polysaccharides also make DNA unsuitable for PCR amplification because they interfere with Taq polymerase activity.

CTAB is a cationic surfactant that disrupts the cell walls and nuclear membrane to expose chromosomes. CTAB precipitates DNA and removes polysaccharides from plant preparations.

Chloroform-isoamyl alcohol is used to remove proteins, lipids and cellular debris. It binds to non-aqueous compounds and separates them from aqueous compounds. Chloroform- isoamyl alcohol treatment followed by centrifugation precipitates proteins, lipids and cellular debris while DNA remains in the aqueous supernatant.

DNA extraction from plant tissues can vary depending upon the source material. However, mechanical shearing of plant cell wall is a necessary step for all types of plant tissues. For this, most commonly followed method is grinding of plant tissue by freezing it in liquid nitrogen (–196°C). This breaks down the cell wall and allows access to DNA while harmful cellular enzymes such as DNases remain inactivated. DNases are the enzymes, which can degrade DNA molecules, but these enzymes are inactivated at extremely low temperatures. Liquid nitrogen is used to flash-freeze cells and break them open by subsequent grinding. Grinding tissue in liquid nitrogen produces fine powder while at the same time, keeps harmful chemicals and cellular enzymes deactivated (Fu *et al.*, 1998).

Once the tissue is sufficiently ground, it can be suspended in a suitable buffer such as CTAB. The insoluble impurities are removed by centrifugation. Mixing with chloroform, followed by centrifugation, separates proteins. Treatment with phenol-chloroform-isoamyl alcohol causes phase separation and DNA is suspended in the upper aqueous phase. DNA is further precipitated by ethanol treatment. The extracted DNA can be stored in TE buffer or sterile water (Fig 4.5).

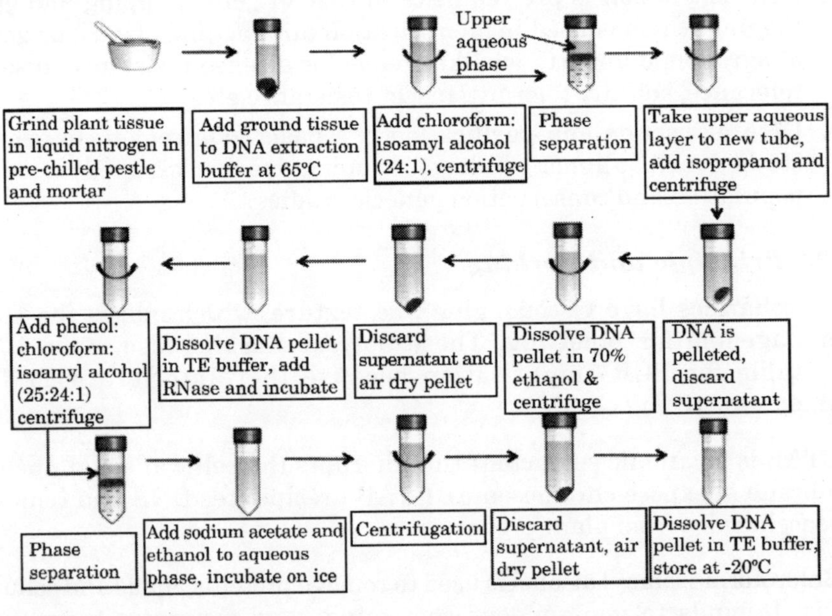

Fig. 4.5: Extraction of DNA from plant tissues

4.7.2.3 *Protocol for Plant DNA Extraction and Purification*

Materials required

- *DNA extraction buffer*

  100 mM Tris-Cl (pH=8.0)

  20 mM EDTA (pH=8.0)

  1.4 M NaCl

  2.0 % (w/v) CTAB

  0.2% (v/v) β-mercaptoethanol
- *Chloroform: isoamyl alcohol 24:1 (v / v)*

  Mix 96 ml of chloroform and 4 ml isoamyl alcohol. Keep at room temperature in a close container.
- *10% (w / v) CTAB (cetyl trimethyl ammonium bromide)*

  Add 10 g of CTAB to approximately 70 ml of water. Dissolve the detergent by warming the solution to 65°C. Adjust volume to 100 ml.
- *0.5 M EDTA (pH=8.0)*

  Add 18.61 g of ethylenediaminetetraacetate.$2H_2O$ to 80 ml of water. Add approximately 2 g of NaOH pellets to adjust the pH to 8.0. (The sodium salt of EDTA will not go into solution until the pH of solution is adjusted to approximately 8.0 by the addition of NaOH). Adjust the volume to 100 ml.
- *5M NaCl*

  Dissolve 29.22 g NaCl in 20 ml water. Adjust the volume to 100 ml.
- *Phenol: chloroform: isoamyl alcohol (25:24:1)*

  Mix 750 µl of buffered phenol (pH=7.8), 720 µl of chloroform and 30 µl of isoamyl alcohol. Mix by vortexing and keep at 4°C in coloured container.
- *1 M Tris*

  Dissolve 12.11 g Tris base in approximately 60 ml water. Adjust the pH to 8.0 by adding HCl. Adjust the volume to 100 ml.
- *3M sodium acetate*

  Dissolve 40.8 g sodium acetate.$2H_2O$ in 60 ml water. Adjust the pH to 5.2 with glacial acetic acid. Adjust the volume to 100 ml with water.

Procedure

1. Grind the plant material in liquid nitrogen to fine powder in a pre-chilled pestle and mortar.
2. Transfer the powder to a 1.5 ml microcentrifuge tube containing 750 µl of prewarmed (65°C) DNA extraction buffer. Mix by gentle inversion.

3. Incubate at 65°C for one hour.
4. Allow the above solution to cool at room temperature for 20 minutes.
5. Add 750 µl chloroform: isoamyl alcohol (24:1, v/v) and mix by inversion to emulsify
6. Centrifuge at 14,000 rpm for 20 minutes at room temperature
7. Pipette out gently the aqueous phase without disturbing the interphase to another tube. Add 2/3rd volume of isopropanol and mix by gentle inversion.
8. Spin down the DNA pellet or take out DNA using bent Pasteur pipette. [DNA is seen as a white thread suspended in the solution]
9. Wash DNA in 70% ethanol and spin at 10,000 rpm for 5-10 minutes at 4°C.
10. Dry the pellet and dissolve in 50-60 µl of sterile water or TE buffer (pH=8.0) depending upon the yield of DNA
11. Add 5 µl of RNase to the dissolved DNA, mix gently and incubate at 37°C for 1 hr
12. Add equal volume of phenol: chloroform: isoamyl alcohol (25:24:1, v/v/v) and mix thoroughly to emulsify
13. Spin at 12,000 rpm for 10 minutes and carefully take out the upper aqueous phase using micropipette without disturbing the interphase
14. Repeat the above step twice
15. Add 1/10 volume of 3M sodium acetate, pH=5.2, and mix thoroughly. Add two volumes of absolute ethanol (95%). Mix by gentle inversion and keep on ice for 20 minutes
16. Spin at 10,000 rpm for 10 minutes to pellet the DNA
17. Wash the DNA pellet with 70% ethanol
18. Spin for 2-3 times
19. Air dry the DNA pellet and dissolve in sterile water or TE buffer depending upon the yield of DNA.
20. Store these aliquots of DNA at –20°C for further use.

4.7.2.4 *Troubleshooting Guide*

1. If there is very low yield of DNA, it could be due to incomplete lysis of samples. To solve this problem, increase the quantity of lysis solution.
2. If you get RNA contamination in the samples, repeat the procedure from 11th step onwards.
3. Note that, DNA purification involving RNase significantly lowers DNA yield.
4. Storage of DNA: The quality of purified DNA in a solution can be hampered by factors such as DNases, extreme pH such as below 6.0

and above 9.0, heavy metals, oxidation by free radicals, high temperature and UV light.

5. EDTA protects nucleic acids against nucleases and heavy metals, therefore it is usually added in the form of TE buffer. Storage of DNA dissolved in TE buffer helps maintain the pH and will not generate free radicals.
6. Low temperature is very important for long-term storage of DNA. Storage of DNA aliquots at 4°C is only recommended for short-term storage such as a few days. Medium duration storage of a few months can be done at –20°C while for long-term storage temperature of –80°C is recommended.

4.7.3 DNA Extraction Protocol- III: Extraction of DNA from Bacteria

4.7.3.1 *Introduction*

Genomic DNA from microorganisms like bacteria is necessary for cloning genes, selecting recombinant constructs, and for taxonomy and diagnostics purposes.

4.7.3.2 *Principle and working*

Various methods for extraction and purification of DNA from bacteria have been developed and studied for over a few decades. These methods include CTAB-SDS-proteinase-K, or Lysozyme-SDS-proteinase-K, or Lysozyme-SDS, or mechanical lysis and sonication.

Although these methods are suitable for DNA extraction from bacteria, they have a few drawbacks including laborious manipulations such as changes of microcentrifuge tubes, incubation, precipitation, elution and drying steps. These methods use SDS, which is a detergent to lyse the cells, and this often poses problems in downstream processing.

In prokaryotes, chromosomal DNA is usually associated with fewer proteins and is usually attached to plasma membrane. It is important to remember that large DNA is sensitive to shearing by mechanical stress that may cause random breaks in DNA strands.

Before lysis of bacterial cells, it is important to add buffer containing EDTA. EDTA acts as a chelating agent and renders Mg^{+2} ions unavailable for biological activity. This inactivates endogenous DNases, which require Mg^{+2} ions for their activity.

There are various methods for extracting DNA from bacteria. All these methods essentially have following steps,

1. Growth of bacterial culture in liquid medium
2. Harvesting of bacterial cells and cell lysis
3. Precipitation of proteins and cell debris followed by removal of cell debris
4. Precipitation of DNA
5. Purification of DNA by RNase treatment

The extracted DNA is likely to be contaminated with cellular RNA. This RNA contamination is removed by addition of RNases to the solution. RNase selectively cleaves RNA molecules that can be separated from DNA by further centrifugation.

4.7.3.3 *Protocol for Bacterial DNA Extraction and Purification*

Growth of bacterial culture:

For extraction of genomic DNA from bacteria, it is important to grow a culture of the bacterium of interest. Freshly grown bacterial cultures give better results in terms of DNA extraction as compared to old cultures. For growing liquid bacterial cultures, take about 100 ml of Luria broth, previously autoclaved. Inoculate the bacterial culture at specified temperature in incubator-shaker. Allow the bacterial cultures to reach log phase.

The protocol given below is adopted from Cheng and Jiang, 2006. In this method, phenol is used for lysing bacterial cells. As phenol is a strong oxidizing agent, it can readily disrupt cell wall. Then the chloroform extraction is used to remove traces of phenol from purified DNA. This method provides DNA samples that are suitable for molecular analysis for PCR, restriction enzyme digestion and nucleic acid hybridization. This method is cost effective and faster than other methods of DNA extraction. This method is also applicable to DNA extraction from yeast and a variety of gram-positive and gram-negative bacteria.

Materials required

- STE buffer (pH=8.0)

NaCl	100 mM
Tris-HCl	10 mM
EDTA	1 mM

- TE buffer (pH=8.0)

Tris-HCl	10 mM
EDTA	1 mM

- Tris- saturated phenol pH=8.0
- Chloroform
- RNase enzyme (10mg/ml)

Procedure

1. To extract DNA, 1 ml of cell suspension is centrifuged at 8000 g for two minutes.
2. Remove the supernatant containing liquid growth medium. Wash the cell pellet in 400 µl of STE buffer
3. Wash the cell pellet again in 400 µl of STE buffer.
4. Centrifuge the cells at 8000 g for 2 minutes.
5. Throw the supernatant. Resuspend the cells in 200 µl of TE buffer.
6. Then add 100 µl of Tris-saturated phenol (pH=8.0). Vortex for 60 seconds, to lyse the cells
7. Centrifuge the samples at 13000 g for 5 minutes at 4°C to separate aqueous phase from non-aqueous phase
8. Transfer the 160 µl of upper aqueous phase carefully to a clean 1.5 ml microcentrifuge tube
9. Add 40 µl of TE buffer to make total volume of 200 µl. Add 100 µl of chloroform. Mix the contents thoroughly.
10. Centrifuge for 5 minutes at 13000 g at 4°C. Extract upper aqueous phase.
11. Repeat chloroform extraction till the white interphase is no longer visible
12. Transfer 160 µl of upper aqueous phase to a clean microcentrifuge tube of 1.5 ml capacity.
13. Add 40 µl of TE buffer and 5 µl of RNase (10 µg/ml) and incubate at 37°C for 10-15 minutes to digest RNA.
14. Add 100 µl of chloroform, mix thoroughly and centrifuge at 13000 g for 5 minutes at 4°C.
15. Transfer carefully, 150 µl of upper aqueous layer to a clean tube of 1.5 ml. This aqueous phase contains purified DNA.
16. Store DNA aliquots at -20°C till further use.

4.7.3.4 *Troubleshooting guide*

1. If growth medium of bacteria is not properly removed, it can cause interference in downstream processing.
2. Proper drying of DNA after ethanol precipitation is important. Residual ethanol can render purified DNA resistant to cleavage by restriction enzymes, thus making it unsuitable for further restriction analysis.
3. Some bacterial strains shed cell wall components into the medium that can inhibit the action of restriction enzymes. This can be overcome by washing the cell pellet in STE solution before lysing the bacteria.

4. If the DNA yield is found to be very low, it could be due to several reasons. Too young or too old bacterial cultures often give low DNA yields. It is important to use the bacterial cultures in late log phase for getting optimum yield of DNA. Also low DNA yields result due to incomplete lysis of bacterial cells. Therefore care should be taken to mix the lysis buffer thoroughly with bacterial cell pellet.
5. Insoluble DNA pellets can result after prolonged drying under vaccum. This causes overdrying of DNA and it might get partially denatured. Such overdried DNA takes longer time to dissolve. Being an acid, DNA is probably more soluble in slightly alkaline solutions like TE buffer (pH=8.0) than sterile water which is at neutral pH. Therefore in such a case, TE buffer should be preferred over sterile water. Also heating the DNA sample for a few minutes at 65°C helps in dissolving DNA.
6. Low temperature is very important for long-term storage of DNA. Storage of DNA aliquots at 4°C is only recommended for short-term storage such as a few days. Medium duration storage of a few months can be done at –20ºC while for long-term storage temperature of –80°C is recommended.

4.7.4 DNA Extraction Protocol- IV: Extraction of Plasmid DNA

4.7.4.1 *Introduction*

The discovery of plasmid DNA inside bacterial cells and the genes of plasmids that provide resistance to antibiotics have become a foundation of genetic engineering. Joshua Lederberg first coined the term 'plasmid' in 1952. The plasmids have an interesting property of transferring themselves from one bacterial cell to another by either conjugation or transformation. This characteristic property of plasmids has made them vectors of gene transfer in the genetic engineering experiments. Plasmids form an important tool in the genetics and biotechnology labs, which are most commonly for gene cloning. In the process of gene cloning the gene of interest is inserted into a plasmid. This plasmid is allowed to replicate itself inside a bacterial cell. As the plasmid replicated itself, it produces multiple copies of the gene of interest along with its own DNA. When the bacterium divides in the growth medium, the number of plasmids or in turn, gene of interest also multiplies.

Apart from gene cloning, plasmids are used for various other studies of genetics such as,

- Plasmids have been used to study the replication process of bacterial DNA.
- Plasmids are used to introduce gene from one organism to another, creating genetically modified organisms.

- They are used to study expression patterns of several genes.
- Antibiotic resistance genes from plasmids are used as selectable marker genes for genetic engineering experiments.
- They are used for production of pharmaceuticals in controlled conditions. For example, production of insulin, growth hormones etc.

There are several methods for extraction and purification of plasmid DNA. Most commonly followed method for plasmid DNA isolation still remains to be alkaline denaturation method (Birnboim and Doly, 1979).

4.7.4.2 *Principle and Working*

Plasmid DNA extraction is slightly more complicated because in this process, genomic DNA needs to be separated from plasmid DNA. Plasmid DNA is smaller molecule as compared to genomic DNA. A rapid alkaline extraction procedure for screening recombinant DNA plasmids was given by Birnboim and Doly in 1979. In this method, the bacterial chromosomal DNA was selectively denatured by alkali treatment, whereas covalently closed circular plasmid DNA remained double stranded, however denatured. On neutralization, a heavy precipitate was formed containing chromosomal DNA, leaving partially purified plasmid DNA in the solution. The methods for plasmid DNA preparation that developed later were mainly based on this method.

In the process of extraction of plasmid DNA, bacterial DNA is denatured by alkali treatment. The plasmid DNA being small covalently closed circular molecule, can readily renature upon neutralization; but genomic DNA is too large to reanneal properly and instead gets tangled. After subsequent centrifugation, genomic DNA gets bound along with proteins and forms heavy precipitate. At this time, plasmid DNA is free and dissolved in aqueous phase. It is important to remember that at this time, the sample should not be mixed together or vortexed.

4.7.4.3 *Protocol for DNA Extraction and Purification*

The following protocol is for extraction of Ti or Ri plasmid from *Agrobacterium* strains. This protocol is adapted from Hayman and Farrand (1990).

Growth of liquid **Agrobacterium** *culture*

For extraction of plasmid DNA from bacteria, it is important to grow a culture of the bacterium of interest. Freshly grown bacterial cultures give better results in terms of DNA extraction as compared to old cultures. For growing liquid *Agrobacterium* cultures, take about 100 ml of Luria broth, previously autoclaved. Inoculate the bacterial culture at temperature of 25-28°C in incubator-shaker. Allow the bacterial cultures to reach log phase.

Materials Required

- Cell wash solution (pH=8.0)

Tris-HCl	50 mM
EDTA	20 mM
NaCl	0.5 M
Sarkosyl	0.05 %

- Cell suspension solution

Tris-HCl	25 mM (pH= 8.0)
EDTA	10 mM
Glucose	50 mM
Lysozyme	2 mg/ml

- Lysis solution

NaOH	0.2 N
SDS	1.0 %

- Tris-HCl 2M (pH= 7.0)
- NaCl 5M
- Phenol buffered with 1.0 M Tris-HCl (pH= 8.0)
- Chloroform: isoamyl alcohol (24:1) (v/v)
- Ice cold ethanol
- TE buffer (pH= 8.0)

Tris-HCl	10mM
EDTA	1mM

Procedure

1. Take previously grown *Agrobacterium* liquid culture. Centrifuge 100 ml of culture at 10000 g for 15 minutes to pellet the cells.
2. Discard the supernatant. Resuspend cell pellet in 15 ml of ice-cold cell wash solution.
3. Repeat centrifugation. Discard the supernatant.
4. Resuspend cells in 15 ml of cold cell resuspension solution. Keep on ice for 5-10 minutes.
5. Add 30 ml of lysis solution and mix gently by inverting the tube for several times. Incubate at room temperature for 10 minutes.
6. Add 7.5 ml of 2M Tris-HCl (pH= 7.0) solution. Mix the contents gently by inverting the tube.

7. Add 75 ml of 5M NaCl and mix well by inversion. Incubate at room temperature for 20 minutes.
8. Add equal volumes of buffered phenol and additional 1.8 ml of 5M NaCl solutions. Mix by repeated inversion. Thorough mixing of the cellular solution and phenol is essential for efficient extraction of proteins and lipids.
9. Centrifuge at 10000 g for 15 minutes at 4°C. At this step, phase separation occurs.
10. Transfer the upper aqueous layer carefully without disturbing the interphase to a clean centrifuge tube.
11. Add equal volumes of phenol: chloroform: isoamyl alcohol (25:24:1) (v/v). Mix thoroughly. Centrifuge at 10000 g for 10 minutes.
12. Take upper aqueous phase to a clean centrifuge tube. Repeat step 11 until the interphase is not visible in the solution.
13. To the upper aqueous phase, add chloroform: isoamyl alcohol (24:1) (v/v). Repeat the centrifugation. Extract upper aqueous phase to a clean tube.
14. Precipitate DNA by addition of two volumes of ice-cold ethanol. Refrigerate at –20°C overnight.
15. Centrifuge the contents of 10000 g for 20 minutes at 4°C. Carefully decant the solution. Air dry the pellet.
16. Dissolve DNA pellet in TE buffer. Store at –20°C until further use.

4.7.4.4 *Troubleshooting Guide*

1. If growth medium of bacteria is not properly removed, it can cause interference in downstream processing.
2. Proper drying of DNA after ethanol precipitation is required. Residual ethanol can render purified DNA resistant to cleavage by restriction enzymes, thus making it unsuitable for further for further restriction analysis.
3. Some bacterial strains shed cell wall components into the medium that can inhibit the action of restriction enzymes. This can be overcome by washing the cell pellet in STE solution before lysing the bacteria.
4. Lysozyme addition is not particularly necessary for plasmid isolation, but it is recommended in glucose-Tris-EDTA solution as it increases the DNA yield.
5. If the DNA yield is found to be very low, it could be due to several reasons. Two young or too old bacterial cultures often give low DNA yields. It is important to use the bacterial cultures in late log phase for getting optimum yield of DNA. Also low DNA yields result due to

incomplete lysis of bacterial cells. Therefore care should be taken to mix the lysis buffer thoroughly with bacterial cell pellet.

6. Insoluble DNA pellets can result after prolonged drying under vaccum. This causes overdrying of DNA and it might get partially denatured. Such overdried DNA takes longer time to dissolve. Being an acid, DNA is probably more soluble in slightly alkaline solutions like TE buffer (pH=8.0) than sterile water which is at neutral pH. Therefore in such a case, TE buffer should be preferred over sterile water. Also heating the DNA sample for a few minutes at 65°C helps in dissolving DNA.
7. Contamination with bacterial chromosomal DNA in plasmid DNA preparations may occur due to mixing or vortexing after alkaline lysis step. Avoid vortexing after this step.

4.8 METHODS FOR DETECTION OF ISOLATED DNA

4.8.1 Gel Electrophoresis and Visualization of DNA Under UV-Transilluminator

DNA is negatively charged and charge per unit length of DNA is constant. Therefore, movement of DNA in a gel is dependent on size and shape. Horizontal agarose gel electrophoresis is useful for the separation of linear DNA molecules of up to 2000 kbp size. The gel is usually stained with ethidium bromide so that the separated DNA bands in the gel can be viewed under UV transilluminator. Polyacrylamide gel electrophoresis (PAGE) may be used to separate smaller DNA fragments.

Conformation of DNA has a significant effect on the rate of migration, response of sample gel preparation and applied voltage. For example, relaxed circular DNA and supercoiled DNA migrate anomalously in agarose gel electrophoresis as compared to linear DNA of similar molecular weight. The mobility of supercoiled DNA is also dependent on the linking number. The mobility increases as the linking number increases.

Procedure for Gel Electrophoresis of DNA

Materials required

- Agarose 0.8%
- Ethidium bromide 4 µl
- Loading dye (0.25% bromophenol blue)
- TBE buffer (10x) (pH=8.0)

Tris	890 mM
Boric acid	890 mM
EDTA	20 mM

To make 1x TBE buffer take one part of 10x TBE buffer and add 9 parts of autoclaved distilled water.

Procedure

1. Prepare 0.8% agarose gel. Dissolve agarose in water completely by heating. Add 4 µl of ethidium bromide and mix well.
2. Pour gelling mixture in gel casting tray with well-placed comb. Allow the gel to polymerize.
3. Take 1x TBE buffer in the electrophoresis apparatus with polymerized gel.
4. Load the DNA samples in wells taking 8 µl of DNA and 4 µl loading dye for each loading mixture.
5. Run the electrophoresis at 65V for two hours.
6. Observe for DNA bands in gel under UV transilluminator.

4.8.2 Detection and Quantitation of DNA Using Spectrophotometer

Purine and pyrimidine bases in DNA absorb UV light in the range of 240-280 nm. Oligonucleotides exhibit strong UV absorption, maximum at approximately 260 nm. This absorption arises almost entirely from complex electronic transitions in the purine and pyrimidine bases. The molar concentrations of oligonucleotides can be estimated in a UV spectrometer by measuring the number of absorbance units at 260 nm and by relating the value obtained to the sum of molar extinction coefficient of the individual nucleotides. Extinction coefficient is a parameter, which shows how strongly a substance absorbs light at a given wavelength per molar concentration.

The concentration and purity of DNA in solution can be measured by spectrophotometer. The concentration of DNA in solution is determined by measuring the absorbance at 260 nm that is A_{260}. When absorbance at A_{260} is 1, the concentration of DNA is about 50 µg/ml. However the actual value depends on the base composition, since the bases differ in extinction coefficient.

To measure the DNA concentration, dilute DNA solution to about 1:50 in TE buffer and measure absorbance at 260 nm (A_{260}) using TE buffer as blank. Then concentration of DNA can be calculated by the following formula (Farrell, 2010),

$$(\text{dsDNA})\ \mu g/ml = A_{260} \times \text{Dilution factor} \times E$$

Where, A_{260} = absorbance in optical density at 260 nm

E= extinction coefficient (Average extinction coefficient of dsDNA is 50 µg/OD_{260})

$$(ssDNA)\ \mu g/ml = A_{260} \times \text{Dilution factor} \times E$$

Where, A_{260} = absorbance in optical density at 260 nm

E= extinction coefficient (Average extinction coefficient of ssDNA is 33 µg/OD_{260})

The ratio of absorbance at 260 and 280 nm (A_{260}/A_{280}) is frequently used to measure purity of DNA. Contaminating salts, organic solvent and proteins can increase the absorbance measured at 260 nm. Normally nucleic acids and proteins have no absorbance at 320 nm. Proteins absorb UV light maximally at 280 nm. A ratio A_{260}/A_{280} of 1.8-2.0 is generally an indication of pure preparation of DNA. A lower ratio is indication of protein contamination.

Procedure for measuring concentration of DNA

Materials required

- DNA sample
- 1 ml quartz cuvette (with 1 cm path length)
- 1x TE buffer
- UV spectrophotometer

Procedure

1. Dilute the DNA solution by adding 20 µl of DNA aliquot to 0.98 ml of TE buffer in a microcentrifuge tube. Mix the contents well.
2. Switch on the UV lamp of spectrophotometer and wait for 15 minutes for the lamp to warm up.
3. Adjust wavelength to 260 nm.
4. Add TE buffer to cuvette to use it as blank and calibrate the spectrophotometer.
5. Now take absorbance reading at 260 nm with diluted DNA sample.
6. Repeat the same procedure and take absorbance reading at 280 nm.
7. Calculate DNA concentration.

Optical Density (OD) of 1 at 260 nm corresponds to approximately 50 µg/ml of double stranded DNA and 40 µg/ml of single stranded DNA.

4.8.3 Troubleshooting Guide

1. The common electrophoresis buffer TBE (Tris Borate EDTA) can inhibit enzymes, while EDTA can interfere with downstream applications.

2. DNase contamination is the possible cause if you get a smear instead of DNA bands in gel. Failure to amplify DNA in PCR is also indicative of nuclease contamination.
3. If you get low A_{260}/A_{280} ratio, it is possible that there is protein contamination. To remove protein contaminants, perform phenol and chloroform extractions once again.
4. Protein contamination often lowers A_{260}/A_{280} ratio. Insufficient centrifugation with respect to speed and time may be a cause of residual protein. At the time of chloroform: isoamyl alcohol precipitation, increase the centrifugal speed and/or time for efficient separation of proteins.
5. If smeared DNA in gel is seen under UV transilluminator, it is possibly due to DNA degradation.
6. If more than one band appears, in a single genomic DNA sample, it may be due to RNA contamination or sheared DNA.
7. Do not overrun the gel or else DNA may come out of the gel. Watch for the migration of bromophenol blue dye in gel as an indicator of the DNA migration.

4.9 REVIEW QUESTIONS

1. What is DNA? Describe the structure of DNA as observed by Watson and Crick.
2. Give in brief the properties of DNA.
3. Which enzyme causes DNA degradation and how?
4. Describe and compare composition of plant bacterial cell walls.
5. Describe the methods for cell wall degradation.
6. Why lysozyme is used in bacterial DNA isolation? What effect would lysozyme have on plant cell wall?
7. Differentiate between cell wall compositions of gram-positive and gram-negative bacteria.
8. Explain roles of the following chemicals in DNA isolation process- SDS, RNase, proteinase K, Sarkosyl, CTAB, TE, EDTA, phenol, ethanol.
9. Describe how will you avoid bacterial chromosomal DNA contamination in plasmid DNA preparations?
10. What is the measure for purity of DNA sample? How will you know if your DNA sample is contaminated?
11. Why would you isolate plant DNA? Give applications where plant DNA can be used.
12. Which enzyme degrades DNA? How will you avoid degradation of DNA in the extraction process?

13. What is the full form of CTAB?
14. What is the effect of liquid nitrogen on plant tissues?

4.10 REFERENCES

Annunziato, A. (2008). DNA packaging: nucleosomes and chromatin. *Nature Education,* 1(1).

Asenjo, J. and Andrews, B. (1990). Enzymatic cell lysis for product release. *Bioprocess Technol.,* **9:** 143–175.

Birnboim, H. and Doly, J. (1979). A rapid alkaline extraction procedure for screening recombinant plasmid DNA. *Nucleic Acids Res.,* **7(6):** 1513–1523.

Birnboim, H. (1992). Extraction of high molecular weight RNA and DNA from cultured mammalian cells. *Methods Enzymol.,* **216:** 154–160.

Campbell, V and Jackson, D. (1980). The effect of divalent cations on the mode of action of DNase I. The initial reaction products produced from covalently closed circular DNA. *J. Biol. Chem.,* **255(8):** 3726–3735.

Casali, N. and Preston, A. (2003). *E. coli* plasmid vectors: methods and applications. *In*: Methods in molecular biology vol.235. Humana Press, Inc. p. 316.

Chaves, A., Vergera, C. and Mayer, J. (1995). Dichloromethane as an economic alternative to chloroform in the extraction of DNA from plant tissues. *Plant Mol. Biol. Rep.,* **13:** 18–25.

Cheng, H. and Jiang, N. (2006). Extremely rapid extraction of DNA from bacteria and yeasts. *Biotechnology Letters,* **28(1):** 55–59.

Cui, L., Veeraraghavan, N., Richter, A., Wall, K., Jansen, R., Leebans-Mack, J., Makalowska, I. and dePamphillis, C. (2006). Chloroplast DB: The chloroplast genome database. *Nucleic Acids Research,* **34:** D692–696.

Dehestani, A. and Tabar, S. (2007). A rapid efficient method for DNA isolation of plants with high levels of secondary metabolites. *Asian J. Plant Sci.,* **6:** 977–981.

Dilworth, E. and Frey, J. (2000). A rapid method for high throughput DNA extraction from plant material for PCR amplification. *Plant Molecular Biology Reporter,* **18:** 61–64.

Dorman, C. (1991). DNA supercoiling and environmental regulation of gene expression in pathogenic bacteria. *Infection and Immunity,* **59(3):** 745–749.

Doyle, J. and Doyle, J. (1987). A rapid DNA isolation procedure for small quantities of fresh leaf tissue. *Phytochem. Bull.,* **19:** 11–15.

Drlica, K. (1984). Biology of bacterial deoxyribonucleic topoisomerases. *Microbiol. Rev.,* **51:** 301–319.

Evans, C. and Aguilera, R. (2003). DNase II: Genes, enzymes and function. *Gene,* **322:** 1–15.

Farrell, R. (2010). RNA methodologies: A laboratory guide for isolation and characterization. 4th Edition. Academic Press. p. 141.

Fischer, H., Scherz, J., Szabo, S., Mildner, M., Bengrafa, C., Torriglia, A., Tschachler, E. and Eckhart, L. (2011). DNase2 is the main DNA-degrading enzyme of the *Stratum corneum. PLoS ONE,* **6(3):** e17581.

Fu, R., Wang, J., Sun, Y. and Shaw, P. (1998). Extraction of genomic DNA suitable for PCR analysis from dried plant rhizomes/roots. *Biotechniques,* **25:** 797–801.

Fulton, T., Chunwongse, J. and Tanksley, S. (1995). Microprep protocol for extraction of DNA from tomato and other herbaceous plants. *Plant Mol. Biol. Rep.,* **13:** 207–209.

Gray, M. (2001). Mitochondrial DNA (mtDNA). *In*: Encyclopedia of genetics. Academic Press pp. 1219–1220.

Green, M. and Thompson, D. (1999). Easy and efficient DNA extraction from woody plants for detection of phytoplasmas by polymerase chain reaction. *Plant Disease,* **83**: 482–485.

Hayman, T. and Ferrand, S. (1990). *Agrobacterium* plasmids encode structurally and functionally different loci for catabolism of agropine- type opines. *Mol. Gen. Genet.,* **223:** 465–473.

Hillen, W., Klein, R. and Wells, R. (1981). Precipitation of milligram amounts of 21 deoxyribonucleic acids and restriction fragments. *Biochem.,* **20:** 3748–3756.

Keegstra, K. (2010). Plant cell walls. *Plant Physiology,* **154:** 483-486.

Kodama, Y., Magaya, S., Shimyo, A. and Kato, K. (2007). Mapping and characterization of DNase I hypersensitive sites in *Arabidopsis* chromatin. *Plant Cell Physiol.,* **48(3):** 459–470.

Kunitz, M. (1948). Isolation of crystalline deoxyribonuclease from beef pancreas. *Science,* **108(2792):** 19–20.

Laschtschenko, P. (1909). Über die keimtötende und entwicklungshemmende Wirkung von Huhnereiweiss. *Z. Hyg. Infekt. Krankheiten,* **64:** 419–426.

Lazarides, E. and Lindberg, U. (1974). Actin is the naturally occurring inhibitor of deoxyribonuclease I. *Proc. Nat. Acad. Sci.,* **71(2):** 4742–4746.

Li, *et al.* (2012). The complete mitochondrial genome and novel gene arrangement of the unique-headed bug *Stenopirates* sp. (Hemiptera: Enicocephalidae). *PLoS ONE,* **7(1):** e29419.

Olins, A. and Olins, D. (1974). Spheroid chromatin units (v bodies). *Science,* **183:** 330–332.

Roizman B and Jenkins F. (1985). Genetic engineering of novel genomes of large DNA viruses. *Science,* **229(4719):** 1208–1214.

Sahastrabuddhe, A. and Deodhar, M. (2010). Standardization of DNA extraction and optimization of RAPD-PCR conditions in *Cinia indica. Int. J. Bot.,* **6:** 293–298.

Santoinni, P. and Rothman, S. (1961). Nucleic acid-splitting enzymes in human epidermis and their possible role in keratinization. *J. Invest. Dermatol.,* **37:** 489–495.

Schaffer, C. and Messner, P. (2005). The structure of secondary cell wall polymers: how gram-positive bacteria stick their cell walls together. *Microbiology,* **151(3):** 643–651.

Schatz, G. (2004). Mitochondrial DNA. In: Encyclopedia of Biological Chemistry. *Elsevier Inc.* **2:** 693.

Shankar, K., Chavan, L., Shinde, S. and Patil, B. (2011). An improved DNA extraction protocol from four *in vitro* banana cultivars. *Asian J. Biotechnol.,* **3:** 84–90.

Sinden, R. (1994). DNA structure and function. Academic Press, Inc. p. 398.

Srivastava, N., Sharma, Y., Kamal, B., Dobrial, A. and Jaden, V. (2010). Polyphenolics free DNA isolation from different types of tissues of *Aconitum heterophyllum* wall-endangered medicinal plant. *J. Plant Sci.,* **5:** 414–419.

Stein, N., Herren, G. and Keller, B. (2001). A new DNA extraction method for high throughput marker analysis in a large genome species such as *Triticum aestivum. Plant Breeding,* **120:** 354–356.

Tabachnik, J. and Freed, R. (1961). Demonstration of nucleases on mammalian skin surface and in saline extract of hair. *Nature,* **190:** 921–922.

Tai, T. and Tanksley, S. (1990). A rapid and inexpensive method for isolation of total DNA from dehydrated plant tissue. *Plant Mol. Biol. Rep.,* **8:** 297–303.

Tiwari, K., Jadhav, S. and Gupta, S. (2012). Modified CTAB technique for the isolation of DNA from some medicinal plants. *Research Journal of Medicinal Plants,* **6:** 65–73.

Tsutsumi *et al.* (2001). DNase is present in the chief cells of human and rat stomachs. *The Histochemical Journal,* **33(9-10):** 531–535.

Weinstein, D., Marsh, J., Glick, M. and Warren, L. (1969). Membranes of animal cells: lipids of the L cell and its surface membrane. *The Journal of Biological Chemistry*, **244(15):** 4103–4111.

Whitchurch, C., Tolke-Neilson, T., Ragas, P. and Mattick, J. (2002). Extracellular DNA required for bacterial biofilm formation. *Science*, **295:** 1487.

5

Ribonucleic Acid and Extraction of RNA

5.1 WHAT IS RNA? ITS STRUCTURE AND PROPERTIES

RNA or ribonucleic acid is a biomolecule that is comparable to DNA in a broad sense. Unlike DNA, ribonucleic acids are genetic material of only RNA viruses. Although RNA is present in cytoplasm of all living cells as well as inside eukaryotic nucleus, it is not the genetic material of living organisms. In fact, RNA is a part of gene expression machinery.

RNA is a polynucleotide molecule that contains ribose sugar, phosphoric acid and bases. Bases present in RNA are adenine, guanine, cytosine and uracil. So RNA contains uracil instead of thymine, which is present in DNA. The bases adenine and guanine are purines and cytosine and uracil are pyrimidines. In RNA, adenine pairs with uracil forming two hydrogen bonds and cytosine pairs with guanine forming three hydrogen bonds. RNA does not form a double helical structure like B-form of DNA. RNA is known to form characteristic secondary structures. RNA has ribose sugar in its sugar-phosphate backbone. Deoxyribose differs from ribose sugar by having hydrogen atom in place of hydroxyl group at second carbon atom. In RNA, phosphate is attached to 3' position of one ribose and 5' position of the next one. This forms the sugar-phosphate backbone of RNA polymer. The bases are attached to 1st carbon on ribose sugar.

When complementary strands of RNA come together, they form a double helix that resembles right-handed A-form of DNA. In this form, there are 11 bases per turn, and length of one turn of helix is 0.24 nm as compared to B-form of DNA where there are 10 bases per turn and 0.34 nm length of one turn of helix. RNA is a polymer of ribonucleoside monophosphates joined together by phosphodiester linkages. Ribonucleoside triphosphates are called NTPs while deoxyribonucleoside triphosphates are called dNTPs.

$$\text{Base (A, C, G or U)} + \text{Ribose sugar} = \text{Ribonucleoside}$$

$$\text{Ribonucleoside} + \text{Phosphate} = \text{Ribonucleotide}$$

$$(\text{Ribonucleotide})_n = \text{RNA polymer}$$

RNA is synthesized by RNA polymerase enzyme by the process of transcription. For transcription, two strands of DNA unwind locally and for each gene, only one strand of the double helical DNA acts as a template for RNA synthesis. This is called sense strand and its complementary strand is called antisense strand. RNA polymerase enzyme then reads the sense strand and incorporates nucleotides to synthesize single stranded DNA. These single stranded RNA molecules form secondary structures by pairing of complementary sequences. The single stranded RNA molecule without any base pairing is said to exhibit primary structure. When complementary bases on ssRNA form base pairs, they are called secondary structures. Many secondary structures such as stem-loop, interior loop, hairpin, multi branched loop etc. are known to form in RNA. The tertiary structure of RNA can be said to be a collection of secondary structures required by RNA molecule to perform biological function.

The phosphoric acid present in the backbone of RNA imparts negative charge to the molecule. The high negative charge of RNA works against its folding into a compact structure. Positive ions promote folding by reducing the repulsion between RNA phosphates. Millimolar Mg^{+2} ion concentrations are known to stabilize RNA tertiary structures that are otherwise only marginally stable in presence of high monovalent cation concentrations.

There are three major types of RNAs found in all living cells. They are messenger RNA (mRNA), transfer RNA (tRNA), and ribosomal RNA (rRNA).

Applications of RNA Extraction

RNA synthesis or transcription forms the first step of central dogma. Characterization of RNA is almost always related to gene expression. RNA extraction and analysis are prerequisites for genetic, molecular and cellular studies. Various fields where RNA extraction is applied are,

1. Gene expression studies. RNA extraction is used to study the nature and number of RNA transcripts produced by a particular gene *in vivo*. This sheds a light on the various functional genes.
2. RNA extraction is used for downstream processing of RNA, PCR, RT-PCR, molecular studies etc.

3. RNA is used in production of RNA probes and for hybridization techniques like northern blotting.
4. Analysis of RNA transcripts for identification of non-coding sequences of eukaryotic genes, introns and exon sequences.
5. Regulation of genes at transcriptional and posttranscriptional events.
6. Production of synthetic oligonucleotides, RNA probes, gene constructs etc.
7. Studies on RNA editing, splicing and alternative splicing events in eukaryotic mRNA.
8. Various RNA technologies like antisense RNA, interference RNA technology etc.
9. For identification of transcriptional start sites (TSS) on genome, mapping of RNA etc.
10. RNA extraction is particularly important in cDNA technology. In this, single stranded mRNA is used to produce double stranded complementary DNA (cDNA) molecules by the process of reverse transcription.
11. Use of mRNA extraction in drug design, *in vitro* translation, understanding protein structure, identification of mutant gene and gene product.

Properties of RNA

RNA molecule is essentially synthesized as a single stranded molecule in the cell. However, double stranded RNA or randomly paired RNA is known to melt at high temperature. This melting of RNA is characteristically similar to DNA melting. The melting of base pairs is a ubiquitous feature of RNA structural transitions, which are widely used to sense and respond to cellular stimuli.

Purified solution of RNA can be quantitated using UV spectrophotometer. RNA absorbs UV light in the range of 240 to 280 nm. RNA has no absorbance at 320 nm where it should give optical density reading of 0. The absorbance below 230 nm has little significance regarding computation of quantity and determination of purity of the RNA sample. RNA gives highest absorbance at 260 nm and as the wavelength increases beyond 260 nm, absorbance drops sharply.

5.2 ENZYMES THAT DEGRADE RNA

Ribonucleases, commonly abbreviated as RNases, are those enzymes, which catalyze degradation of RNA molecules. RNases are present in both prokaryotes and eukaryotes. They are present in every living cell. Human body uses RNases as a defense mechanism against invading bacteria, viruses etc. RNases are secreted in body fluids like tears, saliva, mucus, sweat etc. RNases are ubiquitous in the environment. They can be found in pollen, dust, on skin surface, fingertips and laboratory equipments.

RNA analysis and quantification methods require purified, intact and non-degraded RNA samples to produce optimum results. RNases are difficult to remove or deactivate completely and may become reintroduced during handling. There can be several possible sources of RNases in the laboratory

The presence of RNases in the working environment poses problems, as RNA will be degraded as soon as it is exposed to the enzyme. RNases have a remarkable ability to retain activity even after series of alternate freeze-thawing and even autoclaving. Therefore it is necessary to render the working environment RNase free by treating with RNase inhibiting chemicals. It is mandatory for all sorts of RNA preparations that the apparatus is soaked in DEPC (diethyl pyrocarbonate) treated water and solutions are made with DEPC treated water.

RNase inhibitors can be used to minimize RNA degradation during RNA isolation and handling. Natural proteinaceous RNase inhibitors have been found in a number of tissues like placenta, eye lens, spleen, kidney, lung, bone marrow etc. The most commonly used RNase inhibitors in RNA isolation procedures are SDS (sodium dodecyl sulphate), DEPC and guanidinium thiocyanate. Diethyl pyrocarbonate or DEPC reacts with amino acids of proteins, thereby inactivating them.

Several reagents can be considered for storage of RNA because of their property to inactivate RNase. Formamide can be used to dissolve RNA, which inhibits RNase-A activity even at high concentrations.

RNase inhibitors/inactivators are,

1. DEPC
2. SDS
3. Proteinase K
4. Phenol
5. Bentonite

5.3 TYPES OF RNA

5.3.1 Ribosomal RNA (rRNA)

As the name suggests, ribosomal RNA is found in ribosomes. Ribosomes are the site of protein synthesis in both prokaryotes as well as eukaryotes. The ribosomes are composed of rRNA and proteins. Ribosomal RNA constitutes a major portion of total cellular RNA. It makes up about 80% of total RNA in a cell. The ribosomes of eukaryotes sediment at about 80S, while in prokaryotes like bacteria, they sediment at 70S. Both these types of ribosomes have similar structure and function and are composed of rRNA-protein complex.

The Nobel Prize in Chemistry was awarded jointly to Venkatraman Ramakrishnan, Thomas A. Steitz and Ada E. Yonath in 2009 for studies of the structure and function of the ribosome. They showed what the ribosome looks like and how it functions at the atomic level. All three of the scientists used a method called X-ray crystallography to map the position for each and every one of the hundreds of thousands of atoms that make up the ribosome.

The ribosomes of eukaryotes contain two subunits, called large subunit of 60S and small 40S. In large 60S subunit, 28S rRNA, 5.8S rRNA and 5S rRNA is present while in small 40S subunit, 18S rRNA is present. The prokaryotic ribosomes also contain large (50S) and small (30S) subunits. In prokaryotic ribosome, the large subunit contains 23S rRNA and 5S rRNA, while small subunit contains 16S rRNA. The genes or DNA responsible for the synthesis of rRNA is called rDNA or ribosomal DNA.

The rDNA is closely associated with nucleolus. The eukaryotic rDNA synthesizes 45S initial transcript containing 28S rRNA and 18S rRNA. Thereafter, 45S initial transcript is cleaved into 18S and 32S pieces. During this process, ribosomal proteins also get associated. The 5S rRNA transcript that is not produced in the nucleolus is then incorporated into 32S rRNA to form a large 65S unit. This unit is converted into 60S large subunit of ribosomes.

In biosynthesis of prokaryotic ribosomes, a single initial RNA transcript is made from rDNA. This initial rRNA transcript is further cleaved and processed to give 16S, 23S and 5S rRNA. These rRNA transcripts are then used in the formation of large 50S and small 30S subunit of ribosomes.

All the initial transcripts undergo cleavage and chemical modifications before they become mature rRNA. During cleavage, 16S rRNA sequence is cleaved first. These cleaved fragments are still larger than the final RNAs and are unmethylated. The cleaved fragments are further processed and the bases are methylated. At this time, ribosomal proteins also get associated to form precursor ribosome subunits. Eukaryotic rRNAs are synthesized and processed inside nucleus while prokaryotic rRNAs are synthesized and processed in cytoplasm.

5.3.2 Transfer RNA (tRNA)

Transfer RNA or tRNA is the smallest type of RNA. It has fewer than 100 nucleotide units. These small tRNA molecules of 70-90 nucleotides length, act as adapters to translate the nucleotide sequence of mRNA into protein sequence. Transfer RNA is part of protein synthesizing machinery of cell. The molecular wt. of tRNA is about 25,000 Daltons. The tRNA constitutes about 16% of the total cellular RNA. The universal function of tRNA is to transfer amino acids in the process of protein synthesis.

Transfer RNA carries appropriate amino acid to the ribosome during the process of protein synthesis. The biological role of tRNA is to bind activated amino acid residues and transfer them to ribosomes. The tRNA is a complex, partially double helical structure that is represented in the shape of cloverleaf.

The cells contain more than 60 different types of tRNA molecules. Each molecule of tRNA has specificity for a given amino acid to which it can attach by a covalent bond. Transfer RNA is actually a single polynucleotide chain, which is apparently bent in the middle, and the two arms form loops. All tRNA molecules have –CCA sequence at 3' end and guanine residues at 5' end. The activated amino acids are attached to the 3' end of tRNA. Holley *et al.* first proposed the cloverleaf pattern of alanine transfer RNA in 1965.

All tRNA have common right loop of the right arm that consists of 7 unpaired bases including peudouridine (ψ). It is known as TψC arm. It may be involved in the binding of tRNA to the ribosome. All the tRNA molecules contain 7-15 unusual bases, many of which are methylated or unmethylated derivatives of A, U, G and C. Unusual bases are nucleotides of methylinosine (MeI), peudouridine (ψU), methylated purines A and G, methylated pyrimidines such as thymine or ribothymine, 5-methylcytosine etc. However all of these unusual bases are not present in one tRNA. The nucleotide triplet in the bottom loop of tRNA is called anticodon. The anticodon is a site that is complementary to the triplet codon of the messenger RNA and is the site of codon recognition.

5.3.3 Messenger RNA (mRNA)

In 1960, Francois Jacob and Jacques Manod first gave name 'messenger RNA' to describe template RNA that carried genetic information necessary for protein synthesis. Messenger RNA makes up about 5% of the total cellular RNA. The mRNA has a molecular wt. of about 5,00,000 to 45,00,000 Daltons. It has a sedimentation coefficient of 6 to 23. Messenger RNA is normally attached to the smaller subunits of ribosomes and the length of mRNA is such that it connects a number of ribosomes. Such clusters of ribosomes are called polyribosomes or polysomes. These structures have been observed under electron microscope. The mRNA molecule is a complementary copy of the gene.

Only one strand of DNA molecule acts as a template for mRNA synthesis. Thus, mRNA is essentially single stranded. The main difference between prokaryotic and eukaryotic mRNA is that prokaryotic mRNA is transcribed from several adjacent genes. Such mRNA is called polygenic or polycistronic. In eukaryotes however mRNA is transcribed on a single gene, hence called monogenic or monocistronic.

Examination of mRNA reveals that these molecules are of variable length with wide range of sedimentation coefficients (S values). Unlike prokaryotic mRNA, mRNA of eukaryotes are modified at both 5' and 3' ends. The primary transcript of eukaryotic mRNA is called hnRNA or heterogeneous nuclear RNA. This hnRNA contains exons and introns. Exons are the sequences that code for proteins while introns are those that don't, hence are removed from RNA. Introns are absent in prokaryotic mRNA. In eukaryotic mRNA, a 7-methylguanosine cap (m^7G) is present at 5' end. Messenger RNA without m^7G cap shows poor binding of ribosomes, which suggests its influence in protein synthesis. The 3' end of mRNA is polyadenylated that gives poly-A sequence, which consists of 200-250 adenine residues. The poly-A sequence is added to mRNA in the nucleus before it is transferred to cytoplasm. This modification is popularly called as poly-A tail. This poly-A tail is thought to regulate translation as well as mRNA stability.

All mRNA have initiation codons at the 5' end and stop codon at 3' end. Initiation codons trigger the initiation of protein synthesis while stop codon finishes the process. In both prokaryotes and eukaryotes, stop codons do not code for any amino acid.

5.3.4 Mitochondrial RNA (mtRNA)

The mitochondrial ribosomes are located in the mitochondrial matrix and contain a comparatively low RNA content than cytoplasmic or prokaryotic ribosomes. Although traditionally it has been considered that mitochondrial ribosomes contain only mtDNA-encoded 12S and 16S rRNA, the presence of 5S rRNA species of nuclear origin has been demonstrated inside mitochondria. The mitochondrial biosynthetic machinery has some unusual features including the use of limited set of tRNA. The mitochondrial mRNA lack the typical cap of cytoplasmic mRNA in eukaryotes.

RNA polyadenylation has a different purpose in bacteria and mitochondria than eukaryotic nuclear mRNA. In plant mitochondria, polyadenylation of mRNA triggers its degradation. However, in spite of its prokaryotic origin, human mitochondrial transcripts possess stable 3' poly-A tails.

The human mitochondrial genome consists of a circular chromosome that encodes 2 rRNA (16S and 12S) and 22 tRNA. Plant mitochondrial DNA however encodes a third 5S rRNA. The mitochondria of plants and protozoans differ in utilizing and importing tRNA encoded by mitochondrial as well as nuclear genome, whereas in animal mitochondria, all tRNA are encoded by the same organelle. Many tRNA in prokaryotic and eukaryotic cells are able to recognize more than a single codon in mRNA because of wobble pairing that allows some mismatch between tRNA anticodon and 3rd position of codon. However, at least 30 different tRNA are required to translate the code according to the wobble pairing. Yet human mitochondria successfully translate using 22 tRNA. This is accomplished by an extreme form of wobble in which U in the anticodon of tRNA can pair with any of the four bases in third codon position of mRNA allowing 4 codons to be recognized by a single tRNA. In addition, some codons of mitochondrial mRNA specify different amino acids than the universal code.

5.3.5 Chloroplast RNA (cpRNA)

In plants and algae the components of photosynthetic machinery are located in the thylakoid membrane bound chloroplasts. Chloroplast is a type of plastid that is responsible for conversion of light energy into chemical energy. Just like mitochondria, chloroplasts have their own circular chromosomal DNA. Chloroplasts have their own transcriptional and translational machinery. Chloroplasts are thought to have originated from cyanobacteria like prokaryotes that became incorporated in eukaryotic cells by the process of endosymbiosis. The chloroplast genes encode RNA and proteins that are involved in the gene expression.

The gene expression system of chloroplasts is far more complex than that of their cyanobacterial progenitor. This complexity affects RNA metabolism, transcription and maturation process. Mature chloroplast RNA is generated by nuclear encoded proteins acquired in the process of plant evolution, comprising additional RNA polymerases, sigma factors and sequence specific RNA maturation factors that promote RNA splicing, editing, end formation and translatability.

The ribosomes and tRNA that are used for the translation of chloroplast mRNA are encoded by cpDNA. These include 4 rRNA of 23S, 16S, 5S and 4.5S; and 30 different tRNA. These 30 types of chloroplast tRNAs are sufficient to translate all the mRNA codons. In contrast to nuclear mRNA, polyadenylation of chloroplast mRNA at 3' end causes its degradation by enzyme polynucleotide phosphorylase (PNPase) (Walter *et al.*, 2002).

5.4 POINTS TO REMEMBER WHILE ISOLATING mRNA

When compared to isolation of DNA or total RNA, mRNA extraction is trickier. This is because mRNA is the expression of a gene, complementary sequence of the sense strand. All the genes do not express at the same time. Gene expression *in vivo* is influenced by seasons, physiological conditions of the organism, environmental interactions etc. Also all genes do not have same intensity of expression. The life of mRNA inside the cell is also an important aspect that greatly influences the success of mRNA isolation. Therefore, various things need to be considered before isolating the mRNA of interest. The important points that should be remembered while isolating mRNA are,

1. The sample tissues that are high in nucleases or fatty tissues or samples with large amounts of inhibitors can present problems while isolating mRNA.
2. Deciding the most appropriate method of cell disruption is important for maximizing the yield and quality of mRNA preparation. During sample disruption for RNA isolation, it is crucial that the lytic agents or denaturants are in contact with cellular contents at the time of cell disruption.
3. All the mRNAs produced inside the cell are very much short lived. The mRNA is degraded as soon as the sequences present on it are translated into proteins.
4. The degradation of mRNA is a vital aspect of gene regulation. Turning off of mRNA translation by its degradation ensures that previously translated mRNA does not keep on producing the proteins indefinitely.
5. Polyadenylated mRNA is removed by degradation in eukaryotic cells either by shortening of poly-A tails or removal of 5' cap.
6. It is important to determine the mechanisms of degradation of specific mRNA transcripts under various biological conditions or certain genetic backgrounds.
7. Considering short life of mRNA it is important to select fresh and healthy tissue for its isolation. Long term-stored tissues often give poor results.
8. Some genes are activated only under certain physiological and environmental conditions. This should be taken into account while isolating and analyzing certain mRNA transcripts of gene of interest.
9. Prokaryotic mRNA molecules usually have a short existence of 2-3 minutes while in mammals, average life span of mRNA can range between 10 minutes to 2 days.

5.5 CREATION AND MAINTENANCE OF RNase FREE WORKING ENVIRONMENT

5.5.1 Introduction

Whenever you are working with RNA, it is extremely important to be conscious of the presence of ribonucleases in the surrounding environment.

RNases are very stable and active enzymes. RNases are difficult to inactivate and even minute quantities are sufficient to destroy RNA. RNases are virtually present everywhere so one has to take proper care to set up RNase free working environment for RNA extraction experiment.

5.5.2 Need and Importance of RNase Free Working Environment

Before starting the RNA preparation, one has to ensure RNase free working area in the lab. Laboratories that routinely work with RNA usually maintain a sterile workspace with special chemicals and apparatus. The sources of RNases in the lab can be numerous. Contaminated chemicals, glassware, plasticware, pipette tips, eppendorf tubes, hands, dust particles etc. all add up to degradation of RNA by RNases. Here are a few general routine procedures that anyone working with RNA should follow.

1. Always wear sterile gloves while working. Gloves that have touched bare skin, door handles, pen, pencils etc. should not be considered as RNase free.
2. Whenever possible, use RNase free disposable plasticware for handling RNA. Such materials do not require pretreatment to inactivate RNases.
3. Treat non disposable glassware and plasticware with chemicals or heat to make them RNase free.
4. Remember, sterile doesn't mean RNase free. Most RNases retain their activity even after autoclaving.
5. Make sure pipette tips and eppendorf tubes to be used for RNA preparation are RNase free.
6. Make all solutions and buffers in RNase-free water.
7. Chemicals to be used for RNA isolation and analysis should be reserved and kept separately. They should not be mixed with other chemicals for other uses.
8. Make sure whatever enzymes you use for RNA analysis are RNase free.
9. Clean every surface area, bench surfaces, pipettes, electrophoresis apparatus etc. with RNase inactivating chemicals.
10. Establish a regular cleansing routine for a quick daily/weekly cleaning.
11. Dedicate a workspace, a set of pipettes and other equipment only for RNA work.
12. Always wear gloves and lab coat.
13. Make sure that water which is used to rinse glassware and other equipments are RNase free.

5.5.3 Methods for Deactivation of RNase

5.5.3.1 *Ethanol treatment*

Ethanol is frequently used to create RNase free working area. Bench tops, pipette and other working surfaces are frequently wiped with alcohol. The pestle and mortar used for tissue grinding should be rinsed with ethanol. Electrophoretic apparatus should be cleaned with RNase free water and then rinsed with ethanol. The plasticware that does not withstand baking at high temperatures should be rinsed with ethanol to make them RNase free. However, ethanol treatment is not 100% effective for creating RNase free environment, but it cleans most of the contamination.

5.5.3.2 *DEPC treatment*

DEPC is used as universal inactivator of RNases. Diethylpyrocarbonate or DEPC is most commonly used chemical to control RNase contamination in lab. DEPC is a strong inhibitor of RNase. DEPC destroys enzymatic activity by modifying –NH, –SH and –OH groups in RNase and other proteins. DEPC is commonly used to decontaminate glassware, plasticware, solutions and water. To make 0.1% solution of DEPC, add 0.1 ml DEPC and make final volume of 100 ml by using autoclaved distilled water. Incubate the solution for 5 hours. Then autoclave for 15 minutes to remove any traces of DEPC. Such treated water is free from RNase.

All solutions and buffers should be made in such DEPC treated water. It is important to remove all traces of DEPC from these solutions. Even trace amounts of DEPC can modify purine residues in RNA, making it difficult to hybridize. This can pose problems in case of hybridization experiments like northern blotting.

DEPC is very unstable in water and in presence of certain chemicals. It readily degrades into ethanol and carbon dioxide. The commonly used chemical in buffers is Tris base. Tris is known to react with DEPC, degrading it into ethanol and carbon dioxide. Therefore, DEPC should not be used with Tris buffers. However, DEPC can be used with buffers such as phosphate buffered saline (PBS) and MOPS buffer.

5.5.3.3 *Heat treatment*

RNase enzyme is basically a protein and follows all rules of protein denaturation. Like all proteins, RNase can not withstand extremely high temperatures that denature the proteins completely. This property is used to render glassware RNase free. Laboratory grade glassware can withstand extremely high temperatures. Therefore they can be made RNase free by baking them in oven at 150°C for 3 hrs. However, this treatment is only limited to glassware and metallic ware that can withstand such high temperature.

5.6 PROTOCOLS FOR EXTRACTION AND PURIFICATION OF RNA

5.6.1 RNA Extraction Protocol- I: Extraction of Total RNA from Mammalian Tissues

5.6.1.1 *Introduction*

RNA or ribonucleic acid is present in all types of living cells. Mammalian cells contain three types of RNAs; messenger RNA (mRNA), ribosomal RNA (rRNA) and transfer RNA (tRNA). These three types of RNAs are involved in accurate conversion of genetic information in the form of proteins.

5.6.1.2 *Principle and working*

For RNA extraction process, the technique of cell fractionation by differential centrifugation is used. In this procedure, cell nuclei are first pelleted and the cytoplasmic lysate is used as a source for the RNA extraction. Thereafter, the cell lysate is treated with RNase free DNase enzyme. DNase I degrades DNA and this degraded DNA can be separated from RNA by pelleting.

5.6.1.3 *Protocol for total mammalian RNA extraction and purification*

Materials required

- Cell suspension: about 5 × 10^6 cells
- Phenol: chloroform: isoamyl alcohol (25:24:1)
- 70% and 100% ice-cold ethanol
- TE buffer pH=7.0 (RNase free)

Tris HCl (pH-7.0)	10 mM
EDTA (pH=8.0)	1 mM

- DEPC treated water
- Lysis buffer

Tris-HCl (pH=7.5)	10 mM
NaCl (DEPC treated)	150 mM
$MgCl_2$ (DEPC treated)	1.5 mM
NP40	0.65%

- Extraction buffer

Tris-HCl (pH=7.5)	10 mM
NaCl (DEPC treated)	350 mM
EDTA (DEPC treated)	10 mM
SDS	1%

- DNase I (RNase free)

Procedure

1. Transfer up to 5×10^6 cells in cell suspension to microcentrifuge tubes along with medium.
2. Centrifuge for 5 min at 1000 rpm.
3. Remove the supernatant.
4. Resuspend pellet in 200 µl of ice-cold lysis buffer. Vortex until the pellet is completely dissolved.
5. Centrifuge for 5 min at 1000 rpm at 4°C to pellet the cell nuclei and other cell debris.
6. Transfer supernatant containing cytoplasmic lysate to a fresh microcentrifuge tube containing 200 µl of ice-cold extraction buffer. Add 400 µl of phenol: chloroform: isoamyl alcohol (25:24:1) and vortex briefly.
7. Centrifuge for 5 min at 15000 rpm at 4°C.
8. Carefully transfer the upper aqueous phase to a new microcentrifuge tube.
9. Add 1 ml of 100% ice-cold ethanol and keep on ice for 20 min.
10. Centrifuge at 15000 rpm for 15 min at 4°C.
11. Remove the supernatant carefully without losing the pellet.
12. Add 1 ml of 70% cold ethanol and vortex briefly.
13. Spin at 15000 rpm for 15 minutes. Remove the supernatant carefully.
14. Air-dry the pellet.
15. Resuspend RNA pellet in 200 µl TE buffer.
16. Purification of RNA: Treat the extracted RNA sample with RNase free DNase I to degrade the contaminating genomic DNA. Incubate at room temperature for 3 hrs. After that to denature DNase, add 1% SDS solution. Centrifuge at 10000 g for 15 minutes.
17. Extract supernatant carefully.
18. Perform ethanol precipitation of RNA (step 12 to 14).
19. Dissolve RNA pellet in TE buffer and store at –20°C.

5.6.1.4 *Troubleshooting guide*

1. The glassware should be sterile and free from RNases. The glassware should be autoclaved before use.
2. The plasticware should be rinsed with ethanol for decontamination.
3. RNA is very much susceptible to degradation by enzyme RNase. Therefore, prepare all the solutions using RNase free glassware and plasticware. Use DEPC treated water for preparation of solutions.

4. Wear gloves while carrying out the procedures, as human skin is also known to contain RNases.
5. Work quickly and finish the procedure as fast as possible to prevent RNA degradation.
6. If there is very low yield of RNA, it could be due to incomplete lysis of samples. To solve this problem, increase the quantity of lysis solution.
7. If the final RNA pellet doesn't dissolve completely in TE buffer, try heating at 60°C with intermittent vortexing to dissolve.
8. If you come across RNA degradation, it is possibly due to RNase contamination. Follow the precautions carefully to create RNase-free working environment. Also store RNA at –80°C to avoid degradation upon long term storage.
9. If you get DNA contamination in RNA samples, you might have accidentally taken a portion of interphase along with aqueous phase. It is advisable to leave a small amount of aqueous phase behind, to avoid such error.

5.6.2 RNA Extraction Protocol- II: Extraction of mRNA from Mammalian Tissues

5.6.2.1 *Introduction*

The isolation of pure and intact RNA from cells is important for gene expression studies, measuring transcript levels, production of cDNA etc. RNA isolation procedures need to be frequently performed on various tissue samples and on large number of samples simultaneously. There are various methods available to extract mRNA from cells, but the choice of isolation method depends upon source and nature of the tissue. The primary use of mammalian mRNA extraction is in study of gene expression. The mRNA extraction is of immense importance for identification of a mutant gene. Various studies of mRNA have revealed cellular processes such as RNA editing, mRNA splicing and alternative splicing.

5.6.2.2 *Principle and working*

The effective isolation of pure and intact mRNA is crucial for the study of gene expression. High yield and purity of isolated mRNA is essential especially while studying poorly expressed genes. Various techniques are available to isolate pure RNA. Most of the techniques need total RNA isolation method that disrupts the cells and inactivates RNases simultaneously. The first isolation step is extraction of total cellular RNA from tissue sample. This is followed by selection and separation of polyadenylated mRNA by affinity chromatography using oligo (dT)-cellulose columns. Mammalian mRNA has polyadenylated 3' end, bearing 200 adenine residues. This characteristic property of mammalian mRNA is particularly useful in selectively isolating it from the bulk of total RNA, which contains rRNA and tRNA in addition to mRNA.

The following protocol employs the technique of affinity chromatography for isolation of mRNA. The poly (A) tail of mRNA can bind to oligo (dT)-cellulose by formation of hydrogen bonds between adenine residue of poly (A) tail and thymine residue of oligo (dT)-cellulose which is used in the chromatographic column.

5.6.2.3 *Protocol for mammalian mRNA extraction and purification*

Materials required

- Purified RNA solution containing undegraded total RNA.
- NaOH 0.1M
- Sodium acetate 3M (pH=5.2)
- Sodium lauryl sarcosinate 10% w/v
- 1x column buffer
- 2x column loading buffer
- Elution buffer
- 70% and 100% ice-cold ethanol
- TE buffer pH=7.0 (RNase free)

Tris HCl (pH=7.0)	10 mM
EDTA (pH=8.0)	1 mM

- DEPC treated water
- Oligo (dT)-cellulose beads [oligo (dT)= oligodeoxythymidylate]
- Glass wool, sterile chromatography column
- Sterilized microcentrifuge tubes
- 2x column loading buffer

Tris HCl (pH=7.5)	40 mM
LiCl (DEPC treated)	1 M
EDTA (DEPC treated)	2 mM
Sodium lauryl sarcosinate	0.2%

- Elution buffer

Tris HCl (pH=7.5)	10 mM
EDTA (DEPC treated)	0.1 mM
DS	0.05%

- Preparation of oligo (dT) column: oligo (dT)- cellulose columns are available commercially; or can be prepared by using a sterile syringe. Remove the plunger from the syringe and plug the base with glass wool. Add oligo (dT)- cellulose to the syringe using sterile RNase free

Pasteur pipette. Oligo (dT)- cellulose will collect as a column over glass wool. Loading buffer will escape the glass wool and can be collected.

Procedure

1. Take a sterile chromatography column packed with glass wool at the bottom.
2. Suspend oligo (dT)-cellulose in 0.1N NaOH. Pour the column of oligo (dT)-cellulose in the chromatography column.
3. Equilibrate the oligo (dT) column with 1x loading buffer until pH is less than 8.0
4. Heat RNA in water for 5 min at 65°C. Cool quickly to room temperature. Add equal volume of 2x column loading buffer. [Heating of RNA solution is necessary to disrupt any secondary structures formed with poly (A) tail.]
5. Now add RNA solution to the column. Immediately collect the eluate in a sterile beaker. Up to 10 mg of total RNA can be loaded onto 1 ml of packed oligo (dT)-cellulose column.
6. Wash the column with 1x loading buffer and collect the eluate.
7. Heat all the eluate to 65°C for 5 min and rapidly cool to room temperature. Now apply it to the column. Again collect eluate.
8. Wash the column with 5–10 volumes of 1x column loading buffer.
9. Now elute the poly (A) mRNA with 2–3 column volumes of elution buffer. Collect the fractions of about 1/3rd of column volume.
10. Measure the absorbance of fractions at 260 nm to detect mRNA.
11. Now add 0.1 volumes of 3M of sodium acetate solution to the fractions containing mRNA. Vortex briefly.
12. Add 2.5 volumes of ice-cold 100% ethanol. Mix the contents and keep on ice for 30 min.
13. Centrifuge at 15000 rpm for 15 min at 4°C.
14. Discard the supernatant. Wash the pellet with 70% ethanol.
15. Repeat step 13.
16. Discard supernatant. Air-dry pellet.
17. Resuspend the pellet in TE buffer. Store at –70°C.

5.6.2.4 *Troubleshooting guide*

1. The glassware should be sterile and free from RNases. The glassware should be autoclaved before use.
2. RNA is very much susceptible to degradation by enzyme RNase. Therefore, prepare all the solutions using RNase free glassware and plasticware. Use autoclaved water for preparation of solutions.

3. Wear gloves while carrying out the procedures, as human skin is also known to contain RNases.
4. While packing the column with oligo (dT)-cellulose, take care not to trap any air bubbles, as they will interfere with the process.
5. Test the ability of several milliliters of buffer solution to flow through the oligo (dT)-cellulose column before adding RNA sample. If the column becomes clogged during use, re-suspend the packed resin with gentle mixing and prepare a new column.
6. Remember that applying pressure to pass the buffer or solution through the clogged column will greatly worsen the problem.
7. If you come across RNA degradation, it is possibly due to RNase contamination. Follow the precautions carefully to create RNase-free working environment. Also store RNA at –80°C to avoid degradation upon long storage.
8. Do not autoclave column loading buffer and elution buffer used in mRNA isolation, as they will froth profusely.
9. The mRNA has the shortest life as compared to other forms of RNAs. It is degraded very quickly after it is formed. Sometimes, a particular gene is active only for a short duration in an organism; for example, gene for fruit ripening in plants will be active only after fruiting stage. In such case, if you intend to isolate mRNA of the particular gene, source tissue used for mRNA isolation holds key to its success or failure.

5.6.3 RNA Extraction Protocol- III: Extraction of Total RNA from Plant Tissues

5.6.3.1 *Introduction*

Plant RNA can be isolated and purified from many sources such as leaves, roots, seeds, fruits, *in vitro* cultures etc. Structure of plant cell differs from animal cell significantly. Plant cells have numerous characteristic cell organelles such as chloroplasts, which are absent in animal cells. Plant cells have cell walls, which are made up of cellulose, hemicelluloses and pectins. These toughened cell walls provide definite shape to the cell, as well as structural support to the plant. Plants cells are rich in polysaccharides. They also produce secondary metabolites like phenols, tannins, terpenes etc. All these properties interfere with isolation and purification of RNA.

5.6.3.2 *Principle and working*

The first and foremost step in the process of RNA isolation from plant tissues is, disruption and degradation of plant cell wall. The common practice is to grind the plant tissue to a fine powder in liquid nitrogen

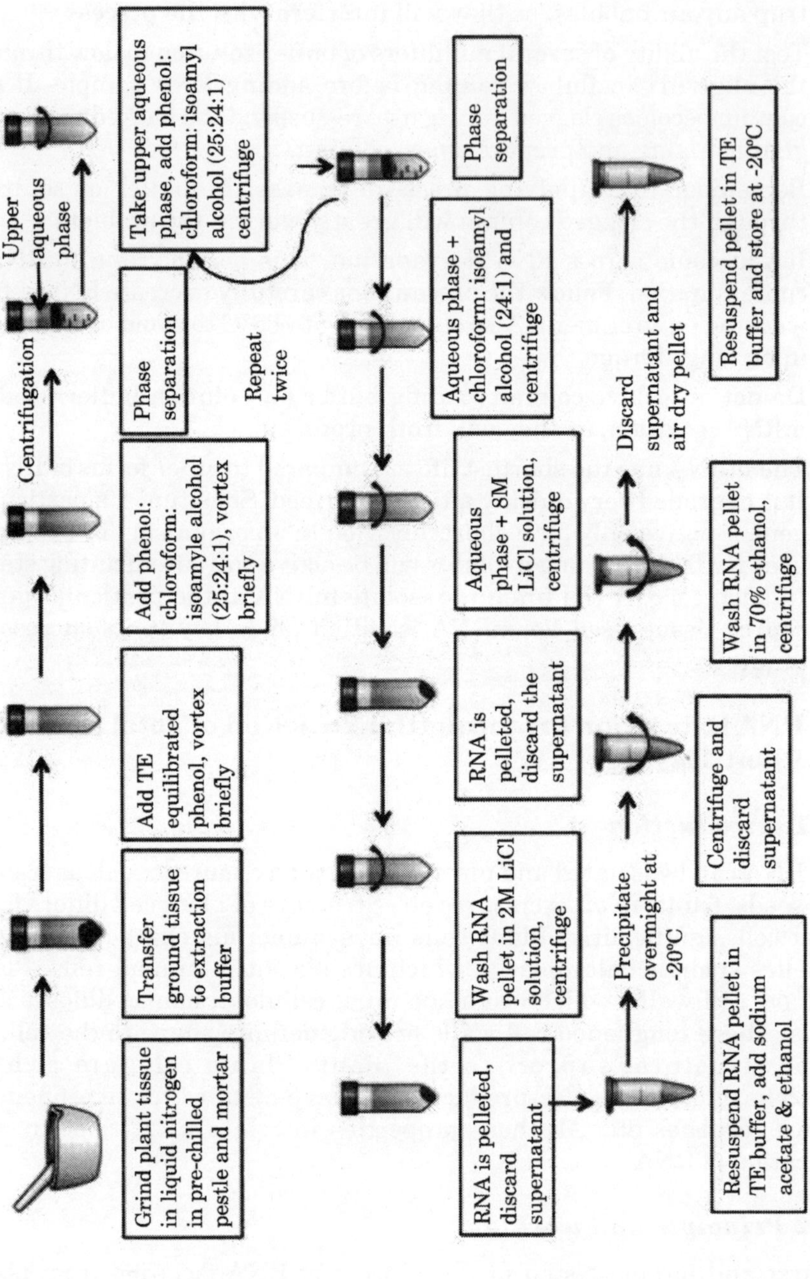

Fig. 5.1: Extraction of total RNA from plant tissues

using pre-chilled pestle and mortar. It is essential that the tissue is kept frozen throughout the grinding process, to ensure that the indigenous RNases remain inactive. Presence of polysaccharides hinders re-suspension of precipitated RNA. Using ethanol with potassium acetate or using 70% ethanol, with 2M NaCl, can effectively remove polysaccharide contamination. Precipitation of RNA in presence of lithium chloride (LiCl) and washing of RNA pellet with sodium acetate is also effective in removing polysaccharide contamination (Fig. 5.1).

5.6.3.3 *Protocol for total plant RNA extraction and purification*

Materials required

- DEPC treated double distilled water
- Liquid nitrogen
- Extraction buffer

Tris (pH=8.0)	100 mM
LiCl	100 mM
EDTA (pH=8.0)	5 mM
NaCl	100 mM
SDS	1% (w/v)

- Phenol equilibrated with TE
- Phenol: chloroform: isoamyl alcohol (25:24:1)
- Chloroform: isoamyl alcohol (24:1)
- TE buffer

Tris (pH=8.0)	10 mM
EDTA (pH=8.0)	1 mM

- LiCl solutions- 2M and 8M
- Sodium acetate solution (pH=5.2) 3M
- 70% and 100% ice-cold ethanol

Procedure

1. Set hot waterbath at 60°C. Preheat 50 ml phenol to 60°C in hot waterbath. Preheat 20 ml extraction buffer to 60°C in the waterbath.
2. Grind about 2 g of plant tissue in liquid nitrogen to fine powder in a pre-chilled pestle and mortar.
3. Quickly transfer ground powder to 20 ml preheated extraction buffer. Mix by inversion for 5 min or vortex.
4. Quickly add 10 ml preheated phenol to the tube containing plant tissue powder in extraction buffer. Vortex for 5–10 min.
5. Add 10 ml of chloroform: isoamyl alcohol (24:1). Vortex briefly.

6. Centrifuge at 2800g at 4°C for 10 minutes. Aqueous phase will be separated and be present as a top layer. The impurities remain in the lower layer.
7. Carefully transfer RNA containing aqueous phase to a fresh sterile polypropylene tube containing 20 ml of phenol: chloroform: isoamyl alcohol (25:24:1). Vortex the mixture for 5-10 minutes.
8. Centrifuge at 2800 g at 4°C for 10 minutes. Recover the upper aqueous phase without disturbing the interphase.
9. Repeat steps 7 and 8 twice, taking upper aqueous layer to a fresh sterile tube every time.
10. Extract the recovered aqueous phase with equal volumes of chloroform: isoamyl alcohol (24:1). Centrifuge for 5–10 minutes at 2800 g. Recover aqueous phase. This step is crucial to remove any traces of phenol, which may interfere with LiCl extraction.
11. To the recovered aqueous phase, add 1/3rd volume of 8M of LiCl solution. Mix the contents by inversion.
12. Pellet RNA by centrifugation at 12000 g at 4°C for 20 minutes. Discard the supernatant.
13. Wash the pellet in 2–5 ml of ice-cold 2M LiCl. Centrifuge at 12000 g for 10 minutes at 4°C. Discard the supernatant.
14. Resuspend pellet in 250 μl of TE buffer and transfer to 1.5 ml microfuge tube. Add 1/10th volume of 3M of sodium acetate and 2.5 volumes of ice-cold 100% ethanol. Precipitate overnight at –20°C.
15. Centrifuge for 15 minutes at 16000 g to pellet RNA at 4°C. Discard supernatant.
16. Wash pellet in 300 μl of 70% cold ethanol. Spin at 16000 g for 5 minutes at 4°C. Recover the pellet. Air-dry the pellet.
17. Resuspend RNA pellet in TE buffer. To determine the yield of total RNA, measure absorbance at 260 nm and 280 nm.

5.6.3.4 *Troubleshooting guide*

1. The glassware should be sterile and free from RNases. The glassware should be autoclaved before use.
2. The plasticware should be rinsed with ethanol for decontamination.
3. RNA is very much susceptible to degradation by enzyme RNase. Therefore, prepare all the solutions using RNase free glassware and plasticware. Use autoclaved water for preparation of solutions.
4. Wear gloves while carrying out the procedures, as human skin is also known to contain RNases. Work quickly and finish the procedure as fast as possible to prevent RNA degradation.
5. Plant cell wall is made up of cellulose, which acts as a barrier to protect the cellular contents. So it needs to be degraded first in order

to isolate RNA. Therefore, plant tissue/sample is ground in liquid nitrogen to fine powder in RNase free pre-chilled pestle and mortar. The rest of the procedure is more or less the same with necessary modifications depending on the plant species or type of samples. The exception is of plant protoplasts. If you have already isolated protoplasts, this preparatory step is not required as plant protoplasts are free of cellulosic cell wall.

6. If the final RNA pellet doesn't dissolve completely in TE buffer, try heating at 60°C with intermittent vortexing to dissolve.
7. If you come across RNA degradation, it is possibly due to RNase contamination. Follow the precautions carefully to create RNase-free working environment. Also store RNA at –80°C to avoid degradation upon long storage.
8. If you get DNA contamination in RNA samples, you might have accidentally taken a portion of interphase along with aqueous phase. It is advisable to leave a small amount of aqueous phase behind, to avoid such error.
9. If the yield of RNA is lower than expected, either based on your previous results or based on reported yields, it could be due to involvement of certain tissue or cell types. RNA yields can vary greatly in different cultured cell types and in different tissues.

5.6.4 RNA Extraction Protocol- IV: Extraction of Mitochondrial RNA from Plant Tissues

5.6.4.1 *Introduction*

In plant cells, mitochondrial RNA constitutes a fraction of total RNA. Its concentration can be about 1-2% of the total RNA. Isolation and purification of mitochondrial RNA (mtRNA) is required for many advanced molecular biology and genetic experiments, such as

- Construction of mitochondrial cDNA library
- Studying gene expression patterns of mitochondrial genes
- Analysis of plant mitochondrial transcription process
- Study of mitochondrial protein synthesis
- Study of cytoplasmic male sterility

5.6.4.2 *Principle and working*

Most procedures for isolating RNA from eukaryotic cells involve lysing and denaturing cells to liberate total nucleic acids. Additional steps are then required to remove DNA.

The process of mtRNA isolation is based on the combination of RNase A-guanidinium thiocyanate-Cesium chloride (CsCl) centrifugation. In this process, mitochondria are first separated from other cellular components such as nuclei and plastids by differential centrifugation of leaf homogenates. The crude mitochondrial suspension is further purified by sucrose gradient centrifugation. To remove the contaminating chloroplast RNA (cpRNA), the purified mitochondria are treated with enzyme RNase A. Thereafter, RNase A is inactivated and adding guanidinium thiocyanate at high concentration lyses mitochondria. Guanidinium thiocyanate is a strong protein denaturant, which inactivates RNase A. Then mtRNA is pelleted by CsCl gradient. Finally co-precipitated single stranded DNA in the CsCl gradient is removed from mtRNA by LiCl precipitation.

5.6.4.3 *Protocol for plant mtRNA extraction and purification*

Materials required

- CsCl solution- 5.7M CsCl, 10 mM EDTA (DEPC treated)
- Denaturation buffer

Formamide	50%
Formaldehyde	12%
MOPS buffer	1x

  (All ingredients are freshly made before use)
- MOPS buffer (pH=7.0)

MOPS	40 mM
Sodium acetate	10 mM
Na_2-EDTA	1 mM
- Extraction buffer

Sorbitol	0.35 M
Tris-HCl (pH=8.0)	50 mM
EDTA	5 mM
BSA	0.1%
Spermine	0.25 mg/ml
Spermidine	0.25 mg/ml

  (Store the buffer at 4°C. Add β-mercaptoethanol 0.2% just before the use)
- Make 4M guanidinium thiocyanate solution in 100mM Tris-HCl (pH=7.5). Store at 4°C. Add β-mercaptoethanol to 1% (final concentration) just before the use.
- 7.5M guanidinium –HCl, 10mM DTT (pH=7.5) store at 4°C.
- 2M and 4M LiCl (DEPC treated). Store at 4°C.
- 2M potassium acetate (pH=5.5) (DEPC treated)
- 2M sodium acetate (pH=7.0) (DEPC treated)

- 5% sodium lauryl sarcosinate
- TE buffer

Tris-HCl (pH=8.0)	10 mM
EDTA (DEPC treated)	1 mM

- Wash buffer

Sorbitol	350 mM
Tris-HCl (pH=8.0)	50 mM
EDTA	20 mM

Procedure

(a) Isolation of mitochondria

Carry out following procedure at 4°C. Keep solutions and bottles in ice bucket.

1. Take about 20 g of fresh green leaves and cut them to small pieces. Add about 200 ml of cold extraction buffer.
2. Homogenize the leaf tissue in blender. Filter homogenate through 2 layers of muslin cloth. Take the strained leaf homogenate in a centrifuge tube.
3. Centrifuge the filtrate at 2000 g for 10 minutes. Carefully transfer the supernatant to a clean tube. (Mitochondria remain in the supernatant while cell debris is pelleted). Centrifuge the supernatant for 20 minutes at 10000 g.
4. Mitochondria will be pelleted this time. Recover the pellet and discard the supernatant. Resuspend pellet in 100 ml extraction buffer. Centrifuge at 10000 g for 20 min.
5. Repeat step no 4. Recover the pellet.
6. Resuspend the mitochondrial pellet in 20 ml of ice-cold wash buffer.
7. Carefully layer each 10 ml mitochondrial suspension on the top of sucrose step gradient (9 ml 0.9 M/11 ml 1.5M/9 ml 1.75 M in wash buffer) and centrifuge for 45 minutes at 80000 g.
8. Collect the mitochondria from 0.9M/1.5M sucrose interphase (which appears as a yellow band) with the help of micropipette to a new centrifuge tube. Dilute it with 5 volumes of wash buffer.
9. Pellet mitochondria by centrifugation at 10000 g for 20 minutes.
10. Resuspend the pellet carefully in 1 ml of ice-cold wash buffer.

(b) Isolation of mtRNA

1. Incubate mitochondria with 20 μg/ml RNase for 60 minutes on ice. [RNase is used here to remove any chloroplast RNA attached to the mitochondrial membrane].

2. Add 5 volumes of 4M guanidinium thiocyanate solution. Add 0.5 volumes of 5% sodium lauryl sarcosinate after 2 minutes at room temperature and mix by vortexing.
3. Centrifuge at 5000 g for 5 minutes.
4. Recover the supernatant. Layer each 3.2 ml of mixture onto 1.1 ml cushion of DEPC treated 5.7M CsCl solution.
5. Carryout ultracentrifugation at 22000 g for 10 hrs.
6. Carefully aspirate the supernatant solution and cut off the top part of centrifuge tube that was in contact with solution. [Avoid direct touching with fingers. Wear gloves.]
7. Completely dissolve the RNA pellet by vortexing in 1 ml of 7.5M guanidinium-HCl solutions.
8. Add 0.05 volumes of 2M-potassium acetate and 0.5 volumes of ethanol.
9. Incubate at –20°C for 4hrs and precipitate mtRNA by centrifugation at 5000 g for 10 minutes.
10. Precipitate mtRNA by adding 0.1 volumes of 2M sodium acetate solution and 2.5 volumes of ethanol. Store at –20°C overnight. Then centrifuge at 10000 g for 30 minutes. Recover the pellet.
11. Add equal volumes of 70% ethanol. Centrifuge at 10000 g for 10 minutes. Discard the supernatant and air-dry the pellet.
12. Dissolve the pellet in TE buffer. To obtain mtRNA free from single stranded DNA, add equal volumes of 4M- LiCl solution and incubate at 4°C overnight. Centrifuge at 10000 g for 20 minutes. Wash the pellet once with 2M LiCl solution and twice with ethanol; pelleting by centrifugation each time.
13. Air-dry the pellet. Dissolve the pellet in RNase free sterile water. Estimate yield of mtRNA by measuring absorbance at 260 nm. The yield of mtRNA will be about 0.2–0.5 µg/g of sample tissue. Store mtRNA at –70°C till further use.

5.6.4.4 *Troubleshooting guide*

1. Cesium chloride extraction has one major drawback that it is a slow and laborious process of RNA isolation. Therefore it is unsuitable for large number of samples. This method however yields very high quality RNA.
2. Residual PEG can interfere with downstream processing of RNA.

5.6.5 RNA Extraction Protocol- V: Extraction of Total RNA from Bacteria

5.6.5.1 *Introduction*

All bacteria have RNAs in the form of rRNA, mRNA and tRNA. In eukaryotic ribosomes, rRNA is present as 25S, 18S and 5S while in prokaryotes; it is present as 23S, 16S and 5S. Unlike mammalian mRNA, bacterial mRNA does not contain 3'-poly (A) tail. Isolation and purification of bacterial RNA is important for molecular biology studies of prokaryotes, such as cDNA synthesis, gene expression studies, RT-PCR etc.

5.6.5.2 *Principle and working*

To isolate RNA from bacterial cells, first the peptidoglycan cell membrane of bacterial cell must be broken down. This is brought about by lysozyme-EDTA treatment. EDTA degrades lipopolysaccharides from outer membrane of gram-negative bacteria and lysozyme degrades the peptidoglycan layer. SDS is subsequently added to inhibit RNase, to remove membrane proteins and also to remove any proteins associated with RNA. Cell wall degradation and RNase inhibition occurs rapidly if SDS and proteinase K are used simultaneously. PEG (polyethylene glycol) is used to precipitate intact viruses and plasmid DNA. The cell debris and DNA can be precipitated by addition of PEG along with potassium phosphate. Phenol treatment with phenol: chloroform: isoamyl alcohol (25:24:1) causes phase separation and RNA is found in the upper aqueous layer while broken cell wall and other debris is trapped in the lower interphase layer (Fig. 5.2).

5.6.5.3 *Protocol for total bacterial RNA extraction and purification*

Materials required

- SDS 5% (w/v)
- Proteinase K 0.5 mg/l, (Store at –20°C)
- NaCl solution 0.85% (w/v). Autoclave and store.
- Polyethylene glycol (PEG)= 12.5% w/v in 1% (w/v) potassium phosphate (pH=5.2)
- PEG 20% (w/v) and 0.75% SDS (w/v) in 7.5% (w/v) potassium phosphate (pH=7.2)
- Sodium phosphate buffer 0.12M (pH=7.2)
- Sodium acetate 3M (pH=5.2)
- TE buffer (pH=8.0)
  Tris-HCl 10 mM
  EDTA 1 mM
- Luria broth= tryptone 10 g/l, yeast extract 5 g/l, NaCl 5 g/l, glucose 1 g/l; autoclave the medium.

Procedure

This protocol is for the isolation of total RNA from *Escherichia coli* culture.

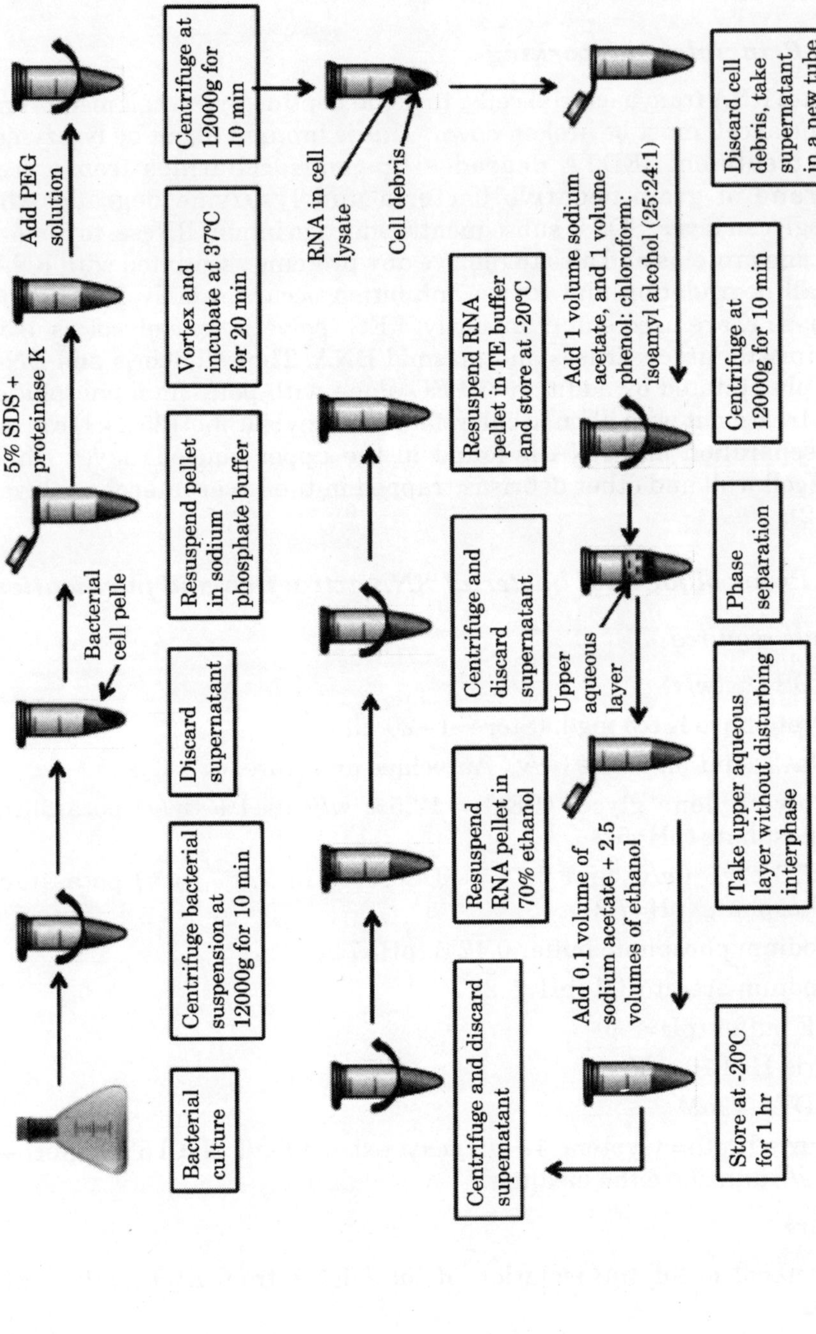

Fig. 5.2: Isolation of total bacterial RNA

(a) Preparation of bacterial culture

1. Inoculate loopful of *E. coli* culture to 50 ml of Luria broth. Incubate overnight at 37°C.
2. Use 5 ml of this culture to inoculate 50 ml of fresh Luria broth. Incubate for 3 hrs on rotary shaker at 37°C. Check for absorbance of 0.45–0.60 at 600 nm.
3. Take 1.5 ml of sample from broth into a sterile eppendorf tube. Centrifuge at 12000 g for 10 minutes. Discard the supernatant.
4. Suspend pellet in 300 µl of saline (NaCl solution). Centrifuge at 12000 g for 10 minutes.
5. Discard the supernatant. Resuspend the pellet in 400 µl of sodium phosphate buffer.

(b) Cell lysis and extraction with PEG

1. Add 50 µl of 5% SDS and 50 µl of proteinase K (0.5 mg/ml) to the above-mentioned bacterial suspension. Vortex briefly and incubate at 37°C for 20 minutes.
2. Add 500 µl of PEG solution. Vortex and centrifuge at 12000 g for 10 minutes.
3. At this stage, cell debris is present in the pellet and RNA remains in the supernatant. Carefully remove the supernatant and put it in a fresh sterile tube.
4. Add 0.1 volumes of (3M) sodium acetate solution. Add 1 volume of phenol: chloroform: isoamyl alcohol (25:24:1). Vortex briefly. Then centrifuge at 12000 g for 10 minutes.
5. Remove the upper aqueous layer and take it in a new sterile centrifuge tube. Repeat the extraction with 1 volume of phenol: chloroform: isoamyl alcohol (25:24:1). Vortex and centrifuge at the same speed. Remove upper aqueous layer and take it in a fresh sterile tube.
6. Add 0.1 volumes of sodium acetate and 2.5 volumes of ethanol. Store at –20°C for 1 hr to precipitate RNA.
7. Centrifuge at 12000 g for 15 minutes to pellet RNA. Discard the supernatant carefully. [The pellet may not be visible at this time].
8. Suspend RNA pellet in 70% ethanol. Centrifuge at 12000 g for 10 minutes. Discard the supernatant and air-dry the pellet.
9. Dissolve RNA pellet in TE buffer at room temperature and store at –20°C. Measure the yield of total RNA by taking absorbance readings at 260 and 280 nm. This protocol yields about 15 µg of total RNA/ml of microbial culture.

5.6.5.4 *Troubleshooting guide*

1. During tissue disruption for RNA isolation, it is crucial that the denaturant is in contact with the cellular contents at the moment

when the cells are disrupted. This can be problematic when using capsulated bacteria or, when samples are numerous, making rapid processing difficult. A common solution to these problems is to freeze the cells in liquid nitrogen. The samples must then be ground with a pre-chilled mortar and pestle into a fine powder and then the denaturing solution should be added.

2. A major problem with bacterial RNA isolation is that the bacterial mRNA has very short life inside the cell.

5.6.6 RNA Extraction Protocol- VI: Extraction of RNA Using Guanidinium Thiocyanate

5.6.6.1 *Introduction*

RNA isolation protocol, which makes use of guanidinium thiocyanate, is now widely used for all sorts of RNA preparations. This protocol gives better and higher yield of undegraded RNA. This protocol is also faster than others, making RNA extraction possible in just 4 hrs. This method provides a pure preparation of undegraded RNA in high yield and can be completed within short period of time. It is particularly useful for processing large numbers of samples and for isolation of RNA from minute quantities of cells or tissue samples.

5.6.6.2 *Principle and working*

The principle at the basis of the method is that RNA is separated from DNA after extraction with an acidic solution containing guanidinium thiocyanate, sodium acetate, phenol and chloroform, followed by centrifugation. Under acidic conditions, total RNA remains in the upper aqueous phase, while most of DNA and proteins remain either in the interphase or in the lower organic phase. Total RNA is then recovered by precipitation with isopropanol and can be used for several applications. By this method, RNA molecules of all sizes are purified and this technique can be easily scaled up or down according to the sample size. This single step method is based on the fact that RNA molecules remain dissolved in the aqueous phase in the guanidinium thiocyanate (4M) solution, in presence of organic phase. At this stage, DNA molecules remain in the organic phase while proteins and other macromolecules are retained in the interphase. Therefore, this process quickly produces a good quality RNA, free from genomic DNA.

5.6.6.3 *Protocol for RNA extraction and purification*

Materials required

- Denaturation solution:

  Guanidinium thiocyanate 4M

Sodium citrate (pH=7.0)	25 mM
N-lauryl sarcosinate (sarcosyl)	0.5% (w/v)
β-mercaptoethanol	0.1M

(Store at room temperature)

- Sodium acetate (pH=4.0) 2M (prepared in glacial acetic acid)
  (Store at room temperature)
- Water saturated phenol
- Chloroform: isoamyl alcohol (49:1)
- Isopropanol
- DEPC treated water
- Sodium dodecyl sulphate

Procedure

1. Take homogenized tissue or cell suspension sample. Add 1ml of denaturation solution per 100 mg of homogenized tissue or 1×10^7 cells of cell suspension. Keep the samples in denaturation solution for 20 minutes.
2. Transfer the cell lysate to a sterile centrifuge tube. Add following reagents to 1 ml of cell lysate: 0.1 ml of sodium acetate (2M). Mix by inversion. Add 1ml of water-saturated phenol. Mix by inversion. Add 0.2 ml of chloroform: isoamyl alcohol (49:1). Shake vigorously or vortex.
3. Cool the samples on ice for 15 minutes.
4. Centrifuge at 10000 g for 20 minutes at 4°C. The acidic pH is critical to ensure the separation of RNA from DNA. Therefore use water saturated phenol and not buffered phenol.
5. The upper aqueous phase contains RNA. Transfer the aqueous phase carefully using sterile RNase-free pipette to sterile eppendorf tube.
6. Add 1ml isopropanol in order to precipitate RNA. Incubate overnight (or at least for 1 hr) at –20°C.
7. Centrifuge for 20 minutes at 10000 g at 4°C. Discard the supernatant. RNA can be seen sticking on the sides of the tube in the form of a gel like pellet.
8. Dissolve RNA pellet in 0.3 ml of denaturation solution. Then add 0.3 ml of isopropanol.
9. Incubate overnight (or at least for 1hr) at –20°C.
10. Centrifuge at 10000 g for 10 minutes at 4°C. Discard the supernatant. This 2nd precipitation step removes DNA and proteins from RNA, but may also reduce the RNA yield.
11. Resuspend RNA pellet in 0.5–1 ml of 70% ethanol, vortex briefly. Incubate samples at 4°C for 15 minutes.

12. Centrifuge for 10 minutes at 10000 g at 4°C. Discard the supernatant.
13. Air-dry the RNA pellet.
14. Dissolve RNA pellet in DEPC treated sterile water and store at –20°C.
15. To quantify RNA, take absorbance readings at 260 nm and 280 nm.

5.6.6.4 *Troubleshooting guide*

1. At the time of phenol-chloroform extraction, the mixture must be spun with sufficient force to ensure adequate separation of organic and aqueous phases. Failing to do this might add protein or genomic DNA contamination in the RNA sample.
2. The interface between aqueous and organic phase is a potential source of protein contamination. While extracting aqueous phase, avoid sucking up of interphase. It is advisable to leave small amount of aqueous layer behind to avoid this mistake.
3. Residual guanidinium salt can interfere with subsequent reactions. Excess salt can be removed by washing RNA pellet in 70% ethanol.

5.7 METHODS FOR DETECTION OF ISOLATED RNA

5.7.1 Gel Electrophoresis and Visualization of RNA Under UV-Transilluminator

RNA is negatively charged due to presence of phosphates and charge per unit length of RNA is constant. Therefore, movement of RNA in a gel is dependent on size and shape. Horizontal agarose gel electrophoresis is commonly used for the separation of RNA molecules of different sizes. The gel is usually stained with ethidium bromide so that the separated RNA bands in the gel can be viewed under UV transilluminator. In pure RNA samples, 28S rRNA and 18S rRNA bands can be distinctly observed in the gel.

Analyzing and detecting RNA by gel electrophoresis is different than DNA analysis. RNA needs to be denatured for optimal electrophoretic separation. As RNA is single stranded, it doesn't incorporate stains as easily as DNA.

Materials required

- Deionized formamide
- 5x MOPS buffer (3-morpholinopropane sulphonic acid)

MOPS (pH= 7.0)	0.2 M
Sodium acetate	0.05 M
EDTA (pH= 8.0)	0.005 M

Autoclave and store at room temperature

- RNA sample buffer

Deionized formamide	10.0 ml
37% formaldehyde	3.5 ml
5x MOPS buffer	2.0 ml

  Store in a tightly sealed container at –20°C
- RNA loading buffer

Glycerol	50%
EDTA	1 mM
Bromophenol blue	0.4%

  Store at –20°C.

Procedure

1. Prepare 1% agarose gel containing ethidium bromide. Dissolve 1 g agarose in 20 ml of 5 x MOPS buffer, 72 ml of DEPC treated water by heating them together.
2. Then gelling mixture is allowed to cool to 55°C. Add 17.6 ml of 37% formaldehyde and 5 µl of 10 mg/ml ethidium bromide under fume hood. Mix the contents gently.
3. Pour the gelling mixture in the gel-casting tray with well placed comb. Allow the gel to polymerize.
4. Prepare RNA loading sample by mixing 1 part RNA sample with 2 parts of RNA loading buffer. Heat the contents to 65°C for 5 minutes and cool on ice for 2 minutes.
5. Then add 2 µl of RNA loading buffer.
6. Place the polymerized gel in the electrophoresis apparatus with 1x MOPS buffer. Load the RNA samples and run the gel at 4–5 V/cm.
7. Continue electrophoresis until the bromophenol blue dye has migrated to about 3/4th of the gel.
8. Visualize RNA migration in gel using UV transilluminator.

5.7.2 Detection and Quatitation of RNA Using Spectrophotometer

To determine purity of RNA sample, spectrophotometric readings at 260 and 280 nm are taken. The ratio of absorbance readings at 260 and 280 nm (A_{260}/A_{280}) provides estimate of purity of RNA sample with respect to contaminants that absorb UV light such as proteins. However it is important to note that A_{260}/A_{280} ratio is influenced considerably by pH. Since absorbance readings are taken in water that is not buffered, the fluctuation in pH can result in changes in A_{260}/A_{280} ratio. Lower pH may result in lower A_{260}/A_{280} ratio and thus can give false reading about protein contamination. Pure RNA has A_{260}/A_{280} ratio in the range of 1.9–2.1.

The concentration and purity of RNA in solution can be measured by spectrophotometer. The concentration of RNA in solution is determined by measuring the absorbance at 260 nm that is A_{260}. When absorbance at A_{260} is 1, the concentration of RNA is about 40 μg/ml. However the actual value depends on the base composition, since the bases differ in extinction coefficient. To measure the RNA concentration, dilute RNA solution to about 1:50 in TE buffer.

Procedure for measuring concentration of RNA

Materials required

- RNA sample
- 1 ml quartz cuvette (with 1 cm path length)
- 1x TE buffer
- UV spectrophotometer

Procedure

1. Dilute the RNA solution by adding 20 μl of RNA aliquot to 0.98 ml of TE buffer in a microcentrifuge tube. Mix the contents well.
2. Switch on the UV lamp of spectrophotometer and wait for 15 minutes for the lamp to warm up.
3. Adjust wavelength to 260 nm.
4. Add TE buffer to cuvette to use it as blank and calibrate the spectrophotometer.
5. Now take absorbance reading at 260 nm with diluted RNA sample.
6. Repeat the same procedure and take absorbance reading at 280 nm.
7. Calculate RNA concentration.

Calculation of concentration of RNA

$$(\text{RNA})\ \mu g/ml = A_{260} \times \text{Dilution factor} \times E$$

Where, A_{260} = absorbance in optical density at 260 nm
E= extinction coefficient (Average extinction coefficient of RNA is 40 $\mu g/OD_{260}$)

5.7.3 Troubleshooting Guide

1. Pure preparations of RNA have ratio A_{260}/A_{280}= 1.9 to 2.1. If there is any protein or phenol contamination, this ratio will be much lower. To remove protein contaminants, perform phenol and chloroform extractions once again.
2. RNase contamination is the possible cause, if you get a smear instead of RNA bands in gel. Failure to amplify RNA in RT-PCR is also indicative of nuclease contamination.

3. A low A_{260}/A_{230} ratio in RNA preparation is indicative of guanidinium salt residue in the RNA sample or some other organic inhibitor. In case of such contamination, repeat the ethanol wash step.
4. Do not overrun the gel or else RNA may come out of the gel. Watch for the migration of bromophenol blue dye in gel as an indicator of the RNA migration.

5.8 REVIEW QUESTIONS

1. Describe the importance of RNases in RNA isolation procedure.
2. What measures would you undertake to create RNase free environment for RNA isolation/extraction?
3. What is the difference between total RNA and mRNA? Give the characteristic property of mammalian mRNA.
4. What is the significance of using oligo (dT)-cellulose for mRNA isolation?
5. Compare DNA isolation with RNA isolation process.
6. Describe in brief the importance and applications of RNA extraction.
7. What is the role of DEPC in RNA extraction process?
8. Compare DNA isolation with RNA isolation procedure.
9. What is mtRNA? How will you isolate mtRNA? Can you isolate mtRNA from animal cells?
10. What is the role of RNase in mtRNA extraction procedure?
11. Give applications of mtRNA isolation.
12. State the differences between a bacterial cell and an animal cell.
13. What are the basic differences between bacterial and mammalian: rRNA and mRNA?
14. What is the role of phenol: chloroform: isoamyl alcohol in the RNA extraction process?

5.9 REFERENCES

Bhat, S., Nagasampagi, B. and Sivakumar, M. (2005). Chemistry of natural products. Narosa Publishing House. p. 419.

Birnboim, H. (1988). Rapid extraction of high molecular weight RNA from cultured cell and granulocytes for northern analysis. *Nucleic Acids Research,* **16(4):** 1487–1497.

Bohmann, K., Hennig, G., Rogel, U., Poremba, C., Mueller, B., Fritz, P., Stoerkel, S. and Schaefer, K. (2009). RNA extraction from archival formalin-fixed paraffin-embedded tissue: A comparison of manual, semi automated and fully automated purification methods. *Clinical Chemistry,* **55(9):** 1719–1727.

Chatterjee, A., Mirer, P., Santamaria, E., Klapperich, C., Sharon, A. and Badge, A. (2010). RNA isolation from mammalian cells using porous polymer monoliths: An approach for high throughput automation. *Analytical Biochemistry,* **82(11):** 4344–4356.

Chomczynski, P. and Sacchi, N. (1987). Single step method of RNA isolation by acid-guanidinium thiocyanate-phenol-chloroform extraction. *Analytical Biochemistry,* **162:** 156–159.

Chomczynski, P. and Sacchi, N. (2006). The single–step method of RNA isolation by acid–guanidinium thiocyanate–phenol–chloroform extraction: Twenty something years on. *Nature Protocols,* **1:** 581–585.

Chomczynski, P. (1992). Solubilization in formamide protects RNA from degradation. *Nucleic Acids Research,* **20:** 3791–3792.

Di Cello, F., Xie, Y., Satyaseela, M. and Kim, K. (2005). Approaches to bacterial RNA isolation and purification for microarray analysis of *Escherichia coli* K1 interaction with human brain microvascular endothelial cells. *Journal of Clinical Microbiology,* **43(8):** 4197–4199.

Draper, D. (2004). A guide to ions and RNA structure. *RNA,* **10:** 335–343.

Farrell, R. (2010). RNA methodologies: A laboratory guide for isolation and characterization. 4th Edition. Academic Press. p. 141.

Garesse, R. and Vallejo, C. (2001). Animal mitochondrial biogenesis and function: A regulatory cross–talk between two genomes. *Gene,* **263:** 1–16.

Gasic, K., Hemandez, A. and Korban, S. (2004). RNA extraction from different apple tissues rich in polyphenols and polysaccharides for cDNA library construction. *Plant Molecular Biology Reports,* **22:** 437a–437g.

Gjerde, D., Hoang, L. and Hornby, D. (2009). RNA purification and analysis: Sample preparation, extraction, chromatography. Wiley–VCH, Weinheim. p. 195.

Hemmrich, K., Denecke, B., Paul, N., Hoffmeister, D. and Pallua, N. (2010). RNA isolation from adipose tissue: An optimized procedure for high RNA yield and integrity. *Lab Medicine,* **41:** 104–106.

Holley, R., Apgar, J., Everette, G., Madison, J., Marquise, M., Merrill, S., Penswick, J. and Zamir, A. (1965). Structure of ribonucleic acid. *Science,* **147(3664):** 1462–1465.

Jacoli, G., Ronald, W. and Lavkulich, L. (1973). Inhibition of ribonuclease activity by Bentonite. *Can. J. Biochem.,* **51:** 1558–1565.

Kreig, P. (1996). A laboratory guide to RNA isolation, analysis and synthesis. Wiley–Liss Inc. USA. p. 445.

Logemann, J., Schell, J. and Willmitzer, L. (1987). Improved method for the isolation of RNA from plant tissues. *Analytical Biochemistry,* **163:** 16–20.

Magalhaes, P., Andreu, A. and Schon, E. (1998). Evidence for the presence of 5S rRNA in mammalian mitochondria. *Mol. Biol. Cell,* **9:** 2375–2382.

Maier, U., Bozarth, A., Funk, H., Zauner, S., Rensing, S., Schmitz–Linnweber, C., Borner, T. and Tillich, M. (2008). Complex chloroplast RNA metabolism: Just debugging the genetic programme? *BMC Biology,* **6:** 36.

Makaroff C and Palmer J. 1987. Extensive mitochondrial specific transcription of *Brassica campestris* mitochondrial genome. *Nucleic Acids Research* **5:** 5141–5156.

Mangan, J., Sole, K., Mitchinson, D. and Butcher, P. (1997). An effective method for RNA extraction from bacteria refractory to disruption including mycobacteria. *Nucleic Acids Research,* **25:** 675–676.

Rapley, R. (2000). The nucleic acid protocols handbook. Humana Press Inc. p. 1050.

Siebert, P. and Chenchik, A. (1993). Modified acid guanidinium thiocyanate–phenol–chloroform RNA extraction method, which greatly reduces DNA contamination. *Nucleic Acids Research,* **21(8):** 2019–2020.

Slomovic, S., Laufer, D., Geiger, D. and Schuster, G. (2005). Polyadenylation and degradation of human mitochondrial RNA: The prokaryotic past leaves mark behind. *Mol. Cell Biol.,* **25(15):** 6427–6435.

Stern, D. and Newton, K. (1986). Isolation of plant mitochondrial RNA. *Methods Enzymol.,* **118:** 488–496.

Tan, S. and Yiap, B. (2009). DNA, RNA and protein extraction: The past and the present. *Journal of Biomedicine and Biotechnology,* **1:** 1–10.

Vasquez, L., Razer, L. and Kerr, R. (2006). Comparison of two total RNA extraction protocols using marine gorgonian coral *Pseudopterogorgia elisabethae* and its symbiont *Symbodinium* sp. *Electronic Journal of Biotechnology,* 9(5).

Walter, M., Kilian, J. and Kudla, J. (2002). PNPase activity determines the efficiency of mRNA 3'–end processing, the degradation of tRNA and the extent of polyadenylation in chloroplasts. *EMBO Journal,* **21:** 6905–6914.

Ye, F., Albaum, M., Markmann–Mullsch, U. and Abel, W. (1993). Improved method for the isolation of mitochondrial RNA from green leaves. *Bio. Techniques,* **14:** 184–185.

6

Radioisotopes, Radiolabelling of Probes and Autoradiography

6.1 INTRODUCTION–USE OF RADIOISOTOPES IN BIOLOGICAL SCIENCES

The discovery of radioactive isotopes and their use in biological sciences have come hand in hand. Henri Becquerel first discovered radioactivity accidentally in 1896. He found that images were developed on the photographic film that was left on top of his desk without any exposure to light. The only source of energy that must have caused these images to develop was the rock containing uranium that was kept in the desk drawer at that time. Therefore he concluded that the uranium inside that rock must have emitted energy that caused the image of the rock to develop on the photographic film. The ability of these radiations to create image on the photographic plate is the basis of almost all diagnostic and research applications of the radioisotopes. In the beginning of 20th century, pioneering work in radioisotopes was done by Marie and Pierre Curie. Marie Curie named the phenomenon of energy transmittance by these elements as 'radioactivity'. Marie and Pierre Curie both took initiatives for the use of radium in research and treatment of cancer.

The hazards of radioactive materials were discovered much later. However, despite the hazards, radioisotopes are used worldwide for numerous diagnostic, research and health measures.

The radiation released by radioisotopes can be detected by placing a photographic film over samples containing radioisotopes or by using an instrument known as scintillation counter. This technique allows researchers to use radioisotopes as tracers in biochemical reactions. In typical experiments, organisms are exposed to chemicals that are labelled with radioisotopes such as ^{14}C or ^{3}H. After the exposure, the biological product in which the tracer appears and the sequence in which they appear reveals the intracellular biochemical reactions. To trace reactions of photosynthesis, Melvin Calvin and his coworkers used radioactive carbon ^{14}C. They grew algal cells in an environment that supplemented radioactive carbon through carbon dioxide. Then they extracted various chemicals at different intervals from algae and exposed them to photographic film. The substances that produced spots on the film due to presence of radioactive carbon in them and the sequence in which they appeared in the cell, allowed the researchers to put together, all reactions of photosynthesis.

Radioisotopes are widely used in medicine to diagnose and cure diseases. Radioactive chemical tracers provide diagnostic information about a person's anatomy and the functioning of specific organs. Radiotherapy also uses radiation in the treatment of some illnesses, such as cancer. More powerful radiation sources like gamma rays are used to sterilize syringes and other medical equipment. The extensive use of radiations in medical field has given rise to a branch of medicine called as 'nuclear medicine' that uses radiation to provide information about the functioning of a person's specific organs or to treat diseases. Today physicians use controlled radiations to make a quick and accurate diagnosis of the patient's illness. The vital organs such as thyroid, bones, heart, liver and many others can be easily imaged and disorders in their function revealed. An example can be given of their use in thyroid related illnesses.

The science of nuclear medicine was developed in the 1950s by physicians who used radioactive iodine (^{131}I) to diagnose and treat thyroid diseases. Thyroid is the only organ in the body that absorbs iodine in major quantity. In this method of diagnosis, the size and shape of the thyroid that reflects its health, is measured by injecting a small amount of radioactive iodine into the patient's bloodstream. After the iodine isotope gets concentrated in the thyroid, the gland is then scanned to produce its image on the photographic film.

The unique advantage that radioisotopes offer for problem solving and research in biological sciences is due to the ease with which it is possible to follow the transfer and transformation of radioisotopes within biological systems. The radioisotopes of elements differ in atomic mass and other physical characteristics, but they essentially have the same chemical properties as their nonradioactive counterparts. Therefore living organisms can take up radioisotopes without changes in their biochemical reactions. Radioactive isotopes of commonly occurring elements such as carbon can serve as tracers because they are chemically very similar to the non-radioactive isotopes and therefore most chemical, biological, and ecological processes treat them in the same identical way. They can thus be used as radiotracers, enabling researchers to follow a biochemical pathway or reaction of interest.

6.2 RADIOACTIVE ISOTOPES

6.2.1 Introduction

The nuclei of some atoms are unstable and disintegrate spontaneously, emitting energy that is carried by nuclear particles like alpha and beta and/or by gamma radiation. The phenomenon of radioactive decay with emission of radiation is called radioactivity. Various radioisotopes of commonly found elements are used in biological research as well as diagnosis and treatment of diseases in nuclear medicine. Most biomedical studies have employed radioactive carbon C^{14} and hydrogen H^3 because they can be easily incorporated into organic molecules. Other radioisotopes used are aluminium (^{26}Al), cadmium (^{41}Ca), iodine (^{131}I), chlorine (^{36}Cl), nickel (^{99}Ni) and (^{63}Ni).

Sources for Acquisition of Radioisotopes in India

Radioisotopes of elements that are not found naturally can be produced in a nuclear reactor by exposing appropriate target material to the neutrons in the reactor, thereby causing a nuclear reaction to occur which leads to the production of desired radioisotope. Bhaba Atomic Research Centre (BARC), Mumbai, India supplies reactor produced radioisotopes and radionuclides for medical and research use. The radioisotopes are processed and supplied by the Board of Radiation & Isotope Technology (BRIT), Mumbai for medical uses across the country, including radiopharmaceuticals, brachy-therapy wires, radio-immunoassay (RIA) kits and various other products, and services. The accelerator at VECC manufactures radioisotopes, which are processed for medical applications whereas; the Regional Radiation Medicine Centre (RRMC) meets the requirements of the eastern region of the country, for radio-diagnosis and therapy.

6.2.2 Radioactivity and Half-Life

The mass of the atomic nuclei can vary slightly for a particular element although the number of electrons remains constant and all the isotopes have the same chemical properties. The nuclei of radioactive isotopes are unstable and they disintegrate to produce new atoms and, at the same time, give off radiations such as electrons (β rays) or radiations (γ rays). Naturally occurring radioisotopes are rare because of their instability, but radioactive atom can be produced in nuclear reactors by bombardment of stable atoms with high-energy particles. The detection methods for radioactivity are extremely sensitive and every radioactive atom that disintegrates can be detected.

All the materials in our surroundings are made of atoms. Atom is the smallest particle to which a matter can be reduced. But even atoms can be further broken down to smaller units. The core of atom is its nucleus, which contains most of its mass, which is surrounded by electrons. The nucleus is made of protons and neutrons. Protons bear positive charge while neutrons contain some mass but are devoid of any charge. The electrons circling the nucleus are negatively charged and are equal in number to the protons in the nucleus. All chemical and biological properties of the atom are due to the electrons, especially their number and the way their orbits around the nucleus are arranged. The number of electrons/protons amounts to the 'atomic number' of that element. The mass of neutrons is same as mass of protons, whereas electrons are almost weightless. The number of protons and neutrons together make up the total mass of the nucleus, which amounts to the 'mass number' of the particular element. Therefore if the number of protons in an atom is kept constant, but the number of neutrons is changed, the chemical properties of the element will remain the same but atomic weight will change. Therefore, atoms of differing weights but identical chemical properties are called 'isotopes'. Many isotopes of elements exist naturally, for *e.g.* Carbon ${}_6C^{12}$ is also found mixed with its isotopes, ${}_6C^{13}$ and ${}_6C^{14}$.

The interaction between a proton and a neutron produces a more stable arrangement of nuclear particles. Therefore, the most stable nuclear configuration has equal number of protons and neutrons. For example, naturally occurring carbon ${}_6C^{12}$ has 6 protons and 6 neutrons, but carbon isotope ${}_6C^{14}$ has 6 protons and 8 neutrons. The nucleus of this isotope has too many neutrons compared to the number of protons, which causes it to disintegrate, thereby giving off radiation. This radioactive isotope of carbon is of great importance in biochemical research. It disintegrates not by loss of neutrons but by transformation of one of the neutrons into proton. Thereby radioactive carbon isotope ${}_6C^{14}$ becomes a stable, nonradioactive isotope of nitrogen ${}_7N^{14}$. As a neutron is composed of one proton and one electron, when a neutron transforms into a proton, it also gives off one electron. The escaping electron is called a β-particle and the radiation is called β-radiation. Similarly there are isotopes that have more protons than neutrons. In this case, the unstable isotope emits positrons in order to achieve a stable configuration. Positron is a weightless particle bearing positive charge.

For biological work, it is important to choose an isotope emitting radiation of the right energy for a particular experiment. For instance, if chromosomes are labelled with two types of radioactive isotopes- carbon isotope ${}_6C^{14}$ and hydrogen isotope H^3, the former emits β radiation of medium energy while the latter emits β radiation of very low energy. When exposed to photographic film, the carbon atom blackens many silver grains of the photographic emulsion with each disintegration, because of the higher energy of its radiation. The energy of H^3 radiation is so low that each disintegrating atom of the isotope causes blackening of one silver grain only. Therefore the result obtained on photographic film by exposure to C^{14} labelled chromosome gives blurred image while resulting image from exposure to H^3 labelled chromosomes appear sharper.

Radioactive Half-life

Each atom of a given radionuclide has the same probability of undergoing radioactive decay per unit time. For a given period of time, some atoms decay while others remain unchanged. In particular, after a period of time, which is called as the radioactive half-life, only half of the atoms originally present remain intact. After a second half-life period, the number of atoms remaining is further halved, *i.e.*, a quarter of the original number, and so on. The decay or disintegration is a continuous process in a radioactive material. The half-life is defined as the time taken when one half of the starting material is decayed. Different materials have different half lives (Refer Table 6.1).

Table 6.1: Various radioactive elements and their half-lives

Radioactive element	*Atomic number*	*Symbol and atomic mass number*	*Half-life*
Hydrogen	1	H^3	12 years
Carbon	6	C^{14}	5000 years
Sodium	11	Na^{24}	14.8 hours
Phosphorus	15	P^{32}	14.3 days
Sulphur	16	S^{35}	87 days
Iodine	53	I^{131}	8 days

6.2.3 Applications of Radioisotopes

Radioisotopes, because of their radiation characteristics and the energy they posses, can be utilized in industry, agriculture, healthcare and research applications.

1. The discipline of cytology, which is the study of cells and their structure, and histology, the study of tissues and their structures can both benefit from autoradiography.
2. Autoradiographs can help scientists learn more about cellular and tissue structures and how radioactive materials are absorbed and distributed.
3. In digital autoradiography, computers and radiation detection instruments work together to provide the radiographic image.
4. Exposure of food items to ionizing radiations like gamma rays can inactivate a spectrum of microorganisms that bring about food

spoilage. Very strong gamma radiation emitting isotopes like cobalt Co^{60} and cesium Ce^{137} are widely used in the food industry for sterilization of packed food items such as milk and meat.

5. Research on chemical receptors, their distribution within tissues and their function can also benefit from autoradiography.
6. In the field of agriculture, radioactive tracers are used to study the action of fungicides and herbicides on physiology of plants.
7. Biochemical agents infused with radioactive substances, called radioligands are injected into a tissue. These agents react with chemical receptors in the tissues, and when an autoradiograph is subsequently taken, it will allow the researcher to determine where these receptors are located, their relative frequency, and density of distribution. This technique is sometimes used to diagnose various medical problems.
8. Autoradiography is used in various detection techniques in molecular biology like southern hybridization, northern hybridization, *in situ* hybridization etc.
9. In the industry, radioactive tracer technique can be used to measure the uniformity of mixture. For the same purpose, short-lived radioisotopes like Na^{24} or Mn^{56} are used. This method is used in the manufacturing of chocolate, cement, paint, soaps, fertilizer, animal feed and medicinal tablets.
10. Radioisotopes are commonly used in radiotherapy for treatment of a number of cancers (Refer Table 6.2). Rapidly dividing cells are particularly sensitive to damage by radiation. For this reason, some cancerous growths can be controlled or eliminated by irradiating the area containing the growth. The different types of cancers that are routinely treated by radioisotopes are as follows,

Table 6.2: Radioisotopes used in radiation therapy for treatment of cancer

Type of cancers	*Radioactive material used*
Thyroid cancer	Iodine (I^{131})
Leukemia	Phosphorus (P^{32})
Brain tumour	Cobalt (Co^{60})
Bone and skeletal cancer	Phosphorus (p^{32})
Neuroendocrine tumours	Lutetium (Lu^{177}), Yttrium (Y^{90})
Pancreatic, ovarian and melanoma cancers	Lead (Pb^{212})
Malignant lymphoma	Cobalt (Co^{60})

11. In physiology, radiotracers are used to study the rate of distribution of elements in the body, study of various metabolic pathways, studying protein synthesis, functions of different cell organelles as well as to measure and monitor the uptake and distribution of nutrients in the body.

12. The mutagenic effects of radioisotopes were found in 1920s and since then, radiation has been used in genetic research as an important means of obtaining new mutations in experimental organisms.
13. Nuclear radiation from radioisotopes has been widely used for inducing mutations in order to develop new crop varieties. This technique is known as mutation breeding. Different types of plant materials like seeds, cuttings, rhizomes, bulbs, tubers, leaves etc. are irradiated with nuclear radiations. In most of the cases, irradiation leads to the development of lethal mutations or undesirable traits in plant tissues. After screening of plants for mutations, useful mutations can be identified and their heritability tested. The stable and desirable mutations can be further used in plant breeding programs for development of varieties.
14. Radiations from certain radioisotopes are used for killing insects, which damage the food grains. Certain seeds and canned food can be stored for longer periods by gently exposing them to radiations.
15. Radioisotopes are also used for producing high yielding crop seeds. The research at Trombay (India), in the field of crop development has led to the development of 23 high yielding varieties of pulses, oilseeds, rice and jute.
16. At BARC several groundnut varieties have been developed using the radioisotopes, which are very popular amongst farmers.
17. Radioisotopes have numerous applications in the industry. These include radiography, gamma scanning of process equipment, use of radiotracers to study sediment transportation at ports and harbours, flow measurements, hydrology and water resource management etc.
18. Using γ-ray photography one can find out wearing of cutting tools and lathes and can locate internal cracks in stones. One can check any non-uniformity in the thickness of a sheet by β or γ-absorption measurements. The sheet is made to run continuously between a radioisotope and a counter like scintillation counter. A change in the counting rate indicates a variation in the thickness of the sheet. The output from the counter may be used to correct the machinery, and thus the thickness is automatically kept constant. This method is used as a thickness control in the manufacture of paper, plastic, metal sheet, etc.

6.3 HEALTH HAZARDS OF RADIOACTIVITY

Radioactive materials that decay essentially produce ionizing radiations. The ionizing radiations such as alpha, beta and gamma have enough energy to destabilize atoms and break chemical bonds. Any living organism coming in contact with the ionizing radiation gets damaged in a unique way. Chronic radiation exposure of low level for a long period of time can cause significant damage to human body, which most probably leads to development of cancer.

Radiation has been a subject of fascination and concern ever since Wilhelm Konrad Röntgen discovered X-rays on 8 November 1895. Over the years, health workers as well as the public have been concerned about medical uses of X-rays, the presence of radon in buildings, radioactive waste from nuclear power stations, fallout from nuclear test explosions, radioactive consumer products, microwave ovens, and many other sources of radiation.

In order to quantify how much radiation we are exposed to in our daily lives and to assess potential health impacts as a result, it is necessary to establish a unit of measurement of radiation exposure. For evaluation of risk of exposure to ionizing radiation to human body, the dose equivalent in 'rem' (Roentgen equivalents in man) can be defined as,

$$\text{rem} = \text{rads} \times Q \times N$$

Where, the radiation dose is given in rads, Q is the quality factor and N is the tissue factor. The value of Q is 1 for X-rays and gamma rays. For alpha particles, value of Q is 20.

The basic unit of radiation dose absorbed in tissue is the gray (Gy), where one gray represents the deposition of one joule of energy per kilogram of tissue. The radiation dose is measured in Rads or Gy, where 1 Gy = 100 Rads.

Other two commonly used units are sievert (Sv) and Rem, where 1 Sv = 100 Rems. Since the sievert is a relatively large value, radiation dose in humans is normally measured in millisieverts (mSv) that is one-thousandth of a sievert.

One gray of beta or gamma radiation has one sievert of biological effect, one gray of alpha particles has 20 Sv effect and one gray of neutrons is equivalent to around 10 Sv depending on their energy. The distinction between low-level and high-level radiation is critical in the evaluation of biological consequences of exposure to ionizing radiations. An acute dose received by persons in a few days above 250-400 Rem is considered to be lethal for at least half of the population exposed. The exposed to doses between 50 Rem and 250 Rem the person will experience 'acute radiation sickness'. The symptoms of such exposure include nausea, vomiting, diarrhea, burns, and hair loss, but may or may not lead to death. Below this level such severe symptoms are rarely observed. For radiation exposure of less than 50 Rem there is the potential for delayed effects such as non-specific life shortening, genetic effects, foetal defects, and cancer.

The Becquerel (Bq) is a unit or measure of actual radioactivity in material, with reference to the number of nuclear disintegrations per second.

$$1 \text{ Bq} = 1 \text{ disintegration/sec.}$$

Quantities of radioactive material are commonly estimated by measuring the amount of intrinsic radioactivity in becquerels – one Bq of radioactive material is that amount which has an average of 1 disintegration per second, *i.e.* an activity of 1 Bq.

Health hazards of ionizing radiation result from alterations of cellular DNA materials leading to genetic mutations. As a reference, an average of 240,000 genetic mutations occur spontaneously every day in the human body. Exposure to radiation of 1 rem intensity adds about 100 more genetic mutations.

One such hazard may come from radioactive iodine. Iodine I^{131} is a radioactive isotope that releases high-energy beta particles. Concentration of iodine I^{131} in the thyroid can cause thyroid cancer. The radioisotope of iodine is volatile and hence it is a major concern if too much iodine I^{131} is accidentally released in the air. This can contaminate food, water, and animal products such as milk. This threat can be brought under control by taking stable iodine pills. These iodine pills contain stable iodine I^{127}, which does not cause cancer. When a person takes these iodine pills his body absorbs the stable iodine to a level that prevents or limits the absorption of I^{131}, which helps to prevent the risk of thyroid cancer. Another fact about radioactive iodine is that its half-life is only about 8 days. This means that in a short duration of time, almost all of the radioactive iodine would get decayed away.

The increased rates of leukemia and breast cancer are indicators of radiation exposure. This is because it is found that bone marrow and breast cells are highly sensitive to radiation.

The use of computed tomography, popularly called as CT scan use has increased dramatically over the past several decades. The total numbers of CT examinations performed annually in the United States have risen from approximately 3 million in 1980 to nearly 70 million in 2007 (Smith-Bindman *et al.*, 2009).

The greater use of CT has resulted in a concurrent increase in the exposure to ionizing radiation for medical reasons. Exposure to ionizing radiation is of major concern, because evidence has linked the exposure to low-level ionizing radiation doses used in medical imaging to the development of cancer. For example, an increased risk of cancer has been identified among long-term survivors of the Hiroshima and Nagasaki atomic bombs, who received exposures of 10 to 100 millisieverts (mSv) (Pierce and Preston, 2000; Brenner *et al.*, 2003; Preston *et al.*, 2007). A single CT scan can deliver an equivalent radiation exposure and patients may receive multiple CT scans in lifetime.

Radiation doses have found to vary significantly between the different types of CT studies. However, it is well known that CT scan delivers much higher radiation doses than conventional diagnostic x-rays. For example, a chest CT scan typically delivers more than 100 times the radiation dose of a routine frontal and lateral chest radiograph. The overall effective doses can range from 2 millisieverts (mSv) for a routine head CT scan to 31 mSv for a multiphase abdomen and pelvis CT scan. It is estimated that 1 in 270 women who underwent CT coronary angiography at age 40 years may develop cancer from that CT scan compared to an estimated 1 in 8100 women who had a routine head CT scan at the same age. For 20-year-old patients, the risks are approximately doubled, and for 60-year-old patients, they are approximately 50% lower (Smith-Bindman *et al*. 2009).

Even though the risk of radiation exposure by CT scan to an individual patient may be small, increasingly large number of people are exposed everyday. This coupled with the increasingly high exposure per examination could translate into many cases of cancer resulting directly from the radiation exposure from CT. Radiation doses from commonly performed diagnostic CT examinations are higher and more variable than generally quoted. This highlights the need for greater standardization across medical institutions. It is important to understand how much radiation medical imaging delivers, so this potential for harm can be balanced against the potential for benefit.

The most prominent health effects of radiation exposure are teratogenic and genetic mutations. Teratogenic mutations result from the exposure of foetuses to radiation. Exposure of foetus in uterus to radiation is of major concern and the risk of damages increases significantly when doses exceed 15 rems. They may result in birth defects such as smaller head or brain size, poorly formed eyes, abnormally slow growth, and mental retardation. Therefore pregnant women are at the maximum risk of damage by exposure to radiation. Genetic defects mostly occur due to mutations arising as a result of DNA damage by radiations. Damage may occur directly as a result of the radiation interacting with cells or formation of toxic substances inside the cells. These effects may be short term or long term. Various studies on animal models have revealed that exposure to radiations results in chromosomal aberrations, genetic mutations due to failure of DNA repair mechanism, chromosomal abnormalities, and protein structure and function alterations.

6.4 PRECAUTIONS AND CARE TO BE TAKEN WHILE HANDLING RADIOISOTOPES

When using radioactive materials for teaching or research, considerations must be given to the physical, chemical and biological hazards that may arise during the procedure. Care should be taken to ensure that the safety measures are undertaken that are necessary for radioisotope handling.

Many radioactive isotopes are dangerous in very small amounts when inhaled or ingested and, of course, all are dangerous in somewhat large quantities through the external effects of their radiations. When such substances are handled without proper precautions serious bodily damage may occur. The chief hazards encountered in handling radioactive isotopes include deposition of radioactive materials in the body, exposure of the whole body to gamma radiation and exposure of the body to beta radiation. Scientists with no previous experience in handling radioactive isotopes should take greater precautions. It is recommended that initial experience should be obtained by working with very small quantities of low energy radioisotopes.

1. While working with radioisotopes, care should be taken to perform all operations under a radioisotope- protection shield such as Plexiglas shield.
2. Protective clothing such as laboratory coat and gloves should be worn in laboratory at all times, especially while handling open radioactive sources.
3. All radioisotope-related work must be done in a confined area. Preferably, a special laboratory should be dedicated only to radioisotope work. This will minimize any accidents by untrained handlers.
4. Activities such as smoking, drinking, or eating food should be strictly prohibited around the radioactive material.
5. Confined areas such as fume cupboards are recommended while using radioactive gases and powders.
6. If heating of solutions containing radioactivity is necessary, use a hotplate with an oil bath or water bath. Radioactive solutions must never be heated directly over a flame.
7. The areas used for radioactive work must be covered with absorbent blotting paper, backed with impervious material for easy and rapid cleaning of any spillage during handling of radioisotopes.
8. All radioactive solutions or liquids must be placed inside double lined containers packed with absorbing material such as vermiculite or cotton. This double containment is recommended to avoid accidents due to spillage of radioactive liquids.
9. All radioactive materials must be stored in a secure area to prevent unauthorized access. All radioactive chemicals must be kept in storage cabinets, refrigerators or freezers that have been designated for this purpose and they should be labelled with the international radiation symbols.
10. Outsides of the gloves should not be touched with bare hands while removing or putting on gloves; especially those that have been used

to handle radioactive material. Disposable gloves can be used in such cases.

11. In case of accidents or spills, senior authority should be informed immediately.
12. Any person handling radioactive material should be familiar with yellow international symbol of radioactive hazard. All radioactive materials and solutions should be labelled and never left unattended in a lab. Any co-worker in the laboratory should be informed about the existence of such materials.
13. Plan your experiment in such a way that mixed waste consisting of hazardous chemical or biologically active combined with radioactivity is not generated.
14. A minimum quantity of radioactive material should be used to satisfy the objective of the experiment.
15. Dispose off radioactive waste properly. Make a reasonable estimate of the amount of radioactivity present in the waste and write it on a radioactive waste tag.
16. Activities such as pipetting by mouth, chewing, applying cosmetics, rubbing eyes or face etc. are strictly prohibited around radioactive material. Use automatic or remote pipetting devices.
17. Portable radiation detection monitors should be used in the laboratory to check spread of radioactive materials.
18. For decontamination, apparatus should be thoroughly washed with detergent. Hands should be washed with soap before leaving the laboratory.
19. A particular sink should be used for the washing of contaminated glassware and equipment. This sink should be clearly labelled with radiation warning signs.
20. Monitoring and contamination control checks must be carried out routinely in the laboratory, within seven days of the usage of radioisotopes at a minimum. Contaminated areas should be cleaned without delay and the cleanliness verified by further contamination control checks.

Disposal of Radioctive Waste

The whole issue of nuclear waste is most feared by the public. The possible dispersion of radionuclides from nuclear waste would result in human exposure mainly though ingestion, inhalation, and external radiation. The measures deployed for radioactive waste management have the purpose of restricting the radioactive material from direct or indirect human contact until all the radioactivity of these materials disappears. The radioactive waste can be classified into different categories depending upon the radioisotopes used, radioactivity levels, half-life and state of the radioactive material. When the radioactivity period is more than 30 years, it is termed as long-lived waste. Depending on the level of radioactivity, following categories are made,

- *Very low-level waste*: This contains very small amounts of radionuclides of about 10-100 Bq/g.
- *Short-lived low and intermediate level waste*: This type contains radioactivity levels in the range of few hundreds to one million Bq/g of which less than 10,000 Bq/g is from long lived radionuclides.
- *Long-lived low-level waste*: This category includes radium bearing waste from extractions of radioactive ores and graphite waste from first generation reactors.
- *Long-lived intermediate level waste*: This mainly comes from spent nuclear fuel and fuel processing accessories from nuclear reactors and from research facilities.
- *High-level waste*: This comes from fission products formed during spent fuel processing in nuclear plants.

For safe disposal of laboratory radioactive waste, following measures should be undertaken:

- Take care to avoid the disposal of non-radioactive material as radioactive waste.
- Do not discard off any radioactive material or a potentially contaminated item into regular trash cans.
- Categorize and separate nuclear waste depending upon the type of radionuclide such as ^{32}P, ^{131}I, ^{14}C, ^{35}S and ^{3}H, and physical characteristics of waste such as solid waste, liquids, scintillation vials, sharp objects etc.
- Package the solid waste, excluding iodine and animal carcasses, in clear plastic bags recommended for nuclear waste disposal. Seal the waste bags properly.
- Place animal carcasses in separate black plastic bags. Seal the waste bags.
- The contaminated sharp objects like broken glass, pipette tips, needles etc. should be placed in plastic bags individually and placed in container for sharp objects.
- To dispose off radioactive liquids, place them in recommended liquid container bottles. Do not mix liquids that might create a hazardous and explosive combination. Place the individual bottles in separate plastic bags.
- The radioactive waste should be periodically removed by the agency that handles radioactive waste disposal.

6.5 RADIOLABELLING OF PROBES

6.5.1 Introduction

Nucleotides labelled with radioactive isotopes are now commonly used in molecular biology and genetic research. Radiolabelling is the incorporation of radioisotope in DNA or RNA molecules. The commonly used radioisotopes for this purpose are hydrogen-H^3, phosphorus-P^{32} and sulphur-S^{35}. Sulphur is medium energy emitter while radioactive phosphorus is high-energy emitter and hence is more hazardous. The application of radiolabelling is in the production of radiolabelled nucleic acid molecules for use in hybridization experiments.

6.5.2 Radiolabelling by Nick Translation

The technique of nick translation depends upon the ability of DNA polymerase I enzyme to translate or move along the DNA molecule a nick created in the double stranded DNA. These nicks or breaks may occur naturally are caused by low concentrations of deoxyribonucleases like DNase I. DNase I removes nucleotide units from DNA strand that causes nicks. These nicks or gaps in the DNA are filled by the action of DNA polymerase I enzyme by incorporating new dNTPs. When these dNTPs supplied are radioactive, it results in the formation of radiolabelled DNA molecule.

Enzyme DNA polymerase I has 5' → 3' exonuclease activity that allows the enzyme to remove nucleotides from double stranded DNA and 5' → 3' polymerase activity that adds the dNTPs in 5' → 3' direction using free 3'-OH group. Therefore, by using radiolabelled dNTPs in the reaction mixture, DNA molecules can be radiolabelled synthetically. The commonly used DNA polymerase enzyme is isolated from *E. coli*. The technique of nick translation is applicable to both linear and covalently closed circular DNA molecules.

6.5.3 Radiolabelling by PCR

The objective of a polymerase chain reaction (PCR) is to amplify a specific DNA segment. Polymerase Chain Reaction technique, popularly know as PCR is an effective method for synthesis and labelling of nucleic acid probes. This technique requires a small amount of template DNA. In the presence of appropriately radiolabelled nucleotide precursors (dNTP), PCR products are labelled as well as amplified simultaneously. Therefore radiolabelled dNTPs are used in the reaction mixture of PCR. PCR has several advantages over other methods of radiolabelling such as higher speed, accuracy and efficiency.

Protocol for Radiolabelling of DNA Oligonucleotide Probes by PCR Reaction

This protocol produces double stranded radiolabelled DNA probes.

1. The reaction mixture consists of 5 µl of 10x amplification buffer, 1.0 µl of 10 mM dNTP solution (dATP, dTTP, dGTP), 1 µl of 0.1 µM dCTP solution, 2.5 µl of 20 µM forward oligonucleotide primer, 2.5 µl of 20 mM reverse oligonucleotide primer, 5-10 µl of template DNA (2-10 ng), 5 µl of 10 mCi/ml $\alpha^{32}P$ labelled dCTP and 48 µl of sterile water. Take this reaction mixture in small PCR tubes.
2. After the above reaction mixture is prepared, add 3 units of thermostable Taq DNA polymerase enzyme to the mixture. Mix the ingredients by gentle tapping of the tube.
3. Overlay the reaction mixture with a drop of mineral oil or bead of paraffin wax. [This step is not required in case of PCR machines with heated lid].
4. Place the tubes in thermal cycler. Amplify the contents using denaturation, annealing and polymerization steps using PCR machine (Table 6.3).

Table 6.3: PCR cycles performed in radiolabelling of oligonucleotide probes

Number of cycles	*Denaturation*	*Annealing*	*Polymerization*
30 cycles	30 to 45 seconds at 94°C	55 to 60 seconds at 55-65°C	1 to 2 minutes at 72°C
Last cycle	1 minute at 94°C	1 minute at 55°C	1 minute at 72°C

The temperature and time required for cycles may vary according to the particular reaction mixture.

5. Remove the tubes from thermal cycler. Remove the mineral oil from tube using micropipette.
6. Extract the reaction mixture using 50 µl of chloroform. Centrifuge for 2 minutes at room temperature to separate the aqueous and organic phases.
7. Remove upper aqueous layer carefully. Add glycogen 5 µg or carrier tRNA and precipitate the DNA using 4M ammonium acetate and 2.5 volumes of ethanol.
8. Incubate the tubes at –20°C for 1 hour. Centrifuge at maximum speed for 5 minutes to precipitate DNA.
9. Discard the supernatant. Dissolve DNA in 20 µl of TE buffer. Store the radiolabelled DNA at –20°C.

6.5.4 End Labeling of Probes

End labeling of nucleic acid probes is an alternative method that generates nucleotide molecules with radioisotopes. Unlike nick translation or PCR method of radiolabelling which radiolabel the nucleotides completely, end labelling only adds radioisotopes at their 3' or 5' ends.

6.5.4.1 *Labeling of 5' ends of DNA or RNA*

In the technique of end labeling of nucleotides, the enzyme T4 polynucleotide kinase is used. This enzyme transfers the terminal phosphate group of ATP onto 5'-hydroxyl termini of nucleic acid molecules. The ATP donor is radioactively labelled with radioactive phosphorus. This technique produces radiolabelled nucleic acids of low specific activity as only termini of each molecule become radioactive. The ATP molecules with radioactive phosphorus at γ position are used in this technique. The T4 polynucleotide kinase enzyme adds radioactive phosphorus to 5-OH position of previously dephosphorylated DNA molecules. The phosphate groups from nucleotide molecules are removed by alkaline phosphatase enzyme, followed by polynucleotide kinase treatment. This treatment replaces all 5' phosphate groups with radioactive phosphate. The kinasing reaction takes about 30 minutes.

This method is highly useful for labelling short oligonucleotides. The 5' end labelling used for generating extremely high specificity probes. It is important to remember that γ positioned P^{32} labelled ATPs are used in this technique; as compared to αP^{32} labelled dATPs that are used in nick translation and PCR.

Protocol for 5' labelling of DNA molecules:

Materials required

- Calf alkaline phosphatase enzyme (1 µg/µl). Store at 4°C.
- Phosphatase buffer
  Tris-HCl (pH=8.0) 50 mM
  EDTA 5 mM
- TE buffer
  Tris-HCl (pH=8.0) 10 mM
  EDTA 1 mM
- Phenol saturated with TE buffer
- Diethyl ether saturated with water.
- Sodium acetate 3 M (pH=5.6)
- $MgCl_2$ 10 mM
- 100% Ethanol

- Kinase buffer (10x)

Tris-HCl (pH=7.6)	0.5 M
EDTA	1 mM
$MgCl_2$	0.1 M
Spermidine	1 mM
Dithiothreitol	50 mM

- Ammonium acetate 2.5 M
- T4 polynucleotide kinase enzyme
- γ^{32}P-labelled ATP

Procedure

1. Dissolve the DNA sample to be labelled in 150 µl of phosphatase buffer and 4-8 units of alkaline phosphatase enzyme.
2. Incubate for 1 hour at 37°C.
3. Inactivate the calf intestinal alkaline phosphatase enzyme by heating the mixture at 65°C for 10 minutes.
4. Add 100 ml of phenol, vortex for 30 seconds. Centrifuge the mixture for 2 minutes.
5. Carefully pipette out the upper aqueous phase containing DNA into a new eppendorf tube. Repeat the phenol extraction for 5 times.
6. Remove the residual phenol by adding 300 µl of diethyl ether followed by centrifugation.
7. After phase separation, upper phase is ether while the lower phase is DNA solution. Repeat the ether extraction 5 times.
8. Remove the residual ether by heating at 65°C for 15 minutes.
9. Add 15 µl of 3M sodium acetate solution and 450 µl of ethanol. Mix the contents carefully.
10. Precipitate the DNA by storing the mixture at –20°C for 20 minutes.
11. Centrifuge for 15 minutes. Decant the supernatant and dry the pellet.
12. Dissolve the DNA pellet in 20 µl of sterile distilled water. Add 5 µl of 10x kinase buffer and desired quantity of γ^{32}P-labelled ATP (about 5–10 ml of aqueous solution).
13. Add sterile distilled water to make volume of 49 µl. Add 1 µl of T4 polynucleotide kinase mixture (about 5 units of enzyme). Mix the contents and incubate at 37°C for 30 minutes.
14. Add 200 µl of 2.5 M ammonium acetate and 1 ml of ice-cold ethanol. Mix the contents and precipitate DNA by storing at –20°C for 20 minutes.
15. Centrifuge for 5 minutes and discard the supernatant. This supernatant is radioactive so dispose it carefully with other radioactive waste.

16. Dissolve the pellet in 300 μl of 0.3 M sodium acetate and add 900 ml of ice-cold ethanol by incubating at –20°C for 20 minutes.
17. Centrifuge for 5 minutes and discard the supernatant. Repeat the precipitation once again.
18. Dissolve the radiolabelled DNA pellet in TE buffer. Store at –20°C.

6.5.4.2 *Labeling of 3' ends of DNA or RNA*

Labelling of 3' end involves addition of radioisotopes to the 3' end of DNA or RNA molecules. The 3' end labelling of DNA is catalyzed by the enzyme terminal transferase. Both single and double stranded DNA molecules can be labelled by the additions of dNTP at 3'OH end. The double stranded DNA fragments with blunt ends and fragments with 3'end overhangs/ sticky ends are most efficiently labelled by this method. The terminal transferase enzyme is non-template dependent enzyme; therefore it does not require any nicks in double stranded molecule for polymerization. The treatment with this enzyme produces 3' protruding overhangs of 4-5 nucleotides.

This process presents drawbacks in case of labelling of short length oligonucleotides. The 3' overhangs increase the length of the molecule, which may change the thermodynamic properties, and specificity of the labelled probes. Therefore use of dATPs is recommended considering lower thermodynamic stability of A:T bonds than G:C pairing.

6.6 TECHNIQUE OF AUTORADIOGRAPHY

6.6.1 Introduction

Autoradiography is the process of taking a type of picture, called an autoradiograph, which shows the relative concentration of radioactive material present within the subject. The subject is usually a biological specimen or human body part. So in other words, autoradiography is the method of detection of radioactive isotopes within biological samples.

6.6.2 Theory and Principle of Autoradiography

The photographic plate or x-ray film is exposed to radioactive emissions from the subject being studied, producing an image. Radiography is the visualization of the pattern of distribution of radiation. In general, the radiation consists of X-rays, gamma or beta rays, and the recording medium is a photographic film. In autoradiography the specimen itself is the source of the radiation, which originates from radioactive material incorporated into it. The first autoradiograph was obtained accidently around 1867 when a blackening was produced on emulsions of silver chloride and iodide by uranium salts.

In autoradiography, each crystal of silver halide in the photographic emulsion is an independent detector, insulated from the rest of the emulsion by a layer of gelatin. Each crystal responds to the charged particle by the formation of a latent image that is made permanent by the process of development. The record provided by the photographic emulsion is cumulative and spatially accurate. It provides information on the localization and distribution of radioactivity within a sample. In biology, this technique may be used to determine the tissue localization of a radioactive substance introduced into a metabolic pathway, bound to a receptor or enzyme, or hybridized to a nucleic acid. The x-ray film or photographic emulsion is exposed to the labeled tissue section to obtain the autoradiograph, which is also called as autoradiogram.

The use of radiolabelled ligands to determine the tissue distributions of receptors is termed either *in vivo* or *in vitro* receptor autoradiography if the ligand is administered into the circulation (with subsequent tissue removal and sectioning) or applied to the tissue sections, respectively. The ligands are generally labeled with ^{3}H or ^{125}I. RNA or DNA probes are usually labeled with ^{32}P or ^{35}S.

An X-ray film catches the radiation that crosses to the other side of the subject. Where it hits on the plate, the radiation excites electrons, or subatomic negatively charged particles. In a photographic plate, the old means of viewing a radiograph, these exposed areas would darken, the less exposed areas would show up gray, and the unexposed areas would remain white. Provided that the film makes close contact with a flat sample, autoradiography provides

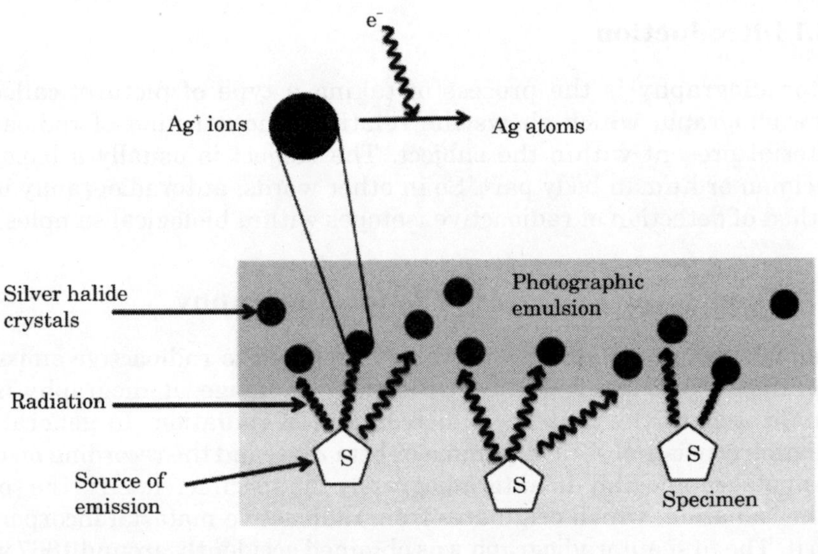

Fig. 6.1: Conversion of silver ions to silver atoms upon exposure to radiation leading to the development of autoradiographic image

excellent spatial resolution and it avoids destruction of the sample. Silver halide crystals in the emulsion respond directly to β-particles and γ-rays emitted from the sample. Each emission converts several silver ions to silver atoms to produce a stable latent image (Fig. 6.1). When the film is subsequently developed these few silver atoms catalyze the reduction of the entire silver halide crystal (grain) to metallic silver to produce an autoradiographic image of the radioisotope distribution. The energy from each β-particle or γ-ray is sufficient to render each grain it hits, to fully develop and many grains may be converted to the developable state by a single β-particle. The local blackening of the film for direct autoradiography is directly proportional to the amount of radiation reaching it.

6.6.3 Procedure of Autoradiography (Used for *In Situ* Hybridization)

Materials required

Prepare Kodak NTB-2 autoradiographic emulsion at 45°C and warm an equal volume of DDW at same temperature. Gently mix together and wrap the container containing the working emulsion solution with aluminium foil.

Procedure

1. Add appropriate volume of working NTB-2 emulsion to a slide dipping chamber.
2. Slowly dip the slides into the dipping chamber and gently pull them out.
3. Place the slides vertically in appropriate rack and air-dry them for 2 hours.
4. Place the slides in slide box containing desiccant and seal the box with black tape. Wrap the box with aluminum foil and allow exposure to proceed at 4°C for 1-10 days depending upon the intensity of the signal. This exposure time depends on the activity of the isotope, the temperature and the background radiation.
5. In the dark room, add sufficient volumes of developer, water and fixer to slide jars and allow the solutions and water to warm up to 20°C.
6. Transfer the slide exposure box from refrigerator to the dark room and allow the box to warm to 20°C.
7. Carefully immerse the slides in the slide jar containing the developer solution for 2-4 minutes and quickly rinse the slides in water 30 seconds.
8. Then immediately place the slides in fixer for 3 minutes.

9. Wash the slides in slowly running tap water for 30 minutes.
10. Remove the emulsion on back of the slide using razor blade and air-dry the slides.
11. One can also carryout counterstaining of the slides in order to enhance the images using various counter stains.
12. The position of the silver grains in the sample is observed by light or electron microscopy Note: the grains are in a different plane of focus in the emulsion overlying the tissue section. Often oil with 100x objective is used for detailed observation with the light microscope.
13. These autoradiographs provide a permanent record. Full details on the batch of emulsion used, dates, exposure time and conditions should be kept for each experiment.

6.6.4 Troubleshooting Guide

1. The slides containing fixed and hybridized tissue specimen must be completely dried prior to exposure because moisture fades the images in autoradiography.
2. The x-ray films used in autoradiography in southern/northern hybridizations often suffer from lack of sensitivity. This can be overcome by using intensifying screens. Intensifying screens enhance the image generated by radioactive molecules on the x-ray film. They are used in conjunction with strong β-emitting isotopes such as ^{32}P or γ-emitting isotopes such as ^{125}I. Emissions from these forms of radiation frequently pass completely through a film, but the intensifying screen fluoresces and exposes the film with multiple photons of light that can absorb them.
3. The x-ray film should be placed between the sample and intensifying screen. For very weak radioactivity, second intensifying screen can be placed on the other side of the radioactive sample.
4. Silver bromide crystals activated by light, β particles or γ rays are highly unstable and quickly revert back to their stable form. To avoid this problem, he film should be exposed at –70°C to stabilize the silver bromide crystals activated by the radioactivity.
5. Among the biggest problems encountered in autoradiography are images that are either too weak or too intense. Varying the exposure time can solve such problems.

6.7 REVIEW QUESTIONS

1. Describe precautions to be taken while handling radioisotopes in the laboratory.
2. What are the health hazards of radioactivity to people?

3. Define autoradiography. How was autoradiography discovered?
4. Describe different types of radiations.
5. Enlist various methods that are used to radiolabel nucleic acid probes. Describe nick translation.
6. Describe the method of 5' end labelling of nucleic acids.
7. What are different types of nuclear wastes? Describe steps of nuclear waste disposal.
8. Define isotopes. What is half-life? Give half-life of different radioisotopes.
9. Give applications of radioisotopes in biological sciences.
10. Give principle and describe how autoradiography works.

6.8 REFERENCES

Annunziata, M. (2007). Radioactivity: introduction and history. Elsevier, p. 609.

Baker, J. (1989). Autoradiography: a comprehensive review. Oxford University Press. p. 110.

Balcar, V. and Nanistos, E. (2006). Autoradiography of [^{3}H] aspartate and glutamate transport in schizophrenia. *Neuropsychopharmacology,* **31:** 685–686.

Billen, D. (1990). Spontaneous DNA damage and its significance for the 'negligible dose' controversy in radiation protection. *Radiat. Res.,* **124:** 242–245.

Brenner, D., Doll, R. and Goodhead, D. (2003). Cancer risks attributable to low doses of ionizing radiation: Assessing what we really know. *Proc. Natl. Acad Sci. USA,* **100(24):** 13761–13766.

Cairns, J. (1962). A minimum estimate for the length of DNA of *Escherichia coli* obtained by autoradiography. *Journal of Molecular Biology,* **4(5):** 407–409.

Dohlman, G., Maunsbach, A., Hammerstrom, L. and Appelgren, L. (1964). Electron microscopic autoradiography: A method for producing uniform monolayers of silver halide crystals using centrifuge sedimentation. *Journal of Ultrastructure Research,* **10(3-4):** 293–303.

Ernst, M., Freed, M. and Zametkin, A. (1998). Health hazards of radiation exposure in the context of brain imaging research: Special consideration for children. *The Journal of Nuclear Medicine,* **39(4):** 689–698.

Farrell, R. (2010). RNA methodologies: A laboratory guide for isolation and characterization. 4th Edition. Academic Press. pp. 274–275.

Ghatak, K. (2011). Techniques and methods in biology. PHI Learning Pvt. Ltd. pp. 338–344.

Goto, A., Takebayashi, Y., Liu, D., Li, L., Saiga, T., Ishikawa, Y., Mori, T., Yamadera, A. and Fukumoto, M. (2002). Microdistribution of alpha particles in pathological sections of tissues from thorotrast patients detected by imaging plate autoradiography. *Radiation Reasearch,* **158(1):** 54–60.

Guerra, A. (2004). Ionizing radiation detectors for medical imaging. World Scientific Publishing Co. Pte. Ltd. pp. 193–233.

Guest, G. (1948). Precautions for workers using radioactive isotopes. *Canadian Journal of Comparative Medicine,* **XII(12):** 323–333.

Laskey, R.A. (1980). The use of intensifying screens or organic scintillators for visualizing radioactive molecules resolved by gel electrophoresis. *Methods Enzymol.,* **65:** 363–371.

Lindell, B. (1987). Radiation and health. *Bull. World health Organ.,* **65(2):** 139–148.

Magaldi, D. and Ranaldi, D. (2010). Using thin section autoradiography to detect the uranium content of Fe/Mn soil nodules. *Geologia Balcanica,* **39(3):** 69–71.

Martin, J., Ellis, M., Keller, P., Biron, K., Lehrman, S., Barry, D. and Furman, P. (1985). Plaque autoradiography assay for the detection and quantitation of thymidine kinase-defecient and thymidine kinase-altered mutants for herpes simplex in clinical isolates. *Antimicrob. Agents Chemother,* **28(2):** 181–187.

Mettler, F.Jr, Thomadsen, B. and Bhargavan, M. (2008). Medical radiation exposure in the US in 2006: Preliminary results. *Health Phys.,* **95(5):** 502–507.

Pierce, D. and Preston, D. (2000). Radiation-related cancer risks at low doses among atomic bomb survivors. *Radiat. Res.,* **154(2):** 178–186.

Preston, D., Ron, E. and Tokuoka, S. (2007). Solid cancer incidence in atomic bomb survivors: 1958-1998. *Radiat. Res.,* **168 (1):** 1–64.

Sahoo, S. and Sahoo, S. (2006). Production and applications of radioisotopes. *Physics Education,* April-June 5-11.

Smith-Bindman, R., Lipson, J. and Marcus, R. (2009). Radiation dose associated with common computed tomography examinations and the associated lifetime attributable risk of cancer. *Arch. Intern. Med.,* **169(22):** 2078–2086.

Walker, J. and Rapley, R. (2008). Molecular Biomethods Handbook. 2nd Edition. Humana Press. pp. 397–410.

7

Immunotechniques

7.1 INTRODUCTION

Immunotechniques used in molecular biology and medicinal research, employ the antigen-antibody specificity and affinity for detection of specific proteins from a complex mixture. The basis of antigen-antibody reaction has been extensively exploited to devise techniques to qualitatively and quantitatively detect the presence of different biomolecules with high sensitivity and specificity.

7.2 THEORY AND PRINCIPLE OF ANTIBODY BASED DETECTION METHODS

The antibodies have the inherent property to bind to the antigen used to elicit antibody response with high avidity and sensitivity. The antigens encompass the biomolecules that are part of bacteria, fungi, protozoan parasites, etc. The process of injection of a foreign antigen into an animal host is called immunization. The immunization involves a series of injections of a given antigen by a suitable route periodically over a period of time. When an antigen is injected into the host, the IgM class of antibodies is the first to develop, followed by gradual development of IgG antibodies. During a natural infection the detection of an antigen specific IgM antibody indicates a recent history of infection.

The ability of the antigen to produce an enhanced and prolonged immune response by formation of specific antibodies (humoral immunity) is achieved by combining or mixing a given antigen with non-specific immuno-stimulatory compounds called adjuvants. The adjuvants that are commonly used include aluminum phosphate, aluminum oxide, Freund's Complete Adjuvant (FCA) and Freund's Incomplete Adjuvants (FIA).

The animals normally develop antibodies against the injected antigen as a part of their natural immune response. The serum derived from these animals is used as the antibody source and tested with reference to specificity, sensitivity and affinity. The specificity of the antibody facilitates the detection of the antigen in the lowest possible concentration in the undiluted body fluids. This is generally referred to as cut off level or detection limit. Sensitivity of the antibody defines a degree to which an assay can distinguish one compound of the same nature from another, which is also the function of the antibody molecules in the serum. Affinity means how strongly the antigen is bound to the antibody.

The animals such as guinea pig, rabbit, horse, mouse and rats are often used to generate specific polyclonal antibodies against the injected antigen. However, recently the use of hybridoma technique developed by Kohler and Milstein (1975) is also used to produce bulk amount of specific monoclonal antibodies *in vitro*. These purified monoclonal antibodies or polyclonal antibodies can be generated for use in highly specific and sensitive techniques like immunoassay, immuno-electrophoresis, ELISA, Western blotting etc.

Theses specific antibodies are then labelled with detection agents like biotin, dioxigenin or fluorochrome for their detection.

Colourimetric Detection

The colourimetric detection is a method of detection of proteins depending upon incubation of the protein-antibody complex with a substrate that reacts with the reporter enzyme such as alkaline phosphatase or horseradish peroxidase that is bound to the antibody (most often secondary antibody). This reaction converts the substrate into a coloured product that can be identified visually. In case of ELISA, the substrate that gives soluble or insoluble coloured product upon reaction with the reporter enzyme can be used for detection. In western blotting, soluble substrate is converted into an insoluble form of a different color that precipitates next to the enzyme and stains the membrane. Development of the blot is then stopped by several washings to remove the soluble substrate. The washing step is important part of western blotting. Therefore the substrates that give soluble coloured product in the reaction are unsuitable for use in the western blot. Compared to other methods, colourimetric or chromogenic detection is rapid and easiest method of detection.

Alkaline Phosphatase (AP) Detection System

Nitroblue Tetrazolium (NBT) and 5-bromo-4-chloro-indolyl phosphate (BCIP) substrate

The NBT and BCIP substrates are very commonly used in colorimetric detection of alkaline phosphatase system in western blotting and ELISA procedures. Upon reduction, NBT yields NBT-formazan, which is colored and water-insoluble product. Hydrolysis of BCIP by alkaline phosphatase results in a blue to purple coloured precipitate that can be deposited on nitrocellulose or nylon membranes. The ideal system for blotting or staining applications with AP is the combination of NBT and BCIP. Together, they yield an intense, black-purple precipitate that provides much greater sensitivity than either substrate alone. NBT/BCIP–alkaline phosphatase system characteristically produces sharp band resolution with little background staining on the membrane.

Fast Red TR/Naphthol AS-MX and TR phosphate (4-Chloro-2-methylbenzenediazonium/ 3- Hydroxy-2-naphthoic acid 2,4-dimethylanilide phosphate) substrate system

This system is used for chromogenic detection in western blotting using alkaline phosphatase enzyme. Fast Red systems produce an insoluble intense red coloured end product. Slides stained with Fast Red TR/Naphthol AS-MX must be covered with coverslip using aqueous mounting media, as the reaction product is alcohol soluble. Fast Red TR/AS-MX substrate is a sensitive method of detection that produces bright red precipitate on the transfer membrane.

p-NPP (p-Nitrophenyl Phosphate) substrate

The p-NPP substrate is widely used for detecting alkaline phosphatase in ELISA applications. This substrate develops the yellow end product p-nitrophenol, when hydrolyzed by alkaline phosphatase enzyme. Since this system forms a soluble end product, it is suitable for ELISA, but not recommended for western blotting. The soluble yellow end product p-nitrophenol absorbs light at 405 nm wavelength. The reaction may be stopped with the addition of 3M sodium hydroxide solution. ELISA applications utilizing p-NPP may be read in assays or stopped with alkaline solutions for delayed readings. The absorbance of the stopped reaction is also read at 405 nm. The substrate p-NPP can be availed from manufacturers either as a crystalline powder, tablets or as a ready-to-use formulation.

Horseradish Peroxidase (HRP) Detection System

TMB (3,3',5,5'-tetramethylbenzidine) substrate

TMB is most often used as a substrate for horseradish peroxidase (HRP) detection system in ELISA. TMB in the presence of HRP and peroxide produces a water-soluble blue product that can be precipitated onto a membrane. The major absorbance of the reaction product is at 370 nm and 652 nm. When sulphuric or phosphoric acids are added, the colour changes to yellow. At this time, it gives maximum absorbance reading at 450 nm. TMB is very sensitive and may produce significant background signal if too much protein or antibody is used. TMB is more quickly oxidized than other HRP substrates, resulting in faster color development. TMB is popularly used in western blotting and immunohistochemistry with HRP system. Precipitation of the product results in dark blue bands where the enzyme is located.

4-CN (4-chloro-1-naphthol) substrate

The 4-CN HR peroxidase substrate produces an insoluble blue coloured reaction product and can be used for development of nitrocellulose blots. This precipitate is not as sensitive or as stable as TMB and DAB, but the alcohol-soluble precipitate photographs well and has a distinct blue-purple color that can be useful in double-staining applications.

DAB (3,3'-diaminobenzidine) substrate

DAB yields a brown precipitate in the presence of HRP and peroxide. The brown, insoluble product can be readily chelated with osmium tetroxide. This property makes DAB ideal for electron microscopy. The brown coloured of the reaction product produced by DAB can be intensified with the addition of metals such as nickel, copper, silver and cobalt that form complexes. The color produced by the metal complexes is darker than the color produced by DAB alone, enhancing the sensitivity in staining applications. DAB is used in immunohistochemistry.

The two substrates 4-CN and DAB can be used together in a single reaction, which combines advantages of both the substrates. The CN-DAB substrate has excellent sensitivity, yielding a dark black precipitate that can be recorded very well. The CN/DAB substrate is used in western blotting and dot blotting applications.

ABTS (2,2'-azino bis [3-ethylbenzothiazoline-6-sulfonic acid]) substrate

ABTS is used to detect HRP in ELISA technique. When ABTS reacts with HRP, yields a water-soluble green coloured reaction product. ABTS is less sensitive than the OPD and TMB substrates for HRP detection. Color development by ABTS takes about 20 minutes.

OPD (o-phenylenediamine dihydrochloride) substrate

OPD is used to detect HRP in ELISA. OPD when reacted with HRP yields a water-soluble yellow-orange reaction product. The reaction product has an absorbance maximum of 492 nm. It takes about 15-20 minutes for the reaction to develop coloured product.

Fluorescence Detection

When fluorescence based detection method is used in ELISA, it is called enzyme linked fluorescent assay (ELFA). ELFA has several advantages over conventional ELISA as improved sensitivity and quicker results. In this assay, the reporter enzyme conjugated to the antibody breaks down the substrate that produces fluorescent end product. This fluorescence from reaction product can be detected by spectrofluorometer. Alternatively, fluorophore is conjugated to the detection antibody instead of the enzyme. The fluorophore can directly detect the presence of protein of interest. The commonly used fluorescent molecules are rhodamine, fluorescein isothiocyanate (FITC) and fluorescein isocyanate. Fluorescence detection is considered to be among the most sensitive detection methods for western blotting analysis.

Fluorescent reagents are popularly used in western blotting also because they offer time saving option compared to chemiluminescent or radioactive detection and produce reduced radioactive and chemical waste. The advancements in imaging technology have provided new fluorescent probes and the reduced cost of fluorescent detection systems. Fluorescent detection technique has the unique advantage of allowing multiple targets to be screened on the same blot at the same time without the need of stripping and reprobing.

There are a number of fluorescent dyes that are commercially available that have different absorption and emission spectra (Refer Table 7.1). Fluorescent dyes are used to detect molecules in a variety of biological

Table 7.1: Various fluorescent labels/dyes used in fluorescent detection system

Fluorescent dye/protein	*Colour*	*Excitation wavelength*	*Emission wavelength*
Rhodamine (TRITC)	Red-orange	552 nm	570 nm
Fluorescein (FITC)	Green	495 nm	525 nm
Texas Red	Red	596 nm	620 nm
Green fluorescent protein (GFP)	Green	395 or 470 nm	509 nm
Enhanced yellow fluorescent protein (EYFP)	Yellow	514 nm	527 nm
Enhanced blue fluorescent protein (EBFP)	Blue	383 nm	445 nm
Enhanced cyan fluorescent protein (ECFP)	Blue-green	434 nm	477 nm
dsRed	Red	558 nm	583 nm

applications. Unlike many visible dyes, fluorescent dyes may be used under various physiological conditions and allow staining of living cells and tissues. The detection can be done by fluorescence microscopy. Certain instruments are specialized for detection of infrared and near-infrared fluorescence, while others can analyze only particular fluorescence in the visible range. For detection of any Western blot, it is desirable to use pre-stained molecular weight markers and transfer them to the membrane along with the protein sample. Unless modified in such a way, molecular weight markers will not be visible during fluorescent imaging since they will not produce any signal.

Similar to enzyme reactions, fluorescent reagents must be optimized for optimal signal: noise ratio. If the degree of fluorescent labeling is too low, the signal will be weak. Both nitrocellulose and PVDF membranes and polyacrylamide gels are known to have fluorescent properties, which can create a significant amount of background noise that reduces signal to noise ratios in fluorescent blotting applications. Fluorophore conjugates should be protected from light during storage and use.

Green Fluorescent Protein (GFP)

The Nobel Prize in Chemistry was awarded jointly to Osamu Shimomura, Martin Chalfie and Roger Y. Tsien in 2008 for the discovery and development of the green fluorescent protein, GFP. Osamu Shimomura first isolated GFP from the jellyfish *Aequorea victoria* that is found off the west coast of North America. He discovered that this protein glowed bright green under ultraviolet light. Today, by using DNA technology, researchers can now tag GFP to other proteins of interest. This glowing marker allows them to watch the movement, positions and interactions of the tagged proteins.

The unique brightly glowing green fluorescent protein, GFP, was first observed in the jellyfish, *Aequorea victoria* in 1962. Since then, this protein became one of the most important tools used in biological sciences. The most studied of the fluorescent proteins is the GFP from the jellyfish *Aequorea victoria* and the sea pansy *Renilla reniformis.* Although the proteins from these organisms are not identical, they are thought to have the same chromophore, which is derived from the primary amino acid sequence of GFP. The differences between them are due to changes in the protein environment of the chromophore. Recent interest in these molecules has arisen from the successful cloning of the *Aequorea gfp* cDNA and the demonstration of its expression outside *Aequorea*, in the absence of other *Aequorea* proteins. This demonstration indicated that GFP could be used as a marker of gene expression and protein localization in living and fixed tissues. Bacterial, plant and animal cells including mammalian, are now known to express GFP, with the aid of genetic engineering. The heterologous expression of the *gfp* cDNA also means that it can be mutated to produce GFP proteins with different fluorescent properties. Variants with more intense fluorescence or alterations in the excitation and emission spectra have been produced.

Mutagenesis of jellyfish GFP encoding gene has yielded proteins that fluoresce from blue to yellowish green, and genetic manipulations have generated GFP variants that are better suited for fluorescence microscopy than wild-type GFP and thus have optimized its usage. For example, the useful cyan fluorescent variant (ECFP) has low absorption and low quantum yield, whereas the yellowish-green fluorescent protein (EYFP) has the highest absorption and quantum yield. The recent cloning of a gene that encodes a red fluorescent protein (dsRed) from the Indo-Pacific sea anemone *Discosoma striata* has provided yet another fluorescent protein that is further red-shifted. This dsRed protein shares only about 25% sequence identity with *Aequorea* GFP. Therefore, other usable GFPs are likely to be discovered in the future.

Chemiluminescence Detection

Chemiluminescence is the generation of light with limited heat emission through a chemical reaction. The reactions are invariably oxidative and are biphasic in nature. There is a rapid initial phase in which the excited state is produced, the rate being dependent on reagent concentration. This is followed by a decay phase, which is characteristic of the chemiluminescent derivative. The basic principle behind any type of light-producing reaction involves the excitation of a molecule from its ground state to an excited state. The chemiluminescence is commonly observed in nature in the form of light produced by firefly, click beetles, glowworm, some types of bacteria, certain mushrooms, marine squid, anglerfish and various crustaceans. This form of luminescence involves chemical reaction occurring inside the organism; hence it is commonly called as bioluminescence. Chemiluminescence principle is used in various commercial products available today, like glow in the dark sticks, glow paints, glow powder, glow rope etc. the phenomenon of chemiluminescence is of immense value in molecular biology research, genetic engineering, medical diagnostics and forensic sciences. Chemiluminescence detection system is commonly used in ELISA and immunoassays.

Chemiluminescent substrates differ from other substrates in a way that the light detected is a transient product of the reaction and is only present while the enzyme-substrate reaction is occurring. This is in contrast to substrates that produce a stable, colored product where colored precipitates remain on the membrane even after the enzyme-substrate reaction has terminated. On a chemiluminescent western blot, the substrate is the limiting reagent in the reaction. As substrate is exhausted, light production decreases and eventually stops. A well-optimized procedure using the proper antibody dilutions will produce a stable production of light for several hours, allowing consistent and sensitive detection of proteins.

For detection of any western blot, it is desirable to use pre-stained molecular weight markers that are transferred to the membrane along with the protein sample. Unless modified in such way, molecular weight markers do not show up on film or the imaging system since they do not produce an output of light. However, molecular weight markers that employ antibody-binding domains from protein A, protein G or biotin can also be used. The antibody capture domains of these proteins are engineered into the molecular weight markers and thus bind to the antibodies used in the western blot and allow a signal to be generated and captured alongside the experimental data. The molecular weight markers with antibody binding domains of protein A or G bind to immunoglobulins and biotinylated molecular markers can be detected by avidin or streptavidin. Many antibodies such as those of the mouse IgG1-subclass however do not bind strongly to protein A or G.

Chemiluminescent Reaction of Luminol

In aqueous solution, the most frequently used chemiluminescent compound is luminol (5-amino-2,3-dihydrophtalazine-1,4-dione) or its derivatives, such as isoluminol. Luminol reacts with H_2O_2 in the presence of a catalyst that can be a metal or metal-containing compound or enzyme. The Oxidation of luminol by hydrogen peroxide in alkaline solution results in creation of an excited state product called 3-aminophthalate, which returns to the ground state with the production of light. This product decays to a lower energy state by releasing photons of light. The intensity of the light signal is low and of short duration. The enhanced chemiluminescent reaction is one of the most sensitive and rapid detection methods in medical and analytical biochemistry.

The reaction of hydrogen peroxide and luminol with immobilized horseradish peroxidase make this a popular western blotting system. As it decays back to the ground state, the 3-aminophthalate emits light at wavelength of 425 nm, which can be captured by imaging devices that detect chemiluminescence. However, this technology has one drawback that it requires a substrate that produces an intense signal of long duration to be captured by the CCD camera. If the film is underexposed, the signal will not be visible and if overexposed, the signal may be lost in the background or bands may become blurred together. Quantitative techniques based on the use of chemiluminescent molecules rely on reproducible measurements of low levels of emitted light.

The luminal chemiluminescence reaction is often used by forensic scientists to detect traces of blood at a crime scene. Luminol powder is mixed with hydrogen peroxide and the mixture is sprayed over the scene. The iron in the hemoglobin protein of RBCs in the blood serves as a catalyst for the chemiluminescent effect.

Chemiluminescence immunoassay definitely offers a viable and safer alternative to a range of analytical procedures based on the use of radioisotopes. Its feasibility has been demonstrated for a range of analytes including both high and low molecular weight species. The stability of the reagents and their analytical performance has important implications in improving and maintaining high quality assay performance. This technology has a potential in the detection and monitoring of infective agents as well as tumour markers.

7.3 EXTRACTION AND PURIFICATION OF PROTEINS

7.3.1 Introduction

An important part of proteomics and biotechnology research is to design or modify proteins with optimized properties for specific industrial applications. In order to do this, one must be able to isolate and purify proteins of interest so their conformations, substrate specificities, reactions with other ligands and specific activities can be studied.

The degree of protein purity required depends on the intended end use of the protein. Each protein purification step usually results in some degree of product loss. Therefore, an ideal protein purification strategy is one in which the highest level of purification is reached in the fewest steps.

7.3.2 Enzymes that Degrade Proteins and their Inhibitors

Protease refers to a group of enzymes whose catalytic function is to hydrolyze (breakdown) peptide bonds of proteins. They are also called proteolytic enzymes or proteinases.

They are enzymes of class 3, the hydrolases, and subclass 3.4, the peptide hydrolases or peptidases. They constitute a large family of enzymes divided into endopeptidases or proteinases and exopeptidases according to the point at which they break the peptide chain. These endopeptidases can be classified further, according to the reactive groups at the active site involved in catalysis, into serine-proteases, cysteine-proteases, aspartic-proteases and metalloproteinases.

Exopeptidases

The exopeptidases act only near the ends of polypeptide chains at the N or C terminus. Those acting at a free N terminus liberate a single amino acid residue are called aminopeptidases. The enzymes that produce free dipetide are called dipeptidyl-peptidases and those producing tripeptide are called tripeptidyl-peptidases.

The exopeptidases acting at a free C-terminus liberate a single amino acid (carboxypeptidases) or a dipeptide (peptidyl-dipeptidases). Some exopeptidases are specific for dipeptides hence called dipeptidases or remove terminal residues that are substituted, cyclized or linked by isopeptide bonds.

Endopeptidases

Endopeptidases act preferentially on the inner regions of peptide chains away from the N and C termini. The presence of free α-amino or α-carboxyl groups has a negative effect on the activity of the enzyme.

Protease inhibitors

Protease inhibitors are the molecules that inhibit the function of proteases. Proteases create much hindrance in extraction and purification of proteins in biochemistry and proteomics studies. Therefore protease inhibitors are often used to avoid degradation of proteins of interest. Also to slowdown unwanted proteolysis, extraction and purification of proteins is carried out at low temperatures. Protease inhibitors are often included in buffers. Different types of protease inhibitors are as follows,

Serine-protease inhibitor

Examples of serine-protease inhibitors are PMSF and Benzamide.

Phenylmethylsulfonyl fluoride (PMSF) 100 mM (100x) stock is prepared in anhydrous 95% ethanol. Store the stock at room temperature. Immediately before use, bring the buffer to 1mM PMSF. Add the stock solution slowly with constant stirring.

Benzamide 0.1 M stock is prepared in ethanol. Dissolve benzamide in anhydrous ethanol together with PMSF.

Cysteine (thiol) protease inhibitors

PHMB is a cysteine-protease inhibitor.

Sodium p-hydroxymercuribenzoate (PHMB) 100 mM stock is prepared in water. Adjust pH=8.0 with NaOH. Final working concentration is 1 mM.

Aspartate (acidic) protease inhibitors

The example of aspartate (acidic) protease inhibitors is DAN.

Diazoacetylnorleucine methyl ester (DAN) 100 mM stock is prepared in methanol. Cu(II) (cupric acetate) is dissolved in water at 100 mM. Each solution is added to 5 mM concentration.

Metalloprotease inhibitors

Commonly used Metalloprotease inhibitor is EGTA.

Ethylene glycol bis (β-aminoethyl ether) N,N,N'N'-tetraacetic acid (EGTA) is an efficient chelator of divalent metal cations other than Mg^{+2}. Prepare 0.5 M stock of EGTA by suspending 0.05 mol in 80 ml water; adjust the pH=8.0 and make final volume to 100 ml. the final working concentration is 10 mM.

7.3.3 Protein Isolation Methods

7.3.3.1 *Trichloroacetic acid- acetone extraction*

Solutions used to resuspend and solubilize proteins generally contain chaotrops, detergents, and reducing agents. Chaotrops unfold proteins by breaking noncovalent bonds. The most commonly used chaotrops are urea and thiourea. TCA precipitation efficiently inhibits protease activity in plant tissues.

The TCE-2ME-Acetone solution inactivates the phenoloxidases and oxidases, preventing phenol oxidation into quinones, which would result in protein binding into insoluble complexes. Acetone alone allows solubilization of the pigments, lipids, and terpenoids possibly present in the tissue. 2ME (2-mercaptoethanol) prevents the formation of disulfide bonds during precipitation.

Materials required

Precipitation solution: 10% TCA (w/v), 0.07% 2ME (v/v) in cold acetone. This solution must be freshly prepared and stored at –20°C until use.

Rinsing solution: 0.07% 2ME (v/v) in cold acetone. This solution can be stored at –20°C for about 1 month.

Solubilization solution-I

5 M urea, 2 M thiourea, 2% 3[3-cholaminopropyl diethylammonio]-1-propane sulfonate (CHAPS) (w/v), 2% N-decyl-N,N-dimethyl- 3-ammonio-1-propane sulfate (SB3-10) (w/v), 20 mM dithiothreitol (DTT), 5 mM phosphine, 0.5% Pharmalyte 4-6.5 (v/v), 0.25% Pharmalyte 3-10 (v/v) in double distilled water. The solution is slightly heated below 30°C to aid urea solubilization. It is aliquoted and can be stored at –80°C for months.

Solubilization solution-II

9.5 M urea, 5 mM K_2CO_3, 1.25% SDS, 5% DTT, 6% Triton X-100, 2% ampholines 3.5 to 9.5 in DDW. $K_2CO_3$is prepared as a 2.8% (w/v) stock solution and SDS as a 10% (w/v) filtered stock solution, Triton X-100 is provided as a 20% solution. Then 3 ml DDW is added to the other components until the urea is solubilized by heating below 30°C. Solutions of 40 ml are usually prepared and then aliquoted and stored at –80°C for months.

Determination of protein concentration

The protein content of the samples (either in solubilization solution I or II) is evaluated using the 2-D Quant Kit from Amersham Biosciences. The method involves a precipitation of proteins from the sample, followed by specific binding of copper ions to proteins as described by the manufacturers instructions.

IPG strip rehydration solution (for solubilization solution II only):

Solution A: 7 M urea, 2 M thiourea, 1.4% CHAPS (w/v), 16 mM DTT, 5 mM phosphine, 0.3% Pharmalyte 3-10 (v/v) in ddH_2O.

Solution B: 7 M urea, 2 M thiourea, 0.5% CHAPS (w/v), 10 mM DTT, 5 mM phosphine, in DDW.

Procedure

Protein precipitation and denaturation

1. Grind the plant tissues in a mortar and pestle in liquid nitrogen to obtain a fine powder.
2. Transfer about 200 µl of powder to a 2-ml Eppendorf tube. Cover with 1.8 ml of the cold TCA-2ME-acetone solution, mix, and then store at –20°C for 1 hour. TCA and acetone denature and precipitate the proteins. 3. Centrifuge for 10 min at 10,000 g in a refrigerated centrifuge, at 4°C.

Rinsing with 2ME-Acetone solution

1. Discard the supernatant and resuspend the pellet in 1.8 ml of cold rinsing solution. Store at –20°C for 1 hour. This step eliminates the acidity caused by TCA that would impede protein recovery.
2. Centrifuge for 15 minutes at 10,000 g in a refrigerated centrifuge, at 4°C.
3. Discard the supernatant. Repeat this step twice.
4. Dry the pellet under vacuum for about 1 hour to eliminate the acetone fully.

Protein solubilization

1. The amount of solubilization solutions II and I for protein solubilization depends on the plant tissue; for instance, 60 µl/mg dry powder for leaf tissue and 50 µl/mg dry powder for seeds

2. Resolubilization is achieved by vortexing briefly for 1 minute. At this stage the sample still contains cellular debris.
3. Centrifuge for 15 minutes at 10,000 g at 25°C, and collect the supernatant in a clean 1.5 ml Eppendorf tube.
4. Centrifuge again for 15 min at 25°C, and transfer the supernatant to a new Eppendorf tube.
5. Samples containing solubilized proteins can then be stored at –50°C or –80°C for months.

7.3.3.2 *Phenol extraction*

Phenol extraction of proteins is an alternative method to classical TCA-acetone extraction. It allows efficient protein recovery and removes nonprotein components in the case of plant tissues rich in polysaccharides, lipids, and phenolic compounds. Phenol extraction of proteins followed by methanol/ammonium acetate precipitation was reported by Hurkman and Tanaka in 1986 for proteomic studies.

Phenol interacts with proteins mainly via hydrogen bonding and causes proteins to become denatured and soluble in the organic phase. The proteins are first extracted in a Tris buffer containing several protecting agents. EDTA inhibits metalloproteases and polyphenol oxidases by chelating metal ions. PMSF irreversibly inhibits serine proteases. β-Mercaptoethanol is a reducing agent that prevents protein oxidation. Moreover, as a precaution against protease activity, the temperature must be kept below 4°C, and samples should be placed on ice during the first step of the extraction process.

Materials Required

Phenol: Tris-HCl saturated, pH 6.6/7.9

Extraction buffer

Prepare a solution of 500 mM Tris-HCl, 50 mM EDTA, 700 mM sucrose, 100 mM KCl and adjust pH to 8.0 with HCl. This solution can be stored for a week at 4°C. Add just before extraction, 2% β-mercaptoethanol and 1 mM phenylmethylsulfonyl fluoride (PMSF).

Precipitation solution

0.1 M ammonium acetate in cold methanol. This solution is stored at –20°C.

Isoelectric focusing buffer

9 M urea, 4% CHAPS, 0.5% Triton X-100, 20 mM DTT, 1.2% Pharmalytes pH 3 to 10. Triton X-100 is provided as a 10% solution. This solution can be aliquoted and stored at –20°C for months.

Determination of protein concentration: protein concentration is evaluated according to a modified Bradford assay using the dye reagent from Bio-Rad.

Procedure

Protein extraction

1. Fresh plant tissue is frozen in liquid nitrogen after harvest and ground to a fine powder within pre-cooled steel cylinders of an automatic cryogenic crusher.
2. Then 1 g of ground tissue is suspended in 3 ml of extraction buffer in a 15 ml Falcon tube, vortexed, and incubated by shaking for 10 minutes on ice.
3. Afterwards, equal volume of Tris-buffered phenol is added, and the solution is incubated on a shaker for 10 minutes at room temperature.
4. To separate insoluble material in the pellet, for aqueous and organic phases, the sample is centrifuged for 10 minutes at 5500 g and 4°C. The phenolic phase, which is on the top of the tube, is recovered carefully to avoid contact with the interphase and poured into a new tube.
5. This phenol phase is then back-extracted with 3 ml of extraction buffer. The sample is shaken for 3 minutes again and vortexed. Centrifugation for phase separation is repeated for 10 minutes at 4°C at 5500 g.
6. The phenol phase still on the top of the tube is carefully recovered and poured into a new tube. Then 4 volumes of precipitation solution are added. The tube is shaken by inverting, and the sample is incubated for at least 4 hours or overnight at –20°C.
7. Proteins are finally pelleted by centrifugation for 10 minutes at 5500 g at 4°C.
8. After centrifugation, the pellet is washed three times with cooled precipitation solution and finally with cooled acetone. After each washing step, the sample is centrifuged for 5 minutes at 5500 g and 4°C.
9. Finally, the pellet is dried under vacuum.

Protein solubilization and quantification

1. The final pellet is resuspended in IEF buffer. For 1 g of fresh tomato fruit tissue, 200 ml of IEF buffer is needed.
2. The sample is incubated for at least 1 hour or more at room temperature under agitation. Do not heat samples; as this would lead to carbamylation of proteins.

3. For quantification, several dilutions of ovalbumin standard are made in IEF buffer (about 8 dilutions from 0 to 60 µg/ml). Then 10 µl of 0.1 N HCl is added to every sample. The final volume is adjusted to 100 µl with water, either for standard curve samples or for tissue sample.
4. Then 3.5 ml of diluted dye reagent are added, and the optical density is read at 595 nm.

7.3.4 Troubleshooting Guide

1. TCA, mercaptoethanol and acetone are toxic chemicals. Hence they should be handled under the fume hood only.
2. The precipitation and rinsing solutions must be cold when used. Always keep an acetone bottle at 4°C to prepare these solutions.
3. Both solubilizing solutions I and II contain high molarity of urea and must not be heated above 30°C because upon heating at high temperatures, isocyanate ions are produced that would result in protein carbamylation.
4. SB3-10 is poorly soluble in urea. This is why the urea concentration is limited to 5 M.
5. A fine powder must be obtained for efficient protein extraction. This may require pre-crushing of hard material (*e.g.* seeds).
6. Up to pellet drying step, it is important to work at a low temperature below 4°C to limit protease action.
7. β-mercaptoethanol and PMSF are toxic compounds. PMSF can be prepared as a stock solution 200 mM in isopropanol, aliquoted, and stored at –20°C.
8. Don't add too much water to solubilize CHAPS and urea powders; for a 25 ml final volume, add only about 10 ml of water. When preparing this solution, avoid heating above 30°C, to prevent protein carbamylation. Solubilization may take time.

7.4 ENZYME LINKED IMMUNOSORBENT ASSAY (ELISA)

7.4.1 Introduction

ELISA (Enzyme-linked Immunosorbent Assay) is a widely used biochemical technique for the detection of an antigen or any other protein in a sample. As the name is immunosorbent assay, it clearly suggests the use of immunoglobulins/antibodies for the detection assay. The purpose of an ELISA is to determine if a particular protein is present in a sample and if possible, to find out its quantity. Other names, such as enzyme immunoassay (EIA), are also used to describe the same technology. Enzyme immunoassay technique was first described by Engvall in 1971. Enzyme immunoassay offers an efficient alternative to the radioactive assay RIA. ELISA is simpler and safer assay of proteins.

There are three main types of ELISA;

1. Direct ELISA
2. Indirect ELISA
3. Sandwich ELISA

Direct ELISA is used for detecting antigens while indirect ELISA is usually used for detection of antibodies. Antigen (Ag) is a substance which when introduced into the body stimulates immune response resulting in the production of an antibody (Ab). Antibodies (also known as immunoglobulins-Ig) are gamma globulin proteins that are found in blood and are used by the immune system to identify and neutralize foreign objects, such as bacteria and viruses. For detection of antigens like virus particles from a blood sample, direct ELISA is used. However, the most powerful ELISA assay type is the sandwich ELISA. This is a type of capture assay called sandwich ELISA, because the analyte or antigen to be measured is bound between two antibodies –the capture antibody and the detection antibody.

ELISA may be run in a qualitative or quantitative format. Qualitative results provide a simple positive or negative result like yes or no for presence of antigen/antibody in a sample. The cutoff between positive and negative is determined by the analyst and may be statistical. In quantitative ELISA, the optical density (OD) of the sample is compared to a standard curve, which is typically a serial dilution of a known-concentration solution of the target molecule. For example, if a test sample returns an OD of 1.0, the point on the standard curve that gave OD = 1.0 must be of the same analyte concentration as your sample.

Interpreting the results of ELISA

After reading the results the standard curve is drawn taking the concentration on the X-axis and the absorbance on the Y-axis.

Assays that are performed for qualitative results like presence or absence of a trait can be performed with brief incubation and with large number of samples in a short duration of time. Such qualitative assays can also be performed outside the laboratory, in the field using squeeze bottles and droppers. The results indicated by colour change, can be seen visually and special optical reader is not required. ELISA is thus an excellent choice for as a confirmation tool. Being non-radioactive assay, it does not required skilled person to handle it.

Applications of ELISA

1. The ELISA has been used as a diagnostic tool in medicine and plant pathology, as well as a quality-control check in various industries.
2. Because the ELISA can be performed to evaluate either the presence of antigen or the presence of antibody in a sample, it is a useful tool for determining serum antibody concentrations such as with the HIV test.
3. It has also found applications in the food industry in detecting potential food allergens.
4. ELISA can also be used in toxicology as a rapid presumptive screen for certain classes of drugs.
5. The ELISA was the first screening test widely used for HIV because of its high sensitivity.

7.4.2 Theory and Principle

Immunoassay in simple words is a laboratory technique that makes use of the binding between an antigen and its homologous antibody in order to identify and quantify the specific antigen or antibody in a sample. In ELISA, an antigen must be immobilized to a solid surface and then complexed with an antibody that is linked to an enzyme. The enzymes are covalently attached to the antibodies in such a way that the enzyme's catalytic property and antibody specificity are unaltered. The enzymes such as peroxidase, alkaline phosphatase, β-galactosidase etc. are most commonly used because they catalyze reactions whose products are coloured and can be detected in small amounts.

ELISA is commonly performed in 96-well plates also called as microtiter plates, which permit high throughput results. The microtiter plates were first described by Takatsy in 1951. Formerly, the predominant use of these microtiter plates was in haemagglutination assay. The bottom of each well is coated with a protein to which the antibody will bind. It is this binding and immobilization of reagents that makes ELISA so easy to design and perform. Having the reactants of the ELISA immobilized to the microplate surface makes it easy to separate bound material from unbound material during the assay.

The unbound material can be washed away in wash buffers. This ability to wash away unbound materials makes the ELISA a powerful tool for measuring specific analytes within a crude preparation. Many different substrates are used in ELISA that produce colourimetric, fluorescent or chemiluminescent end products. ELISA is used for quantitative determination of target proteins. The microtiter plate format allows easy handling of a large number of standards, samples and controls. Most assay protocols can be configured to be completed within a day.

The most common and essential elements of ELISA assay are antigens, antibodies, detection enzyme or other detection systems, their substrates etc.

Antibody

In the ELISA assay, both monoclonal and polyclonal antibodies can be used either as a capture antibody immobilized on the solid support or as detection agent conjugated with enzyme. For better results, antibodies should be highly purified and free from aggregates that may cause erroneous results in the assay. Specificity is of prime importance while using polyclonal antibodies. The polyclonal antibodies are in fact, a complex mixture of antibodies that have specificity for various epitopes on the antigen. Polyclonal antibodies are essentially heterogeneous type of antibodies. Since polyclonal antibodies are elicited against any components used for the immunization of the animals, the purity of the immunogen is a major concern. Polyclonals are also likely to share one or more epitopes with closely related antigens, resulting in higher non-specific signal. If the immunogen is contaminated, the antibodies will be produced against these contaminants also and thus may cause false results in the assay. Polyclonal antibodies can very significantly from batch to batch and must be tested and validated thoroughly.

On the other hand, monoclonal antibodies are superior to polyclonal antibodies as all the immunoglobulins in the preparation are identical. Monoclonal antibodies are homogenous type of antibodies that have specificity for only a particular type of antigen. Therefore, they can be used in all types of ELISA, and most commonly used in sandwich ELISA. The desired specificity is a major factor in the choice of antibodies for ELISA. If closely related antigens exist and their reactivity in the ELISA is unwanted, then monoclonal antibodies should be chosen for the assay. In general, antibodies with high affinity are desirable in ELISA because they rapidly produce the greatest number of stable antigen-antibody complexes and therefore provide the most sensitive detection. The high affinity allows less amount of antibody to be used, which also produces less background.

Antigens

Antigens of small size, less than 100 kDa to large complexes greater than 1000 kDa in size are suitable for ELISA assays. The technique and protocol of assay varies depending upon the size of the antigens to be detected. Smaller antigens that may be as small as 10 kDa require covalent conjugation to a larger carrier molecule to produce a strong immune response. The carrier molecules are such that they do not modify the antigenic sites present on the antigen that help in binding with antibody. This conjugation of small antigen to the carrier molecule helps in better binding of the antigen to the antibody.

Larger antigens do not require binding to the carrier molecule and can be directly immobilized on the walls of the microtiter plate.

Enzymes

Enzymes are used in the various detection systems in ELISA. The important properties of the enzymes that are to be used in ELISA should be, stability after storage and assay, availability, low price, easy conjugation with antibodies, availability of a variety of substrates, easy detection etc. The enzymes should not cause any steric changes in the antigen or antibody when conjugated. The enzyme should have a sufficiently high turnover number for a rapid conversion of substrate. This converted substrate should be stable for a long time and should be quantitatively measurable with the available instrument.

The commonly used enzymes in ELISA are horseradish peroxidase (HRP), alkaline phosphatase (AP), urease, β-galactosidase and glucose oxidase. Out of these, HRP and AP are the most commonly used enzymes in ELISA. HRP has a high turnover and very small molecular weight (40 kDa) and AP is known to have a number of substrates with good sensitivity. These two enzymes have been successfully used in human, agricultural, veterinary and research oriented immunoassays for the past several decades (Brown, 2011). Because of the popularity of these enzymes, a wide range of substrates, are available for colourimetric, fluorescence and chemiluminescent detection systems. However, these enzymes also have some drawbacks that should be considered before selection of the enzyme. The reaction involving HRP is highly dependent on pH, so a slight fluctuation can hamper the results. Some plant tissues are also known to possess endogenous HRP. In such cases, it is advisable to select enzymes other than HRP. In case of AP, its high molecular weight (108 kDa) often results in large conjugates. Also care should be taken to avoid phosphates in buffers and samples while using AP.

Properly prepared antibody-enzyme conjugates can remain stable at 4°C for a few months. HRP is sensitive to light and hence should always be stored in dark coloured bottles, wrapped with aluminium foil. Conjugates of most enzymes can be frozen in 50% glycerol for many years. Many times stabilizers are added to the conjugate stocks.

Substrate

Wide varieties of substrates are available commercially for colourimetric, fluorescence and chemiluminescent detection. The ideal substrate should be stable in the absence of the enzyme, inexpensive, ready to use, nontoxic and should produce readable results. If the substrate is chemiluminescent, it should emit light with constant intensity for a sufficient period of time so that the readings can be taken. However, in ELISA, colourimetric substrates are more popular as they produce stable results, are cheaper and show a high level of sensitivity.

Most of the substrates are somewhat light sensitive. Therefore they should be protected from light during storage as well as during incubation step of the assay. The substrate enzyme reactions should be allowed to run for at least 10 minutes or as per the guidelines from manufacturer.

Stop reagents

Stop reagent is added after completion of the enzyme substrate reaction and conversion of substrate to end product. The enzyme substrate reaction is stopped prior to reading. This step is advantageous in ELISA when the assay includes many plates and therefore time when the reading is taken after incubation is likely to vary.

Adding enzyme inhibitors such as sodium fluoride for HRP, or changing the pH of the reaction mixture stops the reaction. It is required that the enzyme inhibitor should not change the wavelength for the reading of the reaction.

Coating of ELISA plates

The first step is to optimize the plate-coating conditions for the antigen or capture antibody. The choice of plate color depends upon the detection mechanism. Clear polystyrene flat bottom plates are used for colorimetric signals while black or white opaque plates are used for fluorescent and chemiluminescent tests. It is important to inspect plates before use, as imperfections or scratches in the plastic will cause false results when acquiring data from the developed assay.

Plate coating is achieved through adsorption of the protein to the plastic of the assay microplate. This process occurs though hydrophobic interactions between the plastic and non-polar protein residues. Although individual proteins may require specific conditions or pretreatment for optimal binding, the most common method for coating plates involves adding a 2-10 μg/ml solution of protein dissolved in an alkaline buffer such as phosphate-buffer (pH 7.4) or carbonate-bicarbonate buffer (pH 9.4). The plate is left to incubate for several hours to overnight at 37°C. Typically, after removing the coating solution, blocking buffer is added to ensure that all remaining available binding surfaces of the plastic well are covered. Coated plates can be used immediately or dried and stored at 4°C for later use, depending on the stability of the coated protein.

7.4.3 Commonly Used Solutions and Buffers

7.4.3.1 *Blocking buffers*

The binding capacity of microtiter plate wells is typically higher than the amount of protein coated in each well. The remaining surface area must be blocked to prevent antibodies or other proteins from adsorbing to the plate during subsequent steps. A blocking buffer is a solution of irrelevant protein, mixture of proteins, or other compound that passively adsorbs to all remaining binding surfaces of the plate. In addition to blocking, it is essential to perform thorough washes between each step of the ELISA.

Blocking buffer generally contains 2% (w/v) Bovine Serum Albumin (BSA) in wash buffer. Some blocking buffer systems may have surfactant such as Tween-20. Surfactants can help minimize hydrophobic interactions between the blocking proteins and the antigen or the antibodies. Typically a final concentration of 0.05% (v/v) Tween-20 is used. In addition, blocking buffers should be used in sufficient volumes to completely coat the wells.

7.4.3.2 *Wash solutions*

Washing steps are necessary to remove non-bound reagents. Insufficient washing will give false results, while excessive washing might result in decreased sensitivity caused by elution of the antibody and/or antigen from the well. Washing is performed in a buffer such as Tris-buffered saline (TBS) or phosphate-buffered saline (PBS). Usually a detergent such as 0.05% Tween-20 is added to the buffer to help remove nonspecifically bound material. [Phosphate buffered saline (PBS): 3.2 mM Na_2HPO_4, 0.5 mM KH_2PO_4, 1.3 mM KCl, 135 mM NaCl, pH=7.4]

Preparation of standards

Typically a standard curve may be plotted with concentrations ranging from 0 to 1000 pg/ml but may go as high as 3000 pg/ml depending on the predicted amount of antigen in the sample and the amount of standard protein available. Typically two-fold or three-fold dilutions of the standard protein are prepared from the stock solution using the standard diluents. When preparing serial dilutions of a protein standard, use fresh tips after each dilution to avoid contamination.

7.4.4 Direct ELISA

7.4.4.1 *Introduction*

The direct ELISA is technically simpler and requires a smaller sample volume of less than 1 ml. Direct ELISA is used for various purposes like virus indexing, virus detection, detection of disease causing bacteria and parasites from blood samples etc. Direct ELISA is most sensitive method of quantitative detection of viruses but is strain specific and requires highly purified antiserum. Direct detection is generally faster than indirect detection and background signal resulting from cross-reactivity of the secondary antibody with the coating antibody is also eliminated. Direct ELISA however, cannot provide the signal amplification gained from the use of secondary antibody systems. Consequently direct detection is generally less sensitive than indirect detection and is best used only when the target antigen is relatively abundant.

7.4.4.2 *Theory and principle*

In direct ELISA, antibody is first adsorbed on the surface of microtiter wells. This antibody is already labelled with an enzyme. Then sample is added to the microtiter wells, which contains the antigen. This antigen is specifically detected by enzyme-linked antibody. In this way, its specific enzyme-linked antibody directly detects antigen, hence the name given is direct ELISA. The major advantage of a direct ELISA is the ability to use crude or impure samples and still selectively bind any antigen that may be present. Samples are usually considered to be homogeneous or heterogeneous, depending on their complexity. This is essentially equivalent to a purified antigen vs. a crude unpurified mixture. In the simplest case, ELISA samples are diluted in PBS, wash buffer, or other specialty buffers and applied in a final volume of 100 µl.

A complex, heterogeneous protein mixture would be less suitable for coating a plate for direct ELISA detection unless the protein of interest is over-expressed and thus the majority of protein present in the sample.

7.4.4.3 *Procedure of direct ELISA (Fig. 7.1)*

1. Add 50 µl of diluted primary enzyme-linked antibody to each well. The appropriate dilution should be determined using a checkerboard titration prior to testing samples. The amount of antibody used will depend on the individual assay, but if maximal binding is required, use at least 1 µg/well. This is well above the capacity of the well, but the binding will occur more rapidly, and the binding solution can be saved and used again. Allow to incubate for 4 hours at room temperature or 4°C overnight. Wash the wells twice with wash buffer. The antibody solution washes can be removed by flicking the plate over a suitable container.
2. The remaining sites for protein binding on the microtiter plate must be saturated by incubating with blocking buffer. Fill the wells to the top with 3% BSA in PBS with 0.02% sodium azide. Incubate for 2 hrs to overnight in a humid atmosphere at room temperature.
3. Wash wells twice with wash buffer.
4. Add 50 µl of the sample solution to the wells. All dilutions should be done in the blocking buffer (3% BSA in PBS with 0.05% Tween-20). Note: Sodium azide is an inhibitor of horseradish peroxidase. Do not include sodium azide in buffers or wash solutions, if an HRP-labeled conjugate will be used for detection.

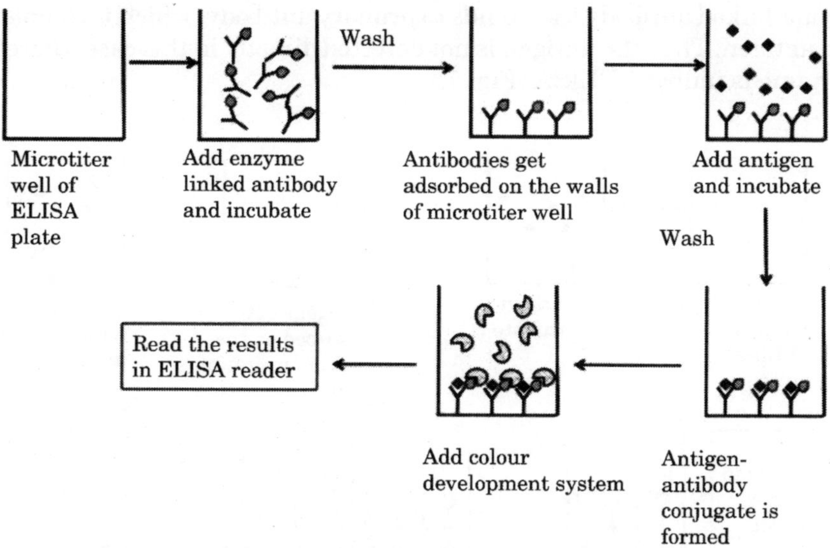

Fig. 7.1: The technique of direct ELISA

5. Wash the plate four times with PBS wash buffer to remove unbound proteins.
6. Add substrate and incubate. After suggested incubation time has elapsed, optical densities at target wavelengths can be measured on an ELISA reader.

7.4.5 Indirect ELISA

7.4.5.1 *Introduction*

Indirect ELISA avoids the limitations of direct ELISA and is a sensitive method. It is commonly used for virus detection from crude samples. A major disadvantage of the indirect ELISA is that the method of antigen immobilization is non-specific. When serum is used as the source of test antigen, all proteins in the sample may stick to the wall of microtiter plate well; so small concentrations of analyte in serum must compete with other serum proteins when binding to the well surface. The sandwich or direct ELISA provides a solution to this problem, by using a 'capture' antibody specific for the test antigen.

7.4.5.2 *Theory and principle*

In indirect ELISA, antigen is not directly detected by enzyme-linked antibody. In fact, two antibodies are used in this process- primary antibody and secondary antibody. Primary antibody binds to the antigen. Secondary enzyme-linked antibody then binds to primary antibody, which is complexed with antigen. Thus the antigen is not detected directly in this case; therefore the name is indirect ELISA (Fig. 7.2).

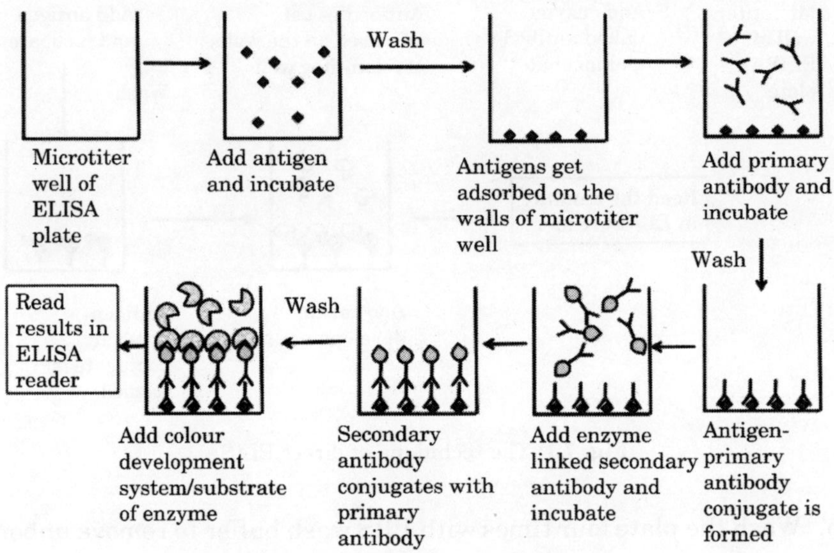

Fig. 7.2: The technique of indirect ELISA

7.4.5.3 *Procedure of indirect ELISA*

1. A buffered solution of the antigen to be tested for is added to each well of a microtiter plate. The plate is incubated and antigen is allowed to bind to the walls of microtiter wells.
2. A blocking solution of non-reacting protein, such as bovine serum albumin or casein, is added to block any plastic surface in the well that remains uncoated by the antigen. This is done to prevent primary antibody binding to the well surface, as it will give false results.
3. Next the primary antibody is added, which binds specifically to the test antigen that is coating the well. This primary antibody could also be in the serum of a donor to be tested for reactivity towards the antigen.
4. Afterwards, a secondary antibody is added, which will bind the primary antibody. This secondary antibody often has an enzyme attached to it, which has a negligible effect on the binding properties of the antibody.
5. A substrate for this enzyme is then added. Often, this substrate changes color upon reaction with the enzyme. The color change shows

that secondary antibody has bound to primary antibody, which strongly implies that the donor has had an immune reaction to the test antigen.

6. Higher the concentration of primary antibody in the serum, stronger is the color change. Often ELISA reader is used to measure quantitative values for color strength.

7.4.6 Sandwich ELISA

7.4.6.1 *Introduction*

In sandwich ELISA, the antigen is sandwiched between the primary antibody and secondary enzyme-linked antibody. Major advantages of this technique are that the antigen does not need to be purified prior to use, and that these assays are very specific.

7.4.6.2 *Theory and principle*

In this technique, one antibody called as capture antibody is purified and bound to a solid phase typically attached to the bottom of a plate well. Antigen is then added and allowed to complex with the bound antibody. Unbound products are then removed with a wash, and labeled second antibody called as detection antibody is allowed to bind to the antigen, thus completing the "sandwich". The assay is then quantitated by measuring the amount of labeled second antibody bound to the matrix, through the use of a colorimetric substrate (Fig. 7.3).

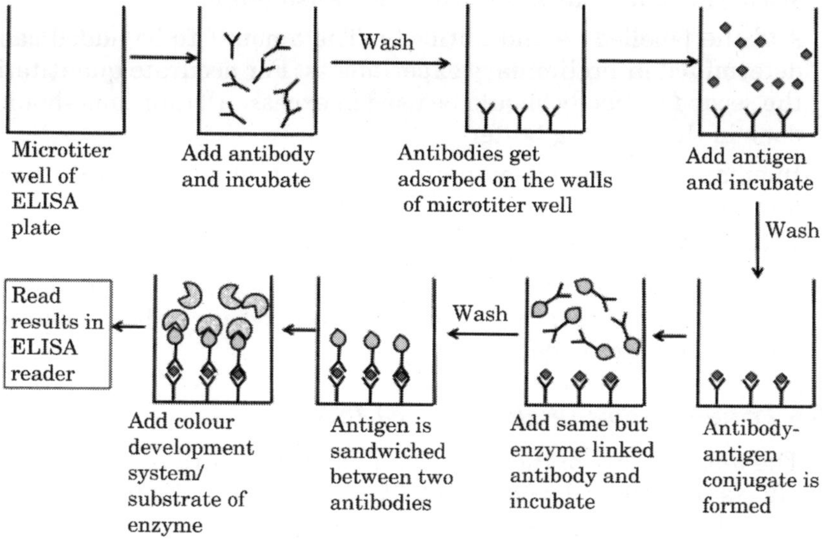

Fig. 7.3: The technique of sandwich ELISA

Before the assay, both antibody preparations should be purified and one antibody must be labeled with enzyme.

7.4.6.3 *Procedure of sandwich ELISA*

1. Bind the unlabeled antibody to the bottom of each well by adding approximately 50 µl of antibody solution to each well. The amount of antibody used will depend on the individual assay, but if maximal binding is required, use at least 1 µg/well. This is well above the capacity of the well, but the binding will occur more rapidly, and the binding solution can be saved and used again
2. Incubate the plate overnight at 4°C, to allow complete binding.
3. Wash the wells twice with wash buffer. The washes can be removed by flicking the plate over a suitable container.
4. The remaining sites for protein binding on the microtiter plate must be saturated by incubating in blocking buffer. Fill the wells to the top with 3% BSA in wash buffer with 0.02% sodium azide. Incubate for 2 hrs to overnight in a humid atmosphere at room temperature. [Note: Sodium azide is an inhibitor of horseradish peroxidase. Do not include sodium azide in buffers or wash solutions if an HRP-labeled antibody will be used for detection.]
5. Wash wells twice with wash buffer.
6. Add 50 µl of the antigen solution to the wells. All dilutions should be done in the blocking buffer (3% BSA in PBS buffer). Incubate for at least 2 hours at room temperature in a humid atmosphere.
7. Wash the plate four times with PBS wash buffer.
8. Add the labelled second antibody. The amount to be added can be determined in preliminary experiments. For accurate quantitation, the second antibody should be used in excess. All dilutions should be done in the blocking buffer.
9. Incubate for 2 hours or more at room temperature in a humid atmosphere.
10. Wash several times with PBS wash buffer.
11. Add substrate and incubate. After suggested incubation time has elapsed, optical densities at target wavelengths can be measured on an ELISA plate reader.

7.4.6.4 *Troubleshooting guide for ELISA*

1. The temperature of the substrate enzyme mixture is important as it affects the colour development. The variation in the temperature of the substrate solution is an important factor in causing differences in the readings between two assays performed with the same technique and reagents.

2. Time of the incubation is very important and should be followed strictly.
3. If the incubation is to be done at room temperature, it usually means at 37°C. However, temperatures fluctuate according to the seasons. Make sure that it is nearly same temperature at any time of the year whenever you are performing the assay.
4. If your assay requires stationary incubation, make sure that plates are kept in a place where no one would disturb them. Avoid shifting, tapping and flicking of the plate at any time during incubation.
5. Always store all the reagents at recommended temperatures.
6. It is advisable to make working dilutions of the enzyme-antibody conjugates fresh and as and when required.
7. Make sure that pipette tips, water, glassware, and plates used for ELISA are sterile and in good condition.
8. If no colour is obtained even after 30 minutes of incubation in HRP system, check whether hydrogen peroxide is added, and check the dilution of hydrogen peroxide.
9. If you find strong colour development in the whole plate, may be conjugate solution is too strong and needs to be diluted.
10. If colour development is very fast, check for dilution of the conjugate. The endogenous enzymes from the tissue samples can also create this problem.
11. If the colour development is very slow, either the conjugate is diluted too much or there is contamination of inhibitory chemicals like sodium azide. Sodium azide is used as preservative for the solutions but it also inhibits the HRP enzyme activity.
12. Insufficient blocking results in nonspecific binding of antibodies and results in high background colour. Follow blocking and washing steps carefully.
13. If the substrate forms precipitate in the wells, it means that too much enzyme may be present. Further dilution of the enzyme conjugate is most likely required along with optimization of antibody dilutions; efficient blocking and additional washing may also be necessary.
14. Do not allow the plates to become dry at any point of time while performing assay. Keep the plates wrapped in aluminium foil always and incubate in humidified chamber.

7.5 RADIOIMMUNOASSAY

7.5.1 Introduction

Radioimmunoassay (RIA) technique was first developed by Salomon Berson and Rosalyn Yalow in 1950s for the determination of insulin from human blood. This was an important discovery in the fight against diabetes. R. Yalow received the Nobel Prize for Physiology or Medicine in 1977 for the development of radioimmunoassay of peptide hormones.

Yalow and Berson, towards the middle 1950's made the surprising discovery that people who had received injections of the polypeptide hormone insulin either for diabetes or treatment of schizophrenia, developed antibodies against the hormone. This discovery conflicted with the prevailing concept that such a small protein like insulin could not be antigenic. It took considerable time before this was accepted. It was also found that the insulin antibodies formed a soluble complex with added insulin labelled with radioactive iodine. Furthermore, when non-labelled insulin was added to this mixture it could displace the labelled insulin bound to the antibody. Therefore percentage binding of labelled insulin to the antibodies is a function of the total insulin concentration in the solution.

This later became the starting point for radioimmunological determination of insulin and other peptide hormones in blood, fluids and tissues. It was accomplished by a combination of immunology, radioisotope research and application of mathematics and physics. RIA technique is so sensitive that it allows determination of insulin in amounts as small as 10-20 picograms and adrenocorticotropic hormone (ACTH) in an amount less than 1 pg per ml of blood plasma. Radioimmunoassay created a revolution in biological and medical research. By using this technique, not only peptide hormones can be detected but also hormones that are not peptides, peptides that are not hormones, enzymes, viruses, antibodies, drugs of the most different kinds can be detected from biological samples. Even today, RIA principles have a wide application in the field of drug analysis, drug monitoring, pharmacokinetic studies and immunodiagnostics in medicine. RIA provides a specific, sensitive, rapid, convenient, reliable and reproducible assay method for screening of biological fluids.

7.5.2 Theory and Principle

Radioimmunoassay is a competitive binding assay that is based on the reversible binding of the labelled antigen to its specific antibody. RIA is based on the principle of antigen-antibody complex formation and the competition between the radiolabelled compound of interest and unlabelled compound in the patient's serum for a limited number of binding sites on the antibody. In this method, known antigen is labelled with a radioactive isotope of iodine I^{125} or I^{131} and the labelled antigen (Ag^*) is mixed with the antibody (Ab). Then a non-labelled antigen (Ag), usually from the patient's blood is added to this mixture at higher concentration to complete with the labelled antigen. Antibody preferentially binds to the non-labelled antigen, displacing the labelled antigen. This mixture is incubated and the equilibrium is reached. Following reactions take place in the process:

$$Ab + Ag^* \Leftrightarrow Ab\text{-}Ag^*$$

$$Ag + Ab\text{-}Ag^* \, Ab \Leftrightarrow Ag^* + Ab\text{-}Ag + Ag^*$$

It is essential that concentration of antibody should be less than that is required to bind all the labelled and unlabelled antigens present in the mixture. In RIA mixture, usually unbound, free, labelled antigen concentration is denoted by F and the labelled antigen-antibody complex ($Ab\text{-}Ag^*$), is called bound antigen and its concentration is denoted by B. for a constant amount of antibody and labelled antigen, the amount of bound antigen will be inversely proportional to the quantity of unlabelled antigen (Ag). This means that if the concentration of unlabelled antigen is increased, the amount of bound antigen will decrease with the concurrent increase in the amount of free antigen (Ag^*) and vice versa.

Thus the ratio of bound to free antigen (B/F) is a function of the concentration of non-labelled antigen in the RIA mixture. This relation is used in the determination of the concentration of the unknown antigen in the sample. Since $B + F = Ag^* =$ constant, measurement of concentration of either B or F is sufficient. Usually, B is measured by capture of the Ag-Ab complex, which can be achieved by a solid-phase secondary antibody or by lattice formation in solution. RIA is a heterogeneous assay that requires complete separation of bound and unbound labelled antigen. The contamination of free antigen with the bound one will increase bound counts per minutes (cpm) and give false low antigen concentration results.

The specific requirements of the RIA can be stated as follows;

1. The labelled and unlabelled antigens must be indistinguishable chemically.
2. The antibody used must be specific to the antigen, which is the compound of interest.
3. The concentration of antibody must be severely deficient to bind all antigens present.

For the preparation of standard curve, a series of samples with increasing but known concentrations of unlabelled antigens and one control (without the unlabelled antigen) are prepared. To each of the samples, constant and known amount of labelled antigen and antibody are added. In the process of RIA, antiserum is incubated with labelled and unlabelled antigens. The time for incubation and the temperature varies with the type of antigen and antibody. After incubation the bound and free fractions of radiolabelled antigens can be separated by one of the following methods:

1. Adsorption of antigen on the solid surface such as charcoal, cellulose and silica.
2. Adsorption of antibodies on a solid surface such as glass or plastic.
3. Gel filtration chromatography.
4. Double antibody system where secondary antibody is added to the mixture that binds to antigen-antibody complex.
5. Precipitation by alcohol, ammonium sulphate or polyethylene glycol.

After separation, the amount of bound and free radiolabelled antigen is found out by measuring radioactivity with scintillation counter or gamma counters. The B/F ratio is calculated for each sample and these values are plotted against logarithms of unlabelled antigen from the sample. This plot is the standard curve of radioimmunoassay for the particular antigen.

7.5.3 Procedure of Radioimmunoassay

The brief general procedure of radioimmunoassay of morphine is described as follows,

1. Preparation of radiolabelled morphine: Morphine is first reacted with sodium-*p*-chloroacetate in absolute alcohol to produce 3-*o*-carboxy-methyl-morphine. The 3-*o*-carboxy-methyl-morphine is then coupled to bovine serum albumin (BSA) by dissolving 3-*o*-carboxy-methyl-morphine in distilled water containing BSA. Keep the pH at 5.5 and add 1-ethyl-3-(3-dimethylaminopropyl) carbodiimide or EDAC. This mixture is then incubated overnight at room temperature and the reaction is allowed to take place. Then resulting product is purified by dialysis. EDAC linked biological substances contain carboxylates and primary amines. EDAC is thus commonly used for cross-linking peptides and formation of immunoconjugates. The resulting product of the reaction is carboxy-methyl-morphine-BSA conjugate. This product is then labelled with tritium.
2. Rabbits are immunized by injecting antigen carboxymethylmorphine-BSA and antibodies from rabbit blood are isolated and purified.
3. This antiserum containing the specific antibody is mixed with a known amount of tritium-labelled antigen carboxymethylmorphine-BSA and incubated.
4. The solution of saturated neutral ammonium sulphate is added to mixture. The mixture is centrifuged at 5000 g for several minutes.
5. The pellet formed is washed twice with 50% ammonium sulphate solution, centrifuging and pelleting each time.
6. The pellet is recovered. This pellet contains antibody bound radiolabelled morphine. The pellet is dissolved in tissue solubilizing

liquids such as NCS. Then radioactivity is measured with the help of either scintillation counters or gamma counters.

7. The tube that contains radioactive morphine and antiserum and no unlabelled morphine should be kept as control, which serves as measure of maximum antibody-bound radioactivity.
8. Then to another tube that has similar contents as 'control', increasing amount of unlabelled morphine is added.
9. The antibody reacts with both radioactive and unlabelled antigens forming both antibody-radioactive morphine and antibody-nonradioactive morphine complexes. The unlabelled morphine competitively binds with the antibody, displacing the radiolabelled morphine.
10. The mixture is centrifuged again and supernatant and pellet are collected separately. The radioactivity in the pellet is measured again. The reduction in the radioactivity corresponds to the amount of morphine present in the unknown sample.
11. The counts obtained from radioactivity measurement are used to find the unknown concentration of morphine in the sample by comparing the readings with the standard curve.

7.5.4 Troubleshooting Guide

1. The method of measurement radioactivity of bound or unbound labelled antigen after the separation depends upon the nature of the radioisotope and the method of separation used to separate bound and unbound fractions.
2. One can determine either the antibody-bound labelled antigen fraction or unbound fraction for measurement of unknown concentration of the antigen from the sample. However, for experimentation purposes, it is advisable to measure both the fractions and compare them to the standard, to make sure that there are no losses unaccounted for.

7.6 WESTERN HYBRIDIZATION

7.6.1 Introduction

The western hybridization is to the analysis of proteins what southern hybridization is to the analysis of DNA. Western blotting is also called immunoblotting because in this technique, an antibody is used to specifically detect its protein antigen. Towbin *et al.* introduced the technique of western blotting in 1979 where they successfully transferred ribosomal proteins from polyacrylamide gel to nitrocellulose membrane. The western hybridization is now a routine technique for protein analysis. The specificity of the antibody-antigen interaction enables a target protein to be identified from purified protein mixture.

Western blotting can produce qualitative and quantitative data about the protein of interest. The electrophoretic western blotting technique produces replicas of proteins separated on polyacrylamide gels with high fidelity. Western blotting, which is also called as protein blotting or immunoblotting, is a powerful and important procedure for the immunodetection of proteins post-electrophoresis, particularly proteins that are of low abundance.

7.6.2 Theory and Principle

The first step in a Western blotting procedure is to separate the proteins using polyacrylamide gel electrophoresis. Proteins are commonly separated using polyacrylamide gel electrophoresis (PAGE) to characterize individual proteins from a mixed sample. After electrophoresis, the separated proteins are transferred or blotted onto a second matrix that is generally a nitrocellulose membrane. Certain type of synthetic membranes like nitrocellulose and PVDF, are able to bind proteins tightly enough so that they can be used for solid support immunoassays. These bound proteins retain their antigenicity and remain accessible for the specific antibody binding (Fig. 7.4).

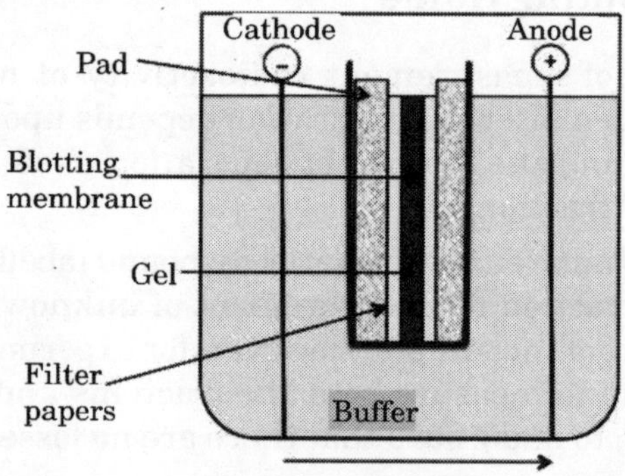

Fig. 7.4: The technique of western blotting

There are a variety of methods that can be used for blotting of proteins to a membrane, such as diffusion transfer, capillary transfer, heat-accelerated convectional transfer, vacuum blotting transfer and electroelution. Transfer efficiency can vary among proteins, depending upon the ability of a protein to migrate out of the gel and to bind to the membrane under a particular set of conditions. The efficiency of transfer depends on factors such as the composition of the gel, complete contact of the gel with the membrane, the transfer time, size and composition of proteins, electric field applied and the presence of detergents and alcohol in the buffer. The transfer of proteins from gel to the membrane is most commonly carried out by electrophoretic transfer method. The proteins from polyacrylamide gel are transferred to the membrane in such a way that a faithful replica of the original gel pattern is obtained.

As the membrane that is used to blot proteins, has very high affinity for all proteins, it may bind to free protein antibodies and give false results. Therefore, nitrocellulose membrane is blocked by blocking solution to prevent any nonspecific binding of antibodies to the surface of the membrane. The blocking buffer contains highly purified proteins that are used to block free sites on the membrane. The blocking buffer should improve the sensitivity of the assay by reducing background interference and improving the signal to noise ratio.

Most commonly, the transferred protein of interest is complexed with an enzyme-labeled antibody. An appropriate substrate is then added to the enzyme and they produce a detectable product such as a chromogenic precipitate on the membrane for colorimetric detection. The most common blocking agent in the blocking buffer is non-fat dry milk. Other blocking agents used are 2% gelatin (usually fish skin gelatin), 2% Bovine Serum Albumin (BSA), 10% horse serum, and 1% casein. The horse serum contains immunoglobulins therefore it is not compatible with some secondary antibodies. It is always important to evaluate which type of primary and secondary antibody is compatible with which blocking agent.

Like many other immunoassay procedures, western blotting procedure consists of a series of incubations with different immunochemical reagents separated by wash steps. Washing steps are necessary to remove unbound reagents from the membrane. However, Insufficient washing will give faulty results, while excessive washing may result in decreased sensitivity caused by elution of the antibody and/or antigen from the blot. The wash buffer formulations may consist of Tris buffered saline (TBS) or phosphate buffered saline (PBS) with or without a detergent such as 0.05% Tween-20. The detergent is added to the buffer to help remove nonspecifically bound material. However, depending on the properties of the protein of interest and antibody used, the amount of detergent in the wash buffer is bound to vary. It is important to note that antibody solutions are often made in wash buffer.

The most sensitive detection methods use a chemiluminescent substrate that, when combined with the enzyme, produces light as a byproduct. The light output can be captured using a film, or a CCD camera that is designed for chemiluminescent detection. Alternatively, fluorescently tagged antibodies can be used, which are directly detected with the aid of a fluorescence imaging system.

The procedures for detection of western blotting vary widely. The western hybridization involves direct or indirect detection of protein. In the direct detection method, the primary antibody that is used to detect an antigen on the blot is labeled with an enzyme or fluorescent dye. In the indirect detection method, a primary antibody is added first to bind to the antigen. This is followed by a labelled secondary antibody that is directed against the primary antibody. The indirect detection amplifies the signal and gives better results. These labels may include biotin, fluorescent probes such as fluorescein or rhodamine, and enzyme conjugates such as horseradish peroxidase or alkaline phosphatase.

Radioisotopes have been used extensively for detection of specific proteins in western hybridization. The radioisotope of iodine ^{125}I is commonly used to radiolabel the antibodies used for detection. However, detection using radioisotopes has several drawbacks such as special handling and care, permit, special laboratory area and radioactive waste disposal. As this technique can be handled by experienced personnel only, the use of radioisotopes in western hybridization is often discouraged.

7.6.3 Procedure of Western Blotting (Hybridization)

Materials required

- Electro-transfer apparatus
- Membrane for blotting- nitrocellulose or PVDF
- Transfer buffer (pH=8.3)

Tris	25 mM
Glycine	192 mM
Methanol	20%

  The buffer can be stored at 4°C for 2-3 months.
- TBS buffer (pH=7.5)

Tris-HCl	0.02 M
NaCl	0.5 M
- TTBS buffer (it is TBS buffer containing 0.05% Tween 20)
- Blocking solution

  TBS containing 5% non-fat dry milk
- Primary antibody (Dilute the primary antibody in TTBS containing 5% non-fat dry milk. The dilutions vary considerably, but are in the range of 1:100 to 1:3000)
- Secondary antibody-enzyme conjugate (alkaline phosphatase or horseradish peroxidase)
- Substrate for the enzyme conjugated to secondary antibody

For alkaline phosphatase- the working substrate solution is prepared by taking 0.15 mg of BCIP and 0.3 mg of NBT per ml of Tris buffer (pH=9.5). The Tris buffer contains 0.1 M of Tris-HCl and 0.5 mM $MgCl_2$.

For horseradish peroxidase- the working substrate solution is prepared by taking 0.015% hydrogen peroxide (H_2O_2) and 0.05% 4-chloro-1-naphthol (4-CN) substrate in TBS buffer containing 16.7% methanol. Protect this solution from light. Alternatively DAB substrate can be used by dissolving 50 mg of DAB (diaminobenzidine) and 100 ml of 3% hydrogen peroxide in 100 ml of TBS.

The substrate working solution should be prepared fresh before use.

Procedure of western hybridization

1. Load the protein mixture onto the PAGE gel and run the gel to separate proteins.
2. Prepare the transfer buffer and electro-transfer apparatus. Make buffer more than that is required to fill the apparatus.
3. Remove the gel and soak it in transfer buffer completely for about 2-3 minutes. Cut off the stacking gel if it was used since the soft gel will stick to the membrane.
4. Soak the filter papers and sponge pads in transfer buffer. Cut the transfer membrane to the size of the gel by using sharp knife or scalpel. Do not touch the membrane by bare hands. Always wear sterile gloves while handling the membrane.
5. Wet the membrane completely with transfer buffer. If PVDF membrane is used, wet it thoroughly with methanol before putting it in aqueous solutions. Now soak the membrane in transfer buffer for about 15 minutes. Do not let the membrane become dry before start of the transfer.
6. Assemble the electro-transfer apparatus. Soak 8-10 filter paper pieces in transfer buffer and place them on top of the anode (+). Take care to eliminate trapped air bubbles. Place pre-wetted membrane on top and roll glass rod or pipette to remove air bubbles. Then place the gel on top of the membrane. Again make sure that no air bubbles are trapped, by rolling pipette/glass rod over each layer. Finally add 6-8 pieces of soaked filter paper on top of the gel. Remove air bubbles. Place cathode (-) over the sandwich.
7. Transfer the assembled cassette in buffer tank with the gel-side facing cathode and membrane-side facing anode. Add more transfer buffer if necessary.
8. Run the transfer for 2-3 hours at 0.8 mA per square cm of the gel.
9. Recover the transfer membrane. Wash the membrane in TBS for 10-15 minutes.

10. Incubate the membrane in TBS containing 5% non-fat dry milk for 30 minutes to 1 hour at room temperature. This is membrane-blocking step. Protein from dry milk blocks the free protein-binding sites on the membrane.
11. Wash the membrane twice for 5 minutes with TTBS buffer.
12. Incubate the membrane for 1-2 hrs at room temperature in primary antibody diluted in TTBS containing 5% non-fat dry milk.
13. Wash the membrane twice for 5 minutes in TTBS buffer.
14. Incubate the membrane for 1-2 hrs at room temperature in secondary antibody-enzyme conjugate diluted in TTBS containing 5% non-fat dry milk (The dilutions vary but range of dilutions is same as in primary antibody).
15. Wash the membrane twice for 5 minutes in TTBS buffer.
16. Wash the membrane in TBS to remove all traces of Tween 20.
17. Incubate the membrane in substrate solution for 30 minutes to 1 hour or till the desired amount of colour intensity is reached.
18. Wash the blot with sterile distilled water. The process of western hybridization is now complete. This washed blot can be dried and stored.
19. Photograph the membrane for the record.

7.6.4 Troubleshooting Guide

1. If you come across problem of non-binding of target protein to the membrane, check the molecular weight of the protein of interest. Proteins with molecular weight less than 12 kDa are retained less efficiently than proteins with high molecular weight.
2. Nitrocellulose membrane is less efficient in binding of low molecular weight proteins as compared to PVDF membrane. Prefer PVDF membrane while working with low molecular weight proteins in the range of 5-12 kDa.
3. The membranes used for transfer are available in various sizes. Membranes with pore size of 0.1 to 0.2 m are recommended for transfer of small proteins.
4. Nitrocellulose membranes have certain drawbacks that should be considered while western blotting. The nitrocellulose membrane is physically fragile and liable to crack and tear when dry. Therefore it is not suitable for stripping and reprobing.
5. PVDF or polyvinylidene difluoride membranes are strong and have high protein binding capacity than nitrocellulose. They are useful for stripping and reprobing. However, they are highly hydrophobic and need to be wetted with methanol before they can be equilibrated with aqueous buffer.

6. While using PVDF, care should be taken so that the membrane does not dry out. If the membrane becomes dry, this might result in uneven blocking and antibody incubation.
7. The washing step should always be done at room temperature and with continuous agitation. The volume of wash buffer should be at least 1 cm deep and the membrane should be completely immersed in wash buffer solution.
8. Remember that wash buffer solution needs to be changed frequently after washing of membrane.
9. Chemicals such as Triton, NP-40 and SDS should not be used in the wash buffer as they may remove membrane-bound antibodies and proteins.
10. All antibody aliquots should be flash frozen in liquid nitrogen and stored at –70°C. They can be stored for few years at such low temperatures.
11. Once the antibody serum is thawed, it should never be frozen again. It should be stored at 4°C. However they can be stored at this temperature only for a short duration. Repeated freezing and thawing of antibody solution results in aggregation and loss of reactivity of antibodies.
12. Before use, antibodies should always be diluted in a buffer containing carrier protein such as BSA.
13. In general, secondary antibodies are less stable than primary antibodies. This is because secondary antibody should retain both the functions of antibody binding and reporter activity. Therefore, stability of antibody chiefly depends upon the stability of reporter groups. Iodinated conjugates are stable for a few weeks while enzyme conjugates are stable for months. Thus it is important to know the type and stability of antibody that you are using.
14. Sodium azide, which is a powerful peroxidase inhibitor, is often used as a buffer preservative. Therefore sodium azide should be washed away if you are using Horseradish Peroxidase (HRP) detection system.
15. Phosphate buffers like PBS should not be used with Alkaline Phosphatase detection system, as the presence of phosphate will inhibit phosphatase reaction. In this case, Tris buffers like TBS should be used.
16. Whenever using antibodies, thumb rule is that 'less is more'. When you are using certain antibodies for the first time, or setting up a new system of western blotting, the detection should be carried out using various dilutions of antibodies (primary or secondary). The highest dilution of the antibody that gives satisfactory results should be used as a working solution. The use of minimum concentrations of primary and secondary antibodies provides the highest specificity with minimum background.

17. If swirls or missing circular patterns occur in the blot, this could be due to poor contact between gel and membrane. This can occur if air bubbles or excess buffer remain between the blot and gel. Carefully move the roller over the membrane in both directions until air bubbles or excess buffer are removed from between gel and membrane, and complete contact is established. Use thicker filter paper in the gel-membrane sandwich.
18. If random staining pattern appears in the blot, this could be due to contaminated or thin fiber pads. The fiber pads deteriorate and compress with time. Replace the fiber pads or thoroughly clean the contaminated pads.
19. If overall high background appears in the blot, it means that blocking was insufficient. Usually PVDF membranes require extensive blocking. Increasing the concentration of blocking agent and/or blocking time can overcome the problem of high background.
20. High background can also result from insufficient washing. Increase the number, duration and stringency of washes.
21. If there is weak or no signal in the blot even after strictly following the procedure, this can happen if amount of protein in the sample was very low. Try more concentrated protein sample that will be loaded in the gel.

7.7 REVIEW QUESTIONS

1. What is the principle underlaying various Immunotechniques?
2. What is ELISA? Explain in brief and give its applications.
3. What is direct ELISA? Explain with a suitable diagram.
4. What is indirect ELISA? Explain with a suitable diagram.
5. What is sandwich ELISA? Explain with a suitable diagram.
6. Differentiate direct and indirect ELISA. Which method would you prefer and why?
7. What is radioimmunoassay? Describe in brief.
8. Describe theory and principle of western hybridization. How does it differ from radioimmunoassay?

7.8 REFERENCES

Bretagne, S., Marmorat–Khuong, A., Kuentz, M., Latge, J., Bart–Delabesse, E. and Cordonnier, C. (1997). Serum *Aspergillus* galactomannan antigen testing by sandwich ELISA: practical use in neutropenic patients. *Journal of Infection,* **35(1):** 7–15.

Brown, M. (2011). Microtiter plate ELISA. *In*: Immunoassays in agricultural biotechnology. John Wiley and Sons Inc. pp. 59–90.

Carpentier, S., Witters, E., Laukens, K., Deckers, P., Swennen, R. and Panis, B. (2005). Preparation of protein extracts from recalcitrant plant tissues: An evaluation of different methods for two–dimensional gel electrophoresis analysis. *Proteomics,* **5:** 2497–2507.

Chegini, S., Erhart–Hoffmann, B., Kaider, A. and Waldhauser, F. (1995). Direct enzyme linked immunosorbent assay and radioimmunoassay for melatonin compared. *Clin. Chem.,* **41(3):** 381–386.

Chalfie, M. (1995). Green fluorescent protein. *Photochemistry and Photobiology* **62(4):** 651–656.

Chi, V. and George, C. (2007). Immunohistochemistry: Paraffin sections using vectastain ABC kit from vector labs. *J. Vis. Exp.,* **8:** 308.

Conway, D., Atkins, N., Lilywhite, J., Bailey, J., Robinson, R., Lindo, J., Bundy, D. and Biunco, A. (1993). Immunodiagnosis of *Strngyloides stercoralis* function: A method for increasing the specificity of the indirect ELISA. *Transactions of The Royal Society of the Tropical Medicine and Hygiene,* **87(2):** 173–176.

Crowther, J. (1995). Methods in molecular biology volume 42: ELISA: theory and practice. Humana Press Inc. p. 223.

Csorba, *et al.* (2011). Development of an ELISA for sensitive and specific detection of IgA autoantibodies against BP180 in pemphigoid diseases. *Orphanet journal of Rare Diseases,* **6:** 31.

Dave, J., Taylor, P., Grange, J. and Gaya, H. (1986). A new enzyme–linked fluorescence assay (ELFA) for use with peroxidase–anti body conjugates: A comparison with ELISA for the quantitation of IgM antibodies to hepatitis B core antigen. *J. Med. Microbiol.,* **21:** 271–274.

Durnez, L., Bortel, W., Denis, L., Roelants, P., Veracx, A., Trung, H., Sochantha, T. and Coosmans, M. (2011). False positive circumsporozoite ELISA: a challenge for the estimation of the entomological inoculation rate of malaria and for the vector incrimination. *Malaria Journal,* **10:** 195–196.

Edwards, M. and Cooper, J. (1985). Plant virus detection using a new form of indirect ELISA. *Journal of Virological Methods,* **11(4):** 309–319.

http://www.nobelprize.org/nobel_prizes/medicine/laureates/1977/press.html

Kemeny, D. and Challancombe, S. (1989). ELISA and other solid phase immunoassays: theoretical and practical aspects. John Wiley and Sons Ltd. p. 357.

Kar, A. (2005). Pharmaceutical drug analysis. 2nd Edition. New Age International.

Kobets, T., Badalova, J., Grekov, I., Havelkova, H., Svobodova, M. and Lipoldova, M. (2010). Leishmania parasite detection and quantification using PCR-ELISA. *Nature Protocols,* **5:** 1074–1080.

Kurien, B. and Scofield, R. (2006). Western Blotting. *Methods,* **38(4):** 283–293.

Ozoslan, M., Caliskan, I., Kilic, I. and Karagoz, I. (2011). Applications of ELISA and HPLC test for detection of aflatoxin in pistachio. *Scientific Research and Essays,* **6(14):** 2913–2917.

Pierangeli, S. and Haris, E. (2008). A protocol for determination of anticardiolipin antibodies by ELISA. *Nature Protocols,* **3:** 840–848.

Ling, K., Zhu, H., Petrovic, N. and Gonsalves, D. (2001). Comparative effectiveness of ELISA and RT–PCR for detecting grapevine leafroll– associated closterovirus–3 in field samples. *Am. J. Enol. Vitic.,* **52(1):** 21–27.

Loset, G., Lunde, E., Bogen, B., Brekke, O. and Sandlie, I. (2007). Functional phage display of two murine α/β T–cell receptors is strongly dependent on fusion format, mode and periplasmic folding assistance. *Protein Engineering, Design and Selection,* **20(9):** 461–472.

Ma, L., Zhang, J., Chen, H., Zhou, J., Ding, Y. and Liu, Y. (2011). An overview on ELISA techniques for FMD. *Virology Journal,* **8:** 419.

Matz, M., Fradkov, A., Labas, Y., Savitsky, A., Zaraisky, A., Markelov, M. and Lukyanov, S. (1999). Fluorescent proteins from nonbioluminescent *Anthozoa* species. *Nat. Biotechnol.*, **17:** 969–973.

Méchin, V., Consoli, L., Le Guilloux, M. and Damerval, C. (2003). An efficient solubilization buffer for plant proteins focused in immobilized pH gradients. *Proteomics,* **3:** 1299–1302.

Messina, J., Hickox, P., Lepow, M., Pollara, B. and Venezia, R. (1985). Modification of a direct enzyme linked immunosorbent assay for the detection of immunoglobulin G and M antibodies to pneumococcal capsular polysaccharide. *Journal of Clinical Microbiology,* **21(3):** 390–394.

Miranda, L., Duval, M., Doherty, H., Seaman, M., Posner, M. and Cavacini, L. (2007). The neutralization properties of a HIV–specific antibody are markedly altered by glycosylation events outside antigen binding domain. *J. Immunol.*, **178:** 7132–7138.

Patterson, G., Day, R. and Piston, D. (2001). Fluorescent protein spectra. *J. Cell. Sci.*, **114:** 837–838.

Riis, P. (2001). Western blotting. *In*: Molecular biology problem solver: A laboratory guide. Wiley–Liss Inc. pp. 373–397.

Saha, G. (2010). Fundamentals of nuclear pharmacy. 6th Edition. Springer–Science Business Media LLC. pp. 249–255.

Towbin, H., Staehelin, T. and Gordon, J. (1979). Electrophoretic transfer of proteins from polyacrylamide gels to nitrocellulose sheets: Procedure and some applications. *Proc. Natl. Acad. Sci. USA,* **76(9):** 4350–4354.

Tsien, R. (1998). The green fluorescent protein. *Annu. Rev. Biochem.*, **67:** 509–544.

Walker, J. and Rapley, R. (2008). Molecular biomethods hadbook. 2nd edition. Humana Press Inc. pp. 657–682.

Woodhead, J. and Weeks, I. (1985). Chemiluminescence immunoassay. *Pure and Appl. Chem.,* **57(3):** 523–529.

Yang, Y. and Ma, H. (2009). Western blotting and ELISA techniques. *Researcher,* **1(2):** 67–86.

Yiber, A., Cetinkaya, F. and Soyutemiz, G. (2011). ELISA screening and liquid chromatography– tandem mass spectrometry confirmation of chloramphenicol residues in chicken muscle and the validation of a confirmatory method by liquid chromatography– tandem mass spectrometry. *Poultry Sci.,* **90(11):** 2619–2626.

8

Techniques of Nucleic Acid Hybridization

8.1 INTRODUCTION

Nucleic acid hybridization involves establishment of a sequence-specific interaction between two complementary strands of nucleic acids. The nucleic acid hybridizations can result in DNA:DNA hybrids or DNA:RNA hybrids or RNA:RNA hybrids. This is the basic technique for identification of specific DNA or RNA sequences from large quantities of genomic DNA or RNA.

Nucleic acid hybridization process can be performed after isolation and purification of nucleic acids from organisms or directly inside the cells by *in situ* approaches. The detection of complementary sequences is done by use of DNA or RNA probes that are labelled with radioactive phosphorus or protein conjugate. The non-radioactive detection approaches include chromogenic detection by enzymatic reaction with substrate that gives coloured product or by chemiluminescence substrates. Fluorimetric detection by use of fluorescent substrates or labels is also very efficient method and is popularly used today.

Several nucleic acids hybridization based approaches such as microarray, and southern or northern blot hybridization have become popular tools for researchers in biochemistry, molecular genetics and biomedicine, and are now in routine use.

8.2 USE OF PROBES IN NUCLEIC ACID HYBRIDIZATION

8.2.1 Introduction to Hybridization Probes

Hybridization probe is a critical component of nucleic acid hybridization system. Success of this technique depends upon the precision of the probe. Nucleic acid probes are fragments of DNA or RNA of known sequence that usually have a reporter group attached and are used to detect complementary sequences from target DNA or RNA.

8.2.2 DNA Probes

Nucleic acid hybridization requires that a probe is complementary to all, or part of the sequence of the mRNA of interest. There are two main forms of hybridization probes used in northern hybridization. The customary approach is to use a complementary DNA (cDNA). Alternatively, anti-sense oligonucleotides that are generally thirty to forty bases in length, can be designed from sequence data and synthesized. Oligonucleotides offer some advantages in terms of simplicity and reduced hybridization time.

DNA probe can be double stranded or single stranded. DNA probes are usually artificially synthesized oligonucleotides measuring about 20 nucleotides in length. A known sequence of DNA can be synthesized with controlled chemical reactions in the automated machines.

Since the oligonucleotide probes are relatively short, they hybridize more efficiently with the target molecules. The time taken for a probe to hybridize with target molecule is dependent upon probe concentration in mg/ml, complexity of the probe and reaction volume in ml.

8.2.3 RNA Probes

RNA probes are essentially single stranded hybridization probes. Antisense RNA probes are created from in vitro transcription of short 200 bp-1.5 kbp PCR products carrying the T7 RNA polymerase recognition sequence at one end. Antisense RNA probes can also be used for microarray hybridization.

Using RNA probes in Northern analysis results in significantly greater sensitivity as compared to double-stranded DNA probes. The 8 to 10-fold increase in sensitivity provided with RNA probe substantially reduces the exposure times and increases the ability to detect rare mRNAs, avoiding the need for poly(A) RNA isolation.

RNA:RNA duplexes have a higher melting point (Tm) and binding affinity than RNA:DNA duplexes. As a result, the hybridization buffer and temperature need to be adjusted to ensure optimal hybridization to target RNA and minimal nonspecific hybridization of probe.

8.2.4 Comparing Probe with Gene

Gene is the unit of heredity in all the living organisms. Gene is a stretch of DNA that codes for certain proteins. Compared to probe, gene is a very large molecule. Hybridization probes are fragments of DNA or RNA of known sequence. On the other hand, sequences of all genes are not necessarily decoded. Hybridization probes can be synthesized complementary to the whole or part of a gene once its sequence is known, for its identification. Basically, probes are artificially synthesized DNA or RNA molecules used to detect sequences complementary to their own.

8.3 COMMONLY USED SOLUTIONS AND BUFFERS IN NUCLEIC ACID HYBRIDIZATION

8.3.1 Pre-Hybridization Solution

Pre-hybridization solution is used to block the nonspecific sites on the membrane. For this purpose, denatured fragmented salmon sperm DNA is invariably included in this solution. The pre-hybridization buffer is also used to wet and condition the microarrays.

In pre-hybridization solutions, Denhardt's reagent is used to block the non-specific attachment of the probe to the surface of solid matrix. Chemicals like dextran sulphate or PEG are used to increase the re-association of the nucleic acids. Denhardt's reagent is used in northern hybridization, and southern hybridization.

8.3.2 Hybridization Buffer

Hybridization buffers are usually made by adding required quantity of labelled probe in pre-hybridization solution. The quantity of probe is very important. Higher concentration of probe can give better hybridization signals but may also increase background due to non-specific binding of probe to the membrane.

8.3.3 Blocking Solutions

Blocking solutions or buffers are used to prevent ligands from sticking to solid support surface. They are used to prevent any non-specific binding of probes.

The most commonly used blocking agents are Denhardt's reagent and BLOTTO (bovine lacto transfer technique optimizer). Heparin is another blocking agent used when the hybridization is carried out in presence of dextran sulphate. Usually the membranes carrying nucleic acid fragments are incubated in blocking agents for 1-2 hours before addition of probe. This considerably reduces background signal. For this reason, pre-hybridization solutions usually contain at least one of these blocking agents.

These blocking agents are included in both pre-hybridization and hybridization solutions when nitrocellulose filters are used. However, when nylon membranes are used, blocking agents can be skipped from hybridization solutions.

8.3.4 Post-Hybridization Wash Solutions

Post-hybridization washes are done to remove unhybridized and non-specifically hybridized probe. Proper post-hybridization washes significantly reduce background. Usually post-hybridization wash solutions are of two types: high stringency wash solutions contain higher concentrations of SSC or SSPE buffers and low stringency wash buffer contains lower concentrations of SSC or SSPE buffers. SDS is also a common component of post-hybridization wash buffers.

8.4 Southern Hybridization

8.4.1 Introduction

Southern blotting was named after Edward M. Southern who developed this procedure at Edinburgh University in the 1970s. Southern blotting is designed to locate a particular sequence of DNA within a complex DNA mixture. Southern Blotting can be used to locate a particular gene within an entire genome. The amount of DNA needed for this technique is dependent on the size and specific activity of the probe. Short probes tend to be more specific.

Any given DNA fragment is unique solely by virtue of its specific nucleotide sequence. The only practical way to find a specific DNA sequence among a vast amount of heterogeneous DNA fragments is to exploit its sequence specificity.

In 1975, Edward Southern invented technique capable of identifying specific DNA fragments. In this method, genomic DNA is digested with one or more restriction enzymes and the resulting fragments are separated according to fragment length on agarose gels. This is a technique of detecting DNA by using a DNA probe. In this technique, DNA is separated by gel electrophoresis and then transferred from the gel to a membrane by blotting. The DNA is detected from the membrane by using a DNA probe complementary to target DNA.

8.4.2 Theory and Principle

Southern blot analysis reveals information about DNA identity, size, and abundance. It is a classic technique that involves separating DNA fragments based on size via electrophoresis, transferring them to a membrane, hybridization with a labeled sequence-specific probe, washing, and finally detection of labeled DNA bands.

In brief, hybridization analysis involves separating the strands of the DNA molecules to be analyzed and then mixing those separated strands with many copies of a single-stranded DNA or RNA molecule, called a probe. The probe contains a sequence complementary to the nucleotide sequence of interest. When a probe is mixed with denatured DNA, hydrogen bonds form between the probe and its complementary sequence in the DNA sample being analyzed. The formation of hydrogen bonds between two complementary strands create a double-stranded complex; this is called hybridization, or annealing. For Southern hybridization analysis, the sample DNA is attached to a solid support, such as a nitrocellulose or nylon membrane.

The amount of genomic DNA needed to generate a detectable hybridization signal depends upon a number of factors including proportion of the genome that is complementary to the probe, size of the probe and the amount of genomic DNA transferred to the filter. Under optimal conditions, this method is sufficiently sensitive to detect less than 0.1 μg of DNA complementary to a labelled probe. As the strength of the signal is proportional to the specific activity of the probe and inversely proportional to its length, southern hybridization has optimum sensitivity when very short probes are used.

The technique of southern hybridization involves several steps. First, the probe and sample DNA are allowed to hybridize under the appropriate conditions. The correct temperature, incubation times, and buffer conditions must be used. Next, the sample DNA is washed using conditions that will remove unhybridized probe but not the hybridized probe. Finally, the sample DNA is tested for the presence of the hybridized probe. The probe can be labelled with a radioactive molecular tag that allows it to be detected following hybridization.

To perform Southern blotting or transfer, the agarose gel is first soaked in a basic solution to denature the DNA fragments. After an additional soaking step to neutralize the base, the gel is placed on a long piece of blotting paper with the ends of the paper suspended in a reservoir of salt solution. A nitrocellulose or nylon membrane is then laid directly on top of the gel. Blotting paper and a stack of dry paper towels are then placed on the membrane. The blotting paper acts like a wick. Driven by capillary action, fluid is drawn from the reservoir up though the gel and into the stack of dry paper.

In the capillary transfer method, DNA fragments are carried from the gel in a flow of liquid and deposited on the surface of the solid support (Fig. 8.1). As the fluid migrates up through the gel, it carries the denatured

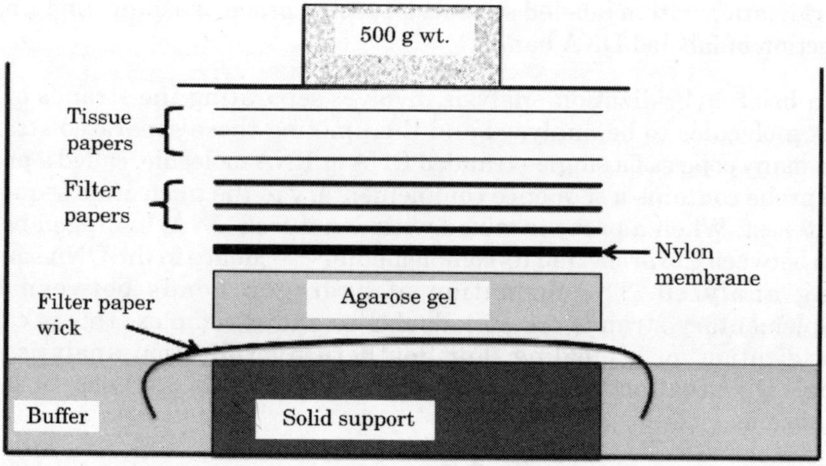

Fig. 8.1: Southern blotting by capillary transfer method

DNA fragments up with it and out of the gel. When the fragments reach the membrane, they stick to the membrane and remain there. Because the wick, gel, membrane, and stack of paper lay directly on top of each other, when the DNA fragments are transferred up onto the membrane, they form the same pattern they formed in the gel.

The rate of the DNA transfer depends on the size of the DNA fragments and the concentration of agarose in the gel. Small fragments less than 1 kb are transferred almost quantitatively from a 0.7% agarose gel within 1 hour, while larger fragments are transferred more slowly. After the transfer is complete, the DNA-containing membrane is rinsed and the denatured DNA molecules are fixed to it through heating or exposure to ultraviolet light. This membrane, with the fixed, single-stranded DNA, is now ready for hybridization with the desired probe.

As a first step in the hybridization procedure, the membrane is immersed in a pre-hybridization buffer that prevents the probe from binding to the membrane in a nonspecific manner. After this pre-hybridization step, the membrane is transferred to the hybridization solution containing the labeled probe. The composition of the hybridization solution and the hybridization conditions vary depending upon the probe used and the DNA sequence you wish to detect. After the hybridization is finished, the membrane is rinsed repeatedly under conditions that will remove unhybridized and nonspecifically bound probe, but that will not disrupt hydrogen bonds between the probe and the target sequence in the sample DNA.

The final step in Southern hybridization is to detect the hybridized probe. In case of radiolabelled probes, the hybridization signal is detected by keeping the nylon membrane in contact with x-ray film for several hours at 4°C. The development of bands on the X-ray film, indicate the position of hybridization of the probe in the gel.

In non-radioactive methods of detection, usually enzymatic procedure is used. To detect biotin-labelled probe, the hybridized membrane is soaked in a solution containing two components bound together. One component is streptavidin, a molecule that binds tightly to biotin. The other component is the enzyme, alkaline phosphatase. The alkaline phosphatase and the attached streptavidin together are called as conjugate. During the time that the membrane is soaked in the solution containing the streptavidin-alkaline phosphatase conjugate, the streptavidin binds tightly to the biotinylated probe. The alkaline phosphatase becomes attached to the probe by virtue of its bond to streptavidin. Once the incubation step to bind the streptavidin-alkaline phosphatase conjugate to the biotin-labeled probe is finished, the membrane is rinsed to remove any unbound conjugate.

Finally, the membrane is placed in a color development solution containing two components: 5-bromo-4-chloro-3-indolyphosphate (BCIP) and nitro-blue tetrazolium (NBT). The alkaline phosphatase portion of the conjugate removes a phosphate group from BCIP; and the resulting product dimerizes to form a dark blue precipitate. This dimerization reaction also releases hybrid ions that reduce the NBT; and the reduced NBT forms a purple precipitate. Since the alkaline phosphatase is bound to the probe via its connection to streptavidin, the precipitates from its reaction with BCIP and NBT form where the probe is bound to the membrane, thereby indicating the location of the DNA fragments hybridized to the probe.

8.4.3 Protocol for Southern Hybridization

The following protocol is for the radioactive detection of specific DNA sequences.

Materials required

- *Denaturing solution*

NaOH	0.4 N
NaCl	1 M

- *Transfer buffer*: 10x SSC or 10x SSPE
- *50x Denhardt's solution*

BSA (bovine serum albumin)	1% (w/v)
Ficoll	1% (w/v)
PVP (polyvinylpyrrolidone)	1% (w/v)

Adjust the final volume with distilled water. Store at –20°C.

^{32}P-Radiolabelled probe

- *Pre-hybridization solution*

  6 x SSC (or 6 x SSPE)

  5 x Denhardt's reagent

  SDS 0.5%

  Denatured fragmented salmon sperm DNA 100 µg/ml

- *Hybridization solution*

  6x SSC (or 6x SSPE)

  SDS 0.5%

  Denatured fragmented salmon sperm DNA 100 µg/ml

  Formamide 50%

Procedure

Capillary transfer of DNA to nylon membrane or Southern Blotting:

1. After electrophoresis is complete, remove the gel carefully from electrophoresis apparatus. Cut the bottom-left corner of the gel to orient it during succeeding process.
2. Denature DNA by soaking the gel in denaturing solution for 15 minutes. The gel should be completely submerged in buffer.
3. Change the solution and continue soaking the gel for 20 minutes with gentle agitation.
4. Cut the nylon membrane 1 mm larger than the size of the gel. Float the membrane on the surface of dish containing deionized water. Wait till the membrane is completely wet from beneath. Then immerse the membrane in denaturing solution for 5 minutes. Then cut off the corner of the membrane, matching that of the gel.
5. Now rinse the gel in deionized water and then neutralize it by soaking it for 30 minutes in several volumes of solution of 1 M Tris (pH=7.4), 1.5 M NaCl at room temperature with constant gentle agitation. Change the neutralization buffer and continue soaking the gel for 15 minutes.
6. While the gel is in the neutralization solution, wrap a piece of 3MM Whatman filter paper around Plexiglas support that is longer and wider than the gel.
7. Place the support in a large glass tray. Fill the tray with transfer buffer 10x SSC. Take care that level of the liquid reaches almost to the top of the support.
8. After the 3MM paper placed on the top of the support is thoroughly wet, smooth out all the air bubbles with a glass rod. This paper serves as a wick that takes up the transfer buffer.

9. Remove the gel from neutralization solution and invert it. Place the inverted gel on the support. Ensure that there are no air bubbles between the 3MM paper and gel.
10. Surround the gel by saran wrap or parafilm. This serves as a barrier that prevents transfer buffer flowing directly onto the paper towels placed on top of the gel.
11. Place the wet nylon membrane on top of the gel so that the cut corners are aligned.
12. Wet 2 pieces of 3MM paper cut to exactly same size as the gel in 2x SSC and place them on top of nylon membrane. Smooth out any air bubbles with a glass rod. Place stack of paper towels 5-8 cm high, just smaller in size than the gel, over 3 MM papers.
13. Put a glass plate on top of the stack and place 500 g weight on top. Allow the binding of DNA fragments to nylon membrane by capillary transfer for 4-24 hours.
14. After the transfer is complete, remove the paper towels and 3 MM papers. Turn over the gel and nylon membrane together and lay them with gel-side up on a dry sheet of 3MM paper.
15. Peel the gel from the membrane. Mark the position of gel side up and mark the position of wells with ball pen.
16. Soak the membrane in 0.5 M Tris-HCl (pH=7.2), 1 M NaCl for 15 minutes at room temperature. This neutralizes the membrane and removes any piece of agarose adhering to it.
17. Remove the membrane from the neutralizing solution and allow any excess fluid to drain away. Place the membrane flat on a paper towel to dry for at least 30 minutes at room temperature.
18. To fix the DNA to the membrane, either place the dried membrane between two pieces of 3MM paper and bake the membrane for 30 minutes to 2 hours at 80°C in oven or, expose the DNA carrying side of the membrane to UV light at 254 nm.

Hybridization of immobilized DNA with probe

1. Float the nylon membrane on surface of tray filled with 6x SSC or 6x SSPE until it is thoroughly wet from beneath. Submerge the filter for 2 minutes.
2. Slip the wet filter into heat-sealable bag. Add pre-hybridization solution 0.2 ml/cm^2 of nylon membrane. Squeeze as much as possible form the bag and seal the open end of the bag.
3. Incubate bag for 1-2 hours submerged in a waterbath set at 42°C temperature.
4. Denature the radiolabelled probe by heating for 5 minutes at 95°C. Chill the probe solution rapidly in ice-cold water.

5. Working quickly, remove the bag containing the filter from the water bath. Cut open the bag. The pre-hybridization solution should be completely removed and immediately replaced with hybridization solution. The probe is then added.
6. Reseal the bag so that no air bubbles are trapped inside. Seal the resealed bag inside a second non-contaminated bag to avoid radioactive contamination of the water bath.
7. Incubate bags in waterbath for the required period of hybridization.
8. After hybridization is complete, remove the bag and cut it open. Pour out the hybridization solution. Cut the bag along the length of three sides.
9. Remove the filter and immediately submerge it in a tray containing excess of 2x SSC and 0.5% SDS at room temperature.
10. After 5 minutes, transfer the membrane to fresh wash solution- 2x SSC and 0.1% SDS and incubate it for 15 minutes at room temperature with gentle agitation.
11. Transfer the membrane to tray containing excess of 0.1x SSC and 0.5 % SDS. Incubate the filter for 30 minutes to 1 hour at 37°C with gentle agitation.
12. Replace the solution with fresh 0.1x SSC and 0.5% SDS and transfer the tray to water bath set at 68°C for the same period of time.
13. Now briefly wash the membrane in 0.1x SSC at room temperature and place the membrane on paper towels.
14. Place the damp membrane on sheet of saran wrap. Now expose the filter to x-ray film to obtain autoradiographic image. The exposure time can vary. In case of low signals, intensifying screens can be used.

8.4.4 Troubleshooting Guide

1. Excessive handling can damage the nylon membrane. Wear gloves and blunt forceps to handle the membrane.
2. If the paper towels placed on top of the membrane are not precisely stacked, they droop over the edge of the gel and may touch the support. This type of short-circuiting is a major cause of insufficient transfer of DNA from gel to the membrane.
3. While capillary transfer, do not remove the membrane once it is placed on top of the gel. Make sure there are no air bubbles between the membrane and the gel.
4. Heating the pre-hybridization solution to 42°C prior to addition to the bag will prevent formation of small bubbles on the surface of the filter as the temperature of the solution increases.

5. Do not allow the nylon membrane to dry out at any stage.
6. If DNA staining is still seen in agarose gel after capillary transfer, it is due to incomplete transfer of DNA. Increase the blotting time.
7. If the blotted membrane shows some trace of bubbles, it is due to bubbles formed between gel and membrane.
8. The absence of any signal after hybridization could be due to lot of reasons, such as incomplete denaturation of probes, incomplete transfer of DNA fragments to nylon membrane, or even absence of complementary DNA sequences in DNA fragments to the probe.
9. If hybridization signals are weak, it could be due to low activity of DNA probe.
10. In case if unexpected bands show up on membrane (in chromogenic detection) or x-ray film (in radioactive detection), it could be due to non-specific binding of probe. To prevent this, increase the blocking time for membrane and increase number of washes after hybridization.

8.5 NORTHERN HYBRIDIZATION

8.5.1 Introduction

The RNA hybridization technique is called northern hybridization because of the fact that it is very similar to the DNA hybridization process commonly known as southern hybridization. Northern hybridization is a variant of southern hybridization. The transfer of RNA fragments from gel to a solid support is called blotting, or more correctly, northern blotting.

Northern blotting has been extensively used to examine the expression of specific genes in nutritional science. Northern blotting is the main method for examining the expression of genes through measurement of mRNA. Tissue specificity and the factors that regulate gene expression can both be determined by northern blotting. A cDNA radioactively labelled with ^{32}P is the most commonly used hybridization probe. Convenient, non-radioactive detection protocols are however, increasingly popular.

Northern hybridization is useful in studying expression patterns of genes. It can be used to determine whether a particular gene is transcribed in all tissues of the organism or only in certain tissues. It can also be used to study gene expression during growth and development. The main advantage of this method is that it is possible to determine total size of a transcript. Since the target material is immobilized on a solid support, it is possible to screen them using different probes simultaneously.

8.5.2 Theory and Principle

The underlying principle of Northern blotting is that RNA are separated by size and detected on a membrane using a hybridization probe with a base sequence complementary to part of the sequence of the target mRNA.

The initial step in northern hybridization process is to extract total RNA from a tissue, using chaotropic agents such as guanidinium isothiocyanate. Such agents disrupt cells and denature proteins including RNases, as well as solubilizing RNA. The extracted RNA species, whether total RNA or poly-A selected, are then separated on the basis of molecular size by agarose-gel electrophoresis. This is followed by blotting or transfer of RNA from gel onto a nylon membrane (Fig. 8.1). Nylon membranes, particularly positively-charged membranes, are normally used in preference to nitrocellulose filters because of their high binding capacity for nucleic acids and their greater robustness on handling. A blot provides a precise reflection of the separated RNA species from a gel.

Following blotting, the RNA must be immobilized on the membrane, either by baking in an oven or by exposure to UV light. This results in the covalent linkage of RNA to the membrane, which prevents the nucleic acid from being washed away during the subsequent processing. The hybridization probe must then be labelled and the probe is hybridized with the RNA fragments on membrane.

This is followed by post-hybridization washes, which ensure that the probe is bound specifically to the target mRNA and that there is no non-specific binding to other mRNA or to the nylon membrane itself.

Detection of hybridization is achieved either through the use of radioactivity or by non-radioactive strategies. Most laboratories using radioactive method of detection use probes labelled with ^{32}P. There are several advantages to using radioactively labelled probes. The procedures are well established and robust, and there is a high level of sensitivity. The disadvantages of radioactive detection method however include safety, instability of probes if the short half-life isotopes are used, and increasing difficulties with the disposal of waste.

In the case of chemiluminescent detection, the breakdown of specific chemiluminescence substrates is catalyzed by enzymes like alkaline phosphatase or horseradish peroxidase, with the emission of light. These enzymes may be conjugated directly to a probe. In more common cases, a probe is labelled with a ligand such as dioxigenin, fluorescein, or biotin; which is then localized by an antibody or avidin or streptavidin (in case of biotin-label) to which alkaline phosphatase or horseradish peroxidase is attached.

X-ray film is the usual means by which hybridization signals are detected, and this can be used for both radioactivity and chemiluminescence. Quantification is achieved by densitometry, and is based on arbitrary units and relative changes between experimental groups and controls.

Pre-hybridization and hybridization is performed in the hybridization oven. The hybridization temperature depends upon the salt concentration and GC content of probe and target DNA. Usually hybridization is done at 37-42°C.

8.5.3 Protocol for Northern Hybridization

The following procedure is for the chemiluminescent detection of specific RNA sequences.

Materials required

- *Transfer buffer*

5x SSC	
NaOH	10 mM

- *Neutralization buffer*

  5x SSC

  pH=7.0

  Sponge, Whatman filter paper, Nylon membrane

- *20x SSPE buffer*

NaCl	3 M
NaH_2PO_4	0.2 M
EDTA	20 mM
pH=7.4	

- *Pre-hybridization solution*

Formamide	50%
SDS	5%
PEG	6%
NaCl	1M

  Salmon sperm DNA 250 mg/ml (denatured at 95°C for 5-10 minutes before adding to heated pre-hybridization buffer)

  This solution is heated to the temperature that is needed for pre-hybridization.

- *Hybridization solution*

  Pre-hybridization solution

  Biotin-labelled probe

The probe is denatured in 0.5 ml of pre-hybridization solution at 95°C for 5 minutes. The probe concentration is kept around 0.1 nM. The denatured probe is added to this solution after 1hour of pre-hybridization.

- *10x PBS buffer*

Na_2HPO_4	0.4M
NaH_2PO_4	0.1M
NaCl	1 M

  pH=7.4
- *X-ray film*
- *Biotin-labelled specific probe*
- *Post-hybridization wash buffers*

  Low stringency wash buffer: 2x SSPE, 0.5% SDS

  High stringency wash buffer: 0.2-0.4x SSPE. 0.5% SDS
- *Blocking buffer*

  1x PBS

  0.5% SDS

  0.1-0.2% casein

  Heat the solution to dissolve casein.
- *Conjugation buffer*

  Dilution to 1:6000 of alkaline phosphatase conjugate in blocking buffer. AP-conjugate should be centrifuged at 4°C for 1 minute at 12000g to remove aggregated proteins from the solution.
- *Wash buffer post-detection*

  1x PBS

  0.5% SDS
- *Assay buffer*

Diethanolamine	0.1 M (pH=10.0)
$MgCl_2$	1 mM

  This buffer can be stored at 4°C for a week when sodium azide 1 mM is added.
- *Substrate buffer*

  Dilution of 1:100 of CSPD (25 mM stock) in assay buffer. This solution is freshly prepared and can be stored at 4°C for only a few hours. (CSPD is a chemiluminescent substrate that is used for the detection of alkaline phosphatase and alkaline phosphatase conjugates. However nylon membranes should be used for blotting while using CSPD if good sensitivity is to be obtained).

Procedure

Capillary transfer of RNA to nylon membrane or Northern blotting

1. The gel containing separated RNA fragments is soaked in transfer buffer for 10 minutes.
2. A sponge is soaked in transfer buffer and placed in a clean container filled with transfer buffer. Remove all the air bubbles from sponge.
3. Place 4-8 transfer buffer soaked Whatman filter papers on top of the sponge. The filter papers should be of the size of the gel.
4. Place the gel upside down on the filter papers. Place a pre-cut Nylon membrane on top of the gel.
5. Place 4-8 pre-soaked Whatman filter papers on top of the membrane. On top of that, place paper towels making about 2 cm thickness.
6. Weight of 500g is placed on the top. This makes assembly for capillary transfer of RNA to nylon membrane.
7. Transfer is allowed to take place for 1-2 hours, depending on the size of the gel.
8. After blotting is complete, membrane is rinsed for 5 minutes in neutralization buffer and air-dried.

Hybridization of immobilized RNA with probe

1. The dried membrane is kept under UV light for fixing the RNA fragments to the membrane.
2. Heat the pre-hybridization solution to the desired temperature.
3. Place the blot into the container/tube and pour pre-warmed pre-hybridization solution over the blot.
4. After 1 hour of incubation, add the denatured probe directly into the pre-hybridization solution. Hybridization is performed overnight at the same temperature of pre-hybridization.
5. After hybridization is complete, wash the membrane in low stringency wash buffer for 10 minutes at room temperature.
6. Then wash the membrane with high stringency wash buffer at 55-65°C for 30-60 minutes. These ashes must be carried out with gentle shaking.

Detection of specific RNA sequences

1. Incubate the membrane in blocking buffer twice for 5 minutes each and again for 30 minutes with gentle shaking.
2. Drain the blocking buffer and add conjugation buffer to the membrane. Perform conjugation for 1 hour with gentle shaking.
3. After conjugation is complete, transfer the membrane to a new container. Allow the excess of conjugation buffer to drain from its surface.

4. Wash twice with blocking buffer for 5 minutes and four times with wash solution.
5. Equilibration to appropriate reaction conditions for alkaline phosphatase is achieved with two washes in assay buffer for 5 minutes each.
6. Remove the excess assay buffer from the membrane. Now add substrate buffer (20 μl/cm^2 of membrane) to empty tray. Place the membrane upside down in the substrate buffer and incubate for 10 minutes.
7. The excess substrate buffer is removed from the blot.
8. Place the moist membrane between two plastic covers. Place the covered membrane into a film-exposure cassette and expose x-ray film for 1-2 hours or as per instructions from commercial supplier for chemiluminescent detection. Usually exposure of 10-15 minutes is sufficient.

8.5.4 Troubleshooting Guide

1. A particular problem in the case of Northern blotting relates to the potential degradation of RNA through the action of RNases. These enzymes are widely distributed in tissues, and environmental contamination is possible at most stages of Northern-blot analysis, with fingers being a major source of contamination. The problem can be overcome by baking, or sterilizing, glassware and solutions, and by employing specific RNase inhibitors such as diethyl pyrocarbonate (DEPC). The use of protective latex gloves is essential.
2. The other important problem is high background on membranes, reducing the signal: noise ratio. Careful adjustment of the post-hybridization wash conditions is important in minimizing background, particularly with chemiluminescence-based detection.
3. Salmon sperm DNA needs to be denatured at 95°C for 5-10 minutes before adding to heated pre-hybridization solution.
4. Pre-hybridization and hybridization solutions can be reused again. They should be stored at 4°C. Before use, the DNA should be denatured at 95°C and cooled to 55°C before adding to the blot.
5. At the time of blotting, make sure that there are no air bubbles trapped between gel and membrane as well as between sheets of filter papers. Remove any air bubbles by rolling glass rod or pipette over it.
6. It is important to remove excess assay buffer before starting the incubation with substrate buffer to avoid dilution of latter buffer.
7. The blot must not be allowed to become dry at any stage of the hybridization procedure.
8. When membrane is immersed in substrate buffer, care should be taken that no air bubbles are trapped under the membrane.

9. Cloudy background can occur in case of insufficient blocking.
10. Spotty background may occur if AP-conjugate is too old or because of high temperature during conjugation or insufficient blocking.

8.6 *IN SITU* HYBRIDIZATION

8.6.1 Introduction

In situ hybridization is a powerful and versatile technique for the analysis of DNA or RNA molecules. This technique is very useful tool for scientists to simultaneously detect and localize specific DNA or RNA sequences with spatial information about their subcellular locations in tissue samples or cells.

In situ hybridization can identify sites of gene expression, tissue distribution of mRNA and identification as well as localization of viral infections. There are two most commonly used procedures for *in situ* hybridization and detection of cellular DNA or RNA. One is radioactive *in situ* hybridization (RISH) that uses a radioactively labelled probe, and other is fluorescent *in situ* hybridization (FISH) that uses fluorescent probes.

8.6.2 Theory and Principle

The foremost step in the *in situ* hybridization process is fixation of cells or tissues onto glass slides. High-quality glass slides play an essential role in preventing cells or tissue sections from falling off the slides during *in situ* hybridization. For this reason, regular glass slides should be pre-coated prior to use for this purpose. Pre-coated glass sides are also commercially available.

Selection of a fixative is very critical to the preservation of cellular or tissue morphology during *in situ* hybridization. Fixation time varies with cell lines ranging from 30 minutes to 10 hours. Therefore the optimal fixation time should be determined empirically for each cell line. Tissue fixation is a key step for the success of *in situ* hybridization and desired fixation time varies with tissues. For good morphology retention and good signal-to-noise ratio following *in situ* hybridization, optimal time has to be determined. The fixation time can vary from 30 minutes to 10 hours depending upon the type of tissue.

After fixation, tissues or cells are embedded in paraffin wax and preserved until further use. Paraffin wax sections must be subjected to complete dewaxing to allow treatments with protease, DNase or RNase as well as penetration of the probe into the tissue. Dewaxing of sections can be completed with xylene. For embedded cultured cells, dewaxing is not necessary.

Protease treatment is applied to tissue sections to be used for *in situ* hybridization of DNA or RNA. Appropriate protease digestion of sections removes some of the proteins from the cellular network to make DNA or RNA more accessible for *in situ* hybridization. It enhances the hybridization signal of interest, while reducing background as well. Protease digestion is a tricky step for the success of *in situ* hybridization. Insufficient protease digestion can result in a diminished or loss of hybridization signal. In case of over digestion tissue morphology may become poor or completely destroyed. In order to obtain a good signal to noise ratio and good preservation of tissue morphology, it is recommended to determine the optimal digestion time for each tissue sample. Different tissues may have different optimal digestion times.

After protease digestion, the sections are subjected to different treatments depending on *in situ* hybridization with cellular DNA or RNA. The DNase treatment serves to remove cellular DNA from tissues. Genomic DNA may compete with RNA of interest in hybridizing with specific probe. Subsequently, DNase digestion can reduce non-specific background for *in situ* hybridization of RNA. RNase treatment is used to degrade cellular RNA in tissue. This is done because RNA can compete with DNA for hybridization of probe. Therefore RNase digestion can enhance the hybridized signal by reducing non-specific background for *in situ* hybridization of DNA.

After treatment with nucleic acid degrading enzymes- RNase or DNase, *in situ* hybridization is performed using a specific probe. Generally, the procedure for DNA hybridization is same as of RNA hybridization except for few differences.

8.6.3 Protocol for *In situ* Hybridization

Materials required

- *4% paraformaldehyde fixative solution*

  Weigh 40 g of paraformaldehyde powder and add to 600 ml of pre-heated DDW at 55-60°C. Allow paraformaldehyde to dissolve at same temperature with stirring. Add one drop of 2N NaOH solution to clear the fixative. Remove solution from heat and add 333.3 ml (1/3rd of total volume) of 3x PBS. Adjust pH to 7.2 and bring final volume to 1 litre with DDW. Filter the solution to remove undissolved particles and cool to room temperature. Store at 4ºC until use.
- *Paraffin wax*
- *Preparation of protease solutions*

  Pepsin solution: 2 mg/ml in DDW. (pH=2.0)

  Trypsin solution: 2 mg/ml in DDW

  RNase-free DNase I enzyme or DNase-free RNase I enzyme, suiting to the requirements. These enzymes are available commercially.

- *Post-DNase / RNase enzyme treatment solution*

Tris-HCl (pH=8.0)	10 mM
NaCl	150 mM
EDTA	20 mM

- *TE buffer: (pH=8.0)*

Tris-HCl	10 mM
EDTA	1 mM

- *PBS buffer: (pH=7.3)*

KCl	2.7 mM
KH_2PO_4	1.5 mM
NaCl	136.9 mM
Na_2HPO_4	15 mM

- *20x SSC: (pH=7.0)*

Na_3-citrate (trisodium citric acid)	0.3 M
NaCl	3 M

Autoclave the buffer.

- *50x Denhardt's solution*

BSA (bovine serum albumin)	1% (w/v)
Ficoll	1% (w/v)
PVP (polyvinylpyrrolidone)	1% (w/v)

Adjust the final volume with distilled water. Store at -20°C.

- *Pre-hybridization solution*

Deionized formamide	50% (v/v)
Tris-HCl (pH=8.0)	15 mM
EDTA (pH=8.0)	2 mM

2x SSC

5x Denhardt's solution

0.2% denatured and sheared salmon sperm DNA for DNA hybridization or 0.2 mg/ml yeast tRNA and 0.2 mg/ml poly(A) RNA for RNA hybridization.

- *Hybridization solution*

Add radiolabelled probe to appropriate volume of freshly prepared pre-hybridization solution.

Procedure

Preparation of glass slides

1. Wash the glass slides in 2N HCl for seven minutes.

2. Rinse the slides in double distilled water (DDW) twice for 1 minute each.
3. Rinse the slides in high-grade acetone for 2 minutes and air-dry.
4. Treat the slides in 2% organosilane solution diluted in high-grade acetone for 1-2 minutes with gentle agitation followed by rinsing in high-grade acetone in 1-2 minutes.
5. Air-dry the slides and store them at room temperature.

Fixation of cultured cells on slides

1. Centrifuge the suspended cultured cells and collect the pellet.
2. Resuspend the cells in serum-free medium at a density of 10^7 cells/ml.
3. Transfer 10-15 µl of the cell solution onto a pre-coated glass slide and allow cells to attach to the slide in a humid chamber for 40 minutes.
4. Place the slides in the slide racks and carefully immerse them in a staining dish, containing sufficient volume of 4% PFA fixative solution.
5. Allow fixation to proceed at room temperature.
6. Carefully transfer the slide rack to a staining dish filled with 3x PBS and incubate for 4 minutes to stop the fixation.
7. Rinse the slides thrice for 4 minutes each with 1x PBS in fresh staining dishes.
8. Carryout complete dehydration by transferring the slide rack through a series of 6 minutes treatments in ethanol dishes such as – 50%, 70%, 95%, 100%, 100%, 100%.
9. Completely dry the slides at room temperature and store at –80°C in a box, containing desiccant until use.

Fixation of tissues

1. Label 10 or 20 ml snap-cap glass vials and their caps according to the samples and fill 2/3rd volume of each vial with freshly prepared 4% PFA.
2. Place freshly dissected tissues, or organs or embryos in the vials and allow fixation to proceed at 4°C for optimal time period.
3. While the samples are being fixed, melt appropriate amount of paraffin wax in a clean glass beaker.
4. After fixation, carryout dehydration of samples by carefully removing the fixative solution from vials and replacing it with 10 ml of 50% ethanol.
5. Perform complete dehydration by transferring the slides every 20 minutes through series of ethanol washes (50%, 70%, 95%, 100%, 100%, 100%) and twice in xylene for 15 minutes each.

Embedding of tissues

1. Carryout partial impregnation of samples by quickly filling each vial with 4 ml of freshly mixed xylene and melted wax (1:1, v/v) using Pasteur pipette.
2. Allow incubation to proceed overnight at room temperature.
3. Melt the mixture at 60°C and immediately replace the xylene-wax mixture with freshly melted wax.
4. Allow impregnation of the samples at 60°C for 1 hour.
5. Continue wax infiltration of tissue samples by quickly replacing with freshly melted wax every 20 minutes using Pasteur pipette.
6. Repeat this step 10 times. Now the samples are ready for embedding.
7. Prepare embedding boats using aluminium foil or moulds that can be obtained commercially.
8. Fill embedding moulds with melted paraffin wax using Pasteur pipette and quickly transfer samples to appropriate moulds.
9. Orient the samples within the mould with Pasteur pipette. Avoid trapping of any air bubbles inside the blocks. Place one sample in each block.
10. Allow the blocks to solidify at room temperature or immerse moulds in the cold water in a tray.
11. Remove the wax blocks from the moulds and store them in a dry place at room temperature until sectioning.

Sectioning of tissues and mounting

1. Using a razor blade carefully cut each of the paraffin wax blocks containing the embedded samples into a trapezoidal shape and continue to trim the section surface until approximately 1-2 mm to the sample.
2. Place the trapezoid block in holding clamp of a microtome with wide edge of the trapezoid surface facing the knife.
3. Set up the thickness of sectioning at about 4-10 μ and start cutting section ribbons.
4. Label pre-coated glass slides and add 1-3 drops of DDW or 0.2 % gelatin solution to the slide.
5. Place the slide on the heating plate at 50°C.
6. Carefully transfer the ribbons of 3-10 sections onto the gelatin or water drop on the slide using a fine brush. Allow the ribbon to stretch to remove any wrinkles until complete drying of the slide.
7. To attach the sections to slides firmly, place the slides at 40°C for 24 hours.
8. Store the slides in a slide box with desiccant at –20°C until use.

Dewaxing of tissue section

1. Transfer sections from freezer to a slide rack and allow them to reach room temperature prior to use.
2. Fill three staining dishes with fresh xylene and 2 dishes with 100% ethanol.
3. Dewax the sections by placing the slide rack into a series of three changes of xylene for 3 minutes each.
4. Wash the sections in 2 changes of 100% ethanol for 3 minutes each.
5. Air-dry the slides and proceed to protease digestion.

Protease digestion

1. Carefully overlay the tissue sections with appropriate volume of protease solution.
2. Carryout digestion of tissues at 37°C for 10 to 30 minutes.
3. After protease digestion is complete, stop proteolysis by rinsing the slides in a staining dish containing PBS (pH=7.4) with 2 mg/ml glycogen.
4. Perform three changes of fresh PBS, 1 minute each.
5. Alternatively wash the slides for 1 minute in a solution of 100 mM Tris-HCl and 100 mM NaCl.
6. Again fix the specimens in freshly prepared 4% PFA for 6 minutes at room temperature.
7. Rinse the slides with PBS thrice, 1 minute each.
8. Rinse the slides in 100% ethanol for 1 minute and air-dry.

DNase treatment for in situ hybridization of RNA

1. Overlay the tissue sections with appropriate volume of RNase free-DNase solution.
2. Allow digestion to proceed for 30 minutes at 37°C.
3. After digestion is complete, inactivate DNase by rinsing the slides in staining dish containing a post DNase treatment solution for 5 minutes.
4. Wash the slides in TE buffer (pH=8.0) thrice for 2 minutes each and 1 minute in 100% ethanol.
5. Air-dry the slides.
6. At this stage the slides are ready for hybridization or can be stored in a slide box with desiccant at –80°C until use.

RNase treatment for in situ hybridization of DNA

1. Overlay the tissue sections with appropriate volume of DNase-free RNase solution.

2. Allow digestion to proceed for 30 minutes at 37°C.
3. After digestion is complete, inactivate RNase by rinsing the slides in dish containing solution of post-RNase treatment solution for 5 minutes.
4. Wash the slides in TE buffer thrice for 2 minutes each and 1 minute in 100% ethanol.
5. Air-dry the slide.
6. Now slides are ready for hybridization. Otherwise, they can be stored in slide box with desiccant at –80°C.

In situ Hybridization

1. For pre-hybridization, overlay tissue specimens with appropriate volume of pre-hybridization solution.
2. Place the slides in humidity chamber and allow pre-hybridization to proceed for 1 hour at 50°C.
3. After pre-hybridization is complete, carefully remove the pre-hybridization solution from the slides carefully using Whatman paper strips or pipette.
4. Immediately overlay each specimen with freshly prepared hybridization solution containing the specific probe. Cover the section with a piece of coverslip slightly larger than the tissue section.
5. Simultaneously denature the probe and target DNA or RNA by placing the slides on plate for 3-4 minutes at 95°C for DNA hybridization and 65-70°C for RNA hybridization.
6. Quickly place the slides in humid chamber and carryout hybridization at 55-65°C for 2-4 hours.
7. After hybridization is complete, quickly place slides in a slide rack and insert the rack into staining dish filled with post-hybridization wash solution containing 2x SSC and 2 mM β-mercaptoethanol.
8. Wash the slides with three changes of fresh wash solution for 15 minutes each at 55°C with gentle shaking.
9. Wash the slides again twice, for 5 minutes each in fresh washing solution containing 0.2x SSC and 2 mM β-mercaptoethanol at room temperature with gentle shaking.
10. Dehydrate the slides by treating with a series ethanol washes- 50%, 70%, 95%, 100%, 100%, 100%, for 3 minutes each.
11. Air-dry the slides.
12. Read the hybridization signal by the process of autoradiography.

8.6.4 Troubleshooting Guide

1. Paraformaldehyde is toxic chemical and should be handled carefully. The preparation of fixative solution should be carried out in a fume-

hood on a heating plate. Temperature of the solution should not be above 65°C and the fixative should not be overheated.

2. While performing fixation of cultured cells, the slides must be completely dried prior to placing them in freezer.
3. Tissue dehydration must be as complete as possible. Otherwise melted paraffin wax may not thoroughly penetrate the tissue resulting in poor paraffin blocks.
4. Xylene is extremely toxic and its handling should be done under a fume hood.
5. Every embedding mould should be clearly labelled for sample identification.
6. It is important to store the paraffin blocks in a dry place. Otherwise moisture may ruin the wax blocks. The paraffin blocks can be stored for years in dry, clean places.
7. The protease solutions once prepared can be stored at –20°C for 2 weeks. It is easy to inactivate and wash pepsin or trypsin by adjusting pH from 2.0 to 7.5.
8. To avoid RNase contamination, prepare all solutions in DEPC treated water.
9. Care should be taken so that air bubbles are not trapped under the coverslip. In case air bubbles are trapped, they should be removed by gently pressing the coverslip.
10. In order to prevent uneven hybridization, slide should be kept levelled in the humid chamber.
11. High background occurs if specimens get dried up during hybridization process.
12. Whenever using radiolabelled probe, all necessary precautions and measures should be followed.
13. If individual sections fall apart during sectioning without formation of ribbons, over-fixation of tissues could be the reason. Reduce the fixation time.
14. If sections are found to crumble during stretching, under-fixation of tissues can be the reason. Increase the fixation time.
15. If tissue sections are found with holes, there might be incomplete dehydration and improper impregnation with wax.
16. If blunt knife is used for cutting sections or if cutting surface of the specimen block is not held parallel to the edge of the knife this may result in sections with uneven thickness or lines throughout.
17. If low signal is found, it could be due to under-digestion of tissue with proteases or activity of the enzyme is low.
18. High background can result from over-fixation of tissue.
19. Tissue over-digestion may result in destruction of tissue morphology.

8.7 COLONY HYBRIDIZATION

8.7.1 Introduction

Colony Hybridization is done for the screening of a gene library with a labelled probe either radioactive or bioluminescent etc. to identify a specific sequence of DNA, RNA, enzyme, protein, or antibody.

Colony hybridization is an efficient technique for screening of bacterial colonies for genes of interest. This technique is dependent upon hybridization of specific labelled DNA probe with the genomic DNA of a particular bacterial strain. Microbes such as bacteria that form colonies when grown artificially on nutrient medium can be screened by this technique. This criterion allows use of great many strains of bacteria. This technique is valuable for screening clones and genetically transformed *E. coli* strains. Therefore colony hybridization is extensively used in genetic engineering experiments.

Advantages of colony hybridization

- This technique allows selection/identification of gene of interest out of large amount of DNA.
- This technique does not require isolation of nucleic acids.
- It is possible to culture and identify microorganisms containing the hybridized sequence.
- This is a time saving method where large number of bacterial strains and clones are to be screened.

Limitations of colony hybridization

- When identifying microorganisms, the procedure will only work if the organisms will grow and form a detectable colony. In case of higher organisms, this method becomes redundant.

8.7.2 Theory and Principle

Colony hybridization is applied to DNA or RNA released from blotted microbial colonies. The microbial colonies are transferred to a nylon membrane. Then the cells are lysed in place by alkali treatment to release the nucleic acids. Then RNA or DNA is fixed to the same filter/membrane after denaturation and hybridized with a labelled probe. The bacterial colonies are replaced by an identical pattern of fixed DNA.

The filter is incubated with a solution of labelled probe, which will hybridize to any bound DNA with complementary sequence to that of probe. Blocking reagent may be added prior to the probe to prevent unspecific binding. Thereafter, filter is thoroughly washed to remove any unbound probe and the hybridization signal is visualized by chromogenic substrate or autoradiography. Colony blot hybridization can be used for screening clones or bacterial isolates. The colonies containing the DNA that is complementary to the probe can be easily identified. Such colonies are taken from the master plate and are grown for further analysis.

In similar way, a very large number of colonies of *Escherichia coli* carrying different hybrid plasmids can be rapidly screened to determine which hybrid plasmids contain a specified DNA sequence or genes. The colonies to be screened are formed on nitrocellulose filters and after a reference set of these colonies has been prepared by replica plating, the cells are lysed and their DNA is denatured and fixed to the membrane *in situ*. The resulting DNA-prints of the colonies are then hybridized to a radioactive probe that defines the sequence or gene of interest, and the result of this hybridization is assayed by autoradiography. Colonies that exhibit positive hybridization signal can then be picked from the replica plate (Fig. 8.2).

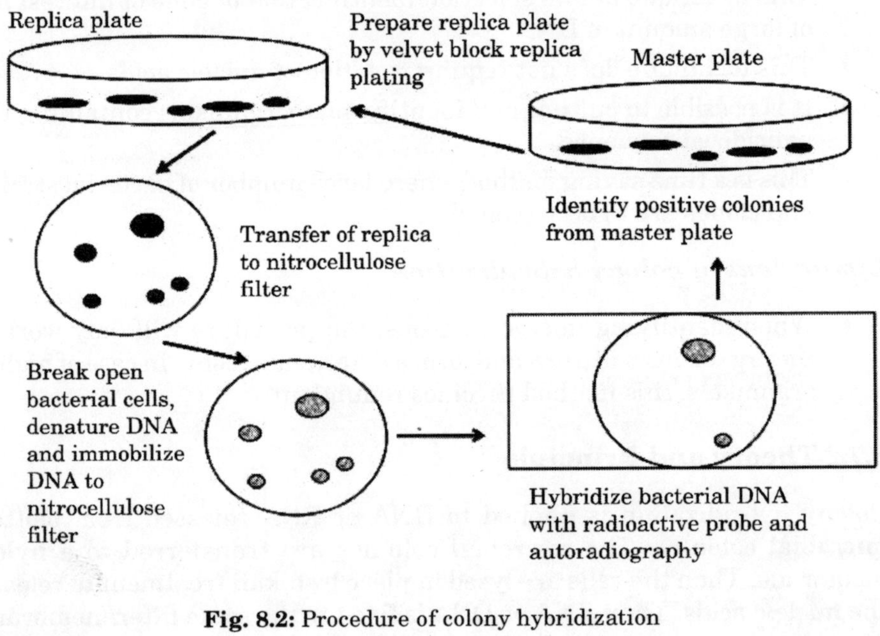

Fig. 8.2: Procedure of colony hybridization

Replica plating

The method of replica plating was developed to achieve the copying of a pattern of microbial growth from one initial agar plate to a series of others. This initial agar plate is called master plate. This method uses velvet or other fabrics to make the transfer without disturbing the shape and size of the colonies.

The principle of replica plating is similar to that of a rubber stamp. In this procedure, a cotton velveteen stamp, is used to stamp a replica of the pattern of cells growing on one plate to one or several other plates. The apparatus consists of a cylindrical holder for the velveteen that is just the right size to fit inside the bottom of a petri plate, and a ring of some sort to hold the velveteen in place. For standard 10 cm plastic plates, the appropriate diameter is approximately 3.3 inches. The holder can be a cylinder of wood, metal, plastic, or any other material. The ring to hold the velveteen can be a metallic or even a large rubber band can be used. The velveteen must be sterilized by autoclaving before use.

This technique may be useful in the detection of biochemical mutants, classification of fermentation reactions, determination of antibiotic sensitivity spectra, and other routine operations requiring repetitive inoculation of several media.

8.7.3 Protocol for Colony Hybridization

Materials required

- *For replica plating*
  Sterile velvet tied to cylindrical wooden block
  Master plate containing microbial colonies.
- *For colony hybridization*
  Solution A: 10% SDS
  Solution B: 0.5 N NaOH, 1.5 M NaCl
  Solution C: 0.5 M Tris-Cl, 1.5 M NaCl (pH 7.4)
  Solution D: 2X SSC
  Nylon-Membrane, cut to size of 100 mm petri dish
  Whatman 3MM filter papers or blotting papers
  15 ×150 mm petri dish

Procedure

Replica plating

1. To replica plate, invert the master plate that contains distinct microbial colonies, over the velveteen on the holder and press the plate firmly against the sterile fabric surface.
2. Then slowly peel the plate off, leaving a replica or print of the cell pattern on the velveteen.

3. Transfer the replica on the velveteen to one or a series of sterile plates by pressing each plate gently onto the velveteen and popping it off.
4. It is possible to make as many as 20 replicas from a single master plate if necessary.

Colony hybridization

1. Grow microbial cells to approximate concentration of 50-100 colonies per plate. Be sure to prepare replica plate before proceeding.
2. Cut a piece of nylon membrane to the size of the petri dish to be probed.
3. Soak 4 of the pieces of blotting paper into one of the 4 solutions A, B, C, and D. Avoid overwetting the paper. It is useful to keep each soaked paper in a 15 × 150 mm petri dish.
4. Gently press the piece of nylon membrane onto the plate containing colonies. Mark 3 spots on both the membrane and plate for orientation.
5. Put the piece of nylon membrane, cells side up, over the 4 blotting papers soaked in their respective solutions in the order, and for the incubation times indicated below:

Incubation	***Incubation time***	***Solution***
1st	3 minutes	Solution A
2nd	5 minutes	Solution B
3rd	5 minutes	Solution C
4th	5 minutes	Solution D

6. Place the membrane with DNA side facing upwards over a dry piece of blotting paper and air dry for 30 minutes.
7. Wrap membrane with saran wrap and crosslink DNA onto membrane by exposing DNA side to UV light for 3 minutes.
8. Hybridize the fixed DNA with labelled probe and follow detection procedure just like standard southern hybridization protocol.
9. Compare the developed blot with replica plate and identify the colonies.

8.7.4 Troubleshooting Guide

1. While transferring microbial colonies to the nylon membrane, the membrane should be pressed gently so that the microbial cells stick to the membrane. Excessive pressing should be avoided.
2. The manner in which you remove the plate from the velveteen controls the amount of yeast transferred from velveteen to plate or from plate

to velveteen. If you peel the master plate off the velveteen with a slow rolling motion, more microbial cells will remain on the velvet than if you pop it off with a quick vertical motion. It is desirable to transfer as few cells as possible and still be able to see the print of the cells on the replica plate.

3. The replica plating procedure should be followed under sterile conditions, in laminar airflow cabinet.

8.8 FLUORESCENT *IN SITU* HYBRIDIZATION

8.8.1 Introduction

Fluorescent *in situ* hybridization (FISH) is a powerful technique used in the detection of chromosomal abnormalities. Being a modification of *in situ* hybridization (ISH), this technique allows detection of target DNA sequences inside cells. This technique skips the process of DNA isolation and has an advantage that one can visualize the location of target DNA in its native state (*in situ*, meaning 'in its native habitat').

FISH involves fluorescent probes, which are used to identify the target DNA sequences. These fluorescent probes can be made by tagging probes with different fluorochromes which give various colours of fluorescence, such as yellow, red, orange etc.

The FISH technique has recently evolved to allow screening of the whole genome simultaneously through multicolour whole-chromosome probes, also called as multicolour FISH or through array based method using comparative genomic hybridization.

FISH is a type of *in situ* hybridization technique which employs fluorescent probes or probes labelled with fluorochromes to detect and visualize target DNA sequence present in chromosome of the cell *in situ*.

Applications of FISH

1. Analysis of chromosomes and chromosomal DNA by FISH is used in many areas of research such as cytogenetic studies and biological as well as clinical research.
2. FISH is technique which is used for detecting DNA sequences in cells, tissues and tumours. It provides a reliable means for studying the genetic composition of cells in mitosis as well as interphase, which is useful in cell biology research.
3. This technique can be used for clinical diagnosis of chromosomal changes in metaphase or interphase nuclei.

4. FISH technique is important in clinical diagnosis of various chromosomal abnormalities including deletions, duplications and translocations.
5. FISH enables cytogeneticists to use comparative genomic hybridization to detect quantitative differences like copy number variations of genes in the chromosomes.
6. FISH can be used for mapping chromosomal genes to study the evolution of genomes analyzing nuclear organization, visualization of chromatin.
7. FISH is used popularly in different chromosomal studies involving somatic hybrids, replication studies, cybrids, detection of chromosomal abnormality studies, chromosome sorting etc.
8. The high specificity and sensitivity of FISH and the speed with which it can be performed has made FISH a pivotal cytogenetic technique. It has provided significant advances in both research and diagnosis of haematological malignancies and tumours.
9. From medical perspective, FISH can be applied to detect genetic abnormalities such as gene fusions, aneuploidy, loss of chromosomal regions or whole chromosomes, or to monitor progression of aberration, therefore serving as a technique which can help diagnose a genetic diseases.
10. FISH can also be applied in identification of oncogenes or genetic aberrations which could lead to various cancers.
11. FISH has established its potential as a diagnostic and research tool in cancer.

8.8.2 Theory and Principle

FISH employs the basic concept of *in situ* hybridization in detection of specific DNA sequences inside the cells. FISH is based on the technique of DNA probe annealing to specific target sequences of specimen DNA. However, it differs from ISH in using fluorescent probes or fluorochromes in visualization of the DNA sequences, where ISH uses radiolabelled probes.

FISH requires fluorescence microscope for visualization of the target DNA sequence and the method is called as fluorescence microscopy. FISH uses fluorescent probes or probes which can be rendered fluorescent later. The probes bind to only those parts of the chromosome with which they show a high degree of similarity. Therefore it is a prior requisite that DNA sequences or a gene of interest which one wants to detect by using FISH should be already sequenced and a probe with complementary sequence should be used.

This modified probe which gives out fluorescence is allowed to hybridize with its complementary sequence of DNA in the chromosomes. Fluorescence

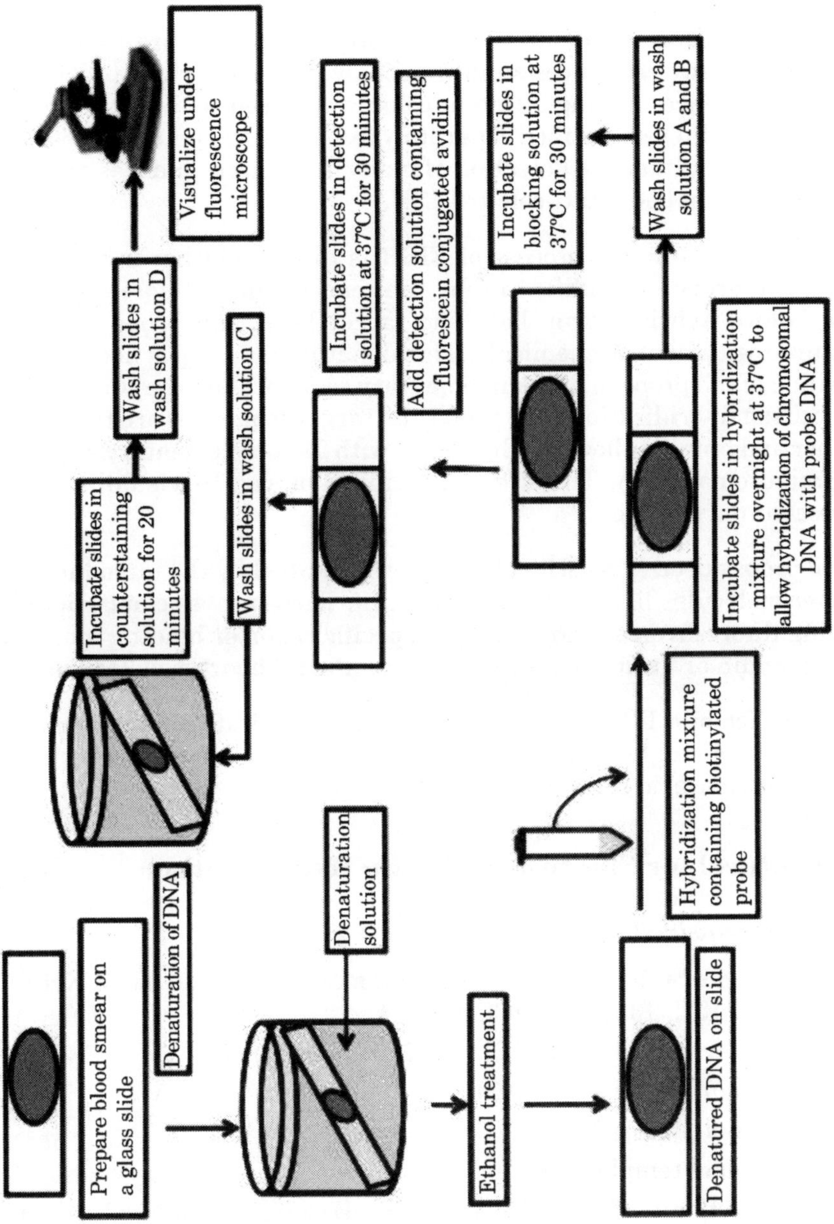

Fig. 8.3: Protocol for fluorescent *in situ* hybridization

microscope is used to examine the specimens containing the above said probes and by using the technique of fluorescence microscopy, the chromosomal sites giving off fluorescence can be detected as the sites of DNA of interest (Fig. 8.3).

Sometimes more than one type of fluorescent probes are used for the detection of many sequences simultaneously. In such a case, probes with different sequences having different colours are used so that their sites of hybridization can be identified separately from their colour of fluorescence. This allows simultaneous detection of many different sequences at a time. This process is called multi-colour FISH.

The first step in the process of FISH is to make either a fluorescent copy of the probe or modified probe that can be made fluorescent later. Then, before hybridization, both the target DNA and probe should be denatured with heat or chemicals. This denaturation step is necessary for annealing of the probe to its complementary sequences on the chromosomal DNA during hybridization. The probe and target sequences are then mixed together and probe is allowed to hybridize with its complementary sequences on chromosomal DNA. The site of hybridization can be visualized using fluorescent microscope.

The probes used in FISH technique are modified by the introduction of reporter molecules like biotin or dioxigenin. These probes can be detected by using fluorochromes conjugated to specific reporter binding molecules such as avidin or antibodies which are specific to reporter molecules.

In multicolour FISH, there is a combined use of different reporter molecules in a single experiment to allow the simultaneous detection of multiple nucleic acid targets.

8.8.3 Protocol for Fluorescent *in situ* Hybridization

Materials required

- Phosphate buffered saline (PBS): 3.2 mM Na_2HPO_4, 0.5 mM KH_2PO_4, 1.3 mM KCl, 135 mM NaCl, pH=7.4.
- TE buffer: 10 mM Tris-HCl, , 1mM EDTA, pH=7.5
- Methanol: Acetic acid (19:1)
- Methanol: Acetic acid (3:1)
- Solution of template DNA
- 10x reaction buffer: 0.5M Tris-HCl (pH=8.0), 50mM $MgCl_2$, 0.5 mg/ml BSA
- 0.1M β-mercaptoethanol: Add 0.1 ml of β-mercaptoethanol to 14.4ml of double distilled water

- 10x nucleotides stock solution: 0.5mM dATP, 0.5mM dCTP, 0.5mM dGTP, 0.5mM biotin-16-dUTP and 0.12 mM dTTP
- *E.coli* DNA polymerase I
- DNase I solution: 3 mg DNase I in 1 ml of 0.15M NaCl, 50% glycerol
- For agarose gel: Agarose, TBE loading buffer (0.089M Tris-borate, 0.089M boric acid, 0.002M EDTA), molecular size marker
- 3M sodium acetate, pH=5.2
- Deionized formamide
- Hybridization buffer: 4x SSC, 20% dextran sulphate. Store at 4°C
- Denaturation solution: 70% deionized formamide, 2x SSC, 50mM sodium phosphate pH=7.0
- Ice cold ethanol: 70%, 90%, 100%

For fluorescent detection

- SSC= saline sodium citrate buffer (pH=7.0): 0.15M sodium citrate, 0.015M NaCl
- Wash solution A: 50% formamide, 2x SSC, pH=7.0
- Wash solution B: 1x SSC-0.1x SSC, pH=7.0
- Wash solution C: 4x SSC, 0.1% tween 20, pH=7.0
- Wash solution D: 2x SSC, 0.05% tween 20
- Blocking solution: 3% BSA (Bovine Serum Albumin), 4x SSC 0.1% tween 20
- Detection buffer: 1% BSA, 4x SSC, 0.1% tween 20
- Detection solution: 5 µg/ml fluorescein conjugated avidin or 6µg/ml rhodamine conjugated anti-dioxigenin
- Counterstaining solution: 2x SSC, 200 ng/ml DAPI (4,6-diamidino-2-phenylindole)

[DAPI is a fluorescent nuclear stain. It is excited by ultraviolet light and shows strong blue fluorescence when bound to DNA. DAPI binds with A=T rich repeats of chromosomes. DAPI is also not visible with regular transmission microscopy but visible under fluorescence microscopy. This stain can be used in living or fixed cells].

Procedure (Fig. 8.3)

Preparation of blood smear slide

1. Incubate blood smeared slide in PBS overnight at 37°C. Then incubate blood smear in TE buffer
2. Incubate in Methanol: Acetic acid (19:1) for 10min at room temperature.

3. Then incubate slides in Methanol: Acetic acid (3:1) at 10 min at room temperature
4. Dehydrate the slides in a series of 70%, 90% and 100% ethanol incubating for 5 minutes each at room temperature
5. Air dry the slides

Procedure for biotinylation of DNA probes by the technique of nick translation

1. Take 2 µg of DNA probe, 10 µl of 10x reaction buffer, 10 µl of β-mercaptoethanol, 10 µl of nucleotides stock solution, 20 units of DNA polymerase I, DNase I in a centrifuge tube. Make final volume of 100 µl with double distilled water.
2. Incubate the reaction mixture for 2 hours at 15°C
3. Place the reaction mixture on ice.
4. Now remove 10 µl of reaction mixture and add to gel loading buffer. Denature it by boiling in waterbath for 2-3 minutes.
5. Transfer the mixture to ice for 3 minutes
6. Now load the mixture on standard 1-2% agarose minigel. Run the gel for 30 minutes
7. For visualization of DNA, stain the gel in 0.5 µg/ml ethidium bromide and visualize the gel in UV transilluminator
8. Check the size of the probe using molecular size marker
9. Now, in order to inactivate DNase I in the reaction mixture, add 2 µl of 0.5M EDTA and 1 µl of 10% SDS. Heat the reaction mixture in waterbath for 10 minutes at 68°C.

Labeling of probes with dioxigenin

Labeling of DNA probe with dioxigenin is carried out in exactly the same way as in biotinylation of probe with a minor change that;

The 10x nucleotide stock solution contains- 0.5 mM dATP, 0.5 mM dCTP, 0.5 mM dGTP, 0.125 mM dioxigenin-11-dUTP, 0.375 mM dTTP.

Hybridization

1. Combine 20 to 60 ng of single copy DNA and 3-5 µg of sheared salmon sperm DNA
2. Mix well and incubate at –70°C for 30 minutes. Then thaw the mixture and centrifuge at 12,000 rpm for 10 min at 4°C.
3. Discard the supernatant and wash the DNA pellet by adding 50 µl of 70% ethanol. Centrifuge again at 12,000 rpm for 10 min at 4°C.
4. Discard the supernatant. Resuspend the pellet in 5 µl of deionized formamide and vortex for few minutes at room temperature.

5. Add 5 µl of hybridization buffer and vortex for 5-10 min
6. Denature the DNA probe at 75°C for 5 minutes followed by incubation on ice for 5 minutes. Now the probe is ready for hybridization.
7. Now take the previously prepared blood smear slides and incubate them in oven at 60°C. This is done to avoid lowering of temperature of denaturing solution when added to the slides.
8. Fill denaturation solution in a suitable container and put in waterbath heated to 70°C. Check the temperature of the denaturation solution with the help of thermometer. It should reach 70°C.
9. Transfer the prewarmed slides to the container of denaturation solution. Incubate the slides for 2 minutes. The solution denatures DNA present in the blood smear.
10. Immediately transfer the slides to a jar containing 70% ethanol on ice. Incubate the slides in series of 70%, 90% and 100% ethanol for 5 minutes each.
11. Air dry the slides
12. Add 10 µl of hybridization mixture containing denatured probe on the denatured DNA of blood smear.
13. Place a coverslip. Take care to avoid trapping of any air bubbles.
14. Seal the edges of coverslip with wax or rubber cement. Incubate the slides overnight at 37°C to allow hybridization of blood smear DNA with probe.

Detection of hybridization signal

1. Prewarm wash solutions A and B at 42°C and 60°C respectively in a waterbath.
2. Remove rubber cement from slides carefully. Place the slides in prewarmed wash solution A (42°C) and keep in shaking waterbath for 10 min.
3. Remove coverslips carefully and place the slides in fresh wash solution A (42°C). Keep on shaking waterbath for 10 minutes.
4. Transfer the slides to prewarmed solution B (60°C) and wash for 5 minutes. Change the solution twice and wash for 5 minutes each time.
5. Remove the slides. Drain the wash solution and apply 200 µl of blocking solution. Cover slides with coverslips and incubate at 37°C for 30 minutes.
6. Remove the coverslips. Drain excess fluid and add 200 µl of detection solution containing fluorochrome conjugated reporter/detection agent.
7. Incubate the slides in humidified chamber at 37°C for 30 minutes.
8. Remove coverslips. Transfer slides to wash solution C and wash for 3 times for 5 minutes at 42°C in shaking waterbath.

9. Place slides in a jar containing counterstaining solution and agitate at room temperature for 20 minutes.
10. Transfer the slides to wash solution D. Incubate for 1-2 minutes at room temperature.
11. Remove the slides and cover with coverslip. These slides can be stored at 4°C or can be used for immediate visualization under fluorescence microscope.

Procedure for multicolour FISH

FISH technique also allows simultaneous detection of multiple target sequences on chromosomes. This is referred to as multicolour FISH. These target sequences are marked with differently labelled probes, which are detected by different fluorochromes. For multicolour FISH, the basic procedure is the same, except for the following steps;

1. Instead of a single probe, several probes with different detection agents are used.
2. The appropriate detection agent should be added to detection buffer. Each detection agent should be coupled with fluorochrome, which should be spectrally distinguishable from other fluorochromes used in the experiment.

8.8.4 Troubleshooting Guide

1. In order to prevent uneven hybridization, slides should be kept levelled in the humid chamber.
2. Care should be taken so that air bubbles are not trapped under the coverslip. In case air bubbles are trapped, they should be removed by gently pressing the coverslip.
3. If no hybridization signal is detected, it is due to absence of probes or target sequences may not be denatured properly prior to hybridization. Make sure to denature probe and target DNA sequences completely before hybridization.
4. If the intensity of fluorescence signal in specimens is too bright with low signal-to-noise ratio, it could be due to overloading of fluorescent probes or detection solutions. Reduce the amount of probe or detection solution.

8.9 DNA MICROARRAY

8.9.1 Introduction

Microarray technology plays a major role in genomic and proteomic profiling. With DNA microarray or gene chip technology, it is possible to analyze thousands of genes simultaneously combining numerous parameters within a single experiment. This technology increases speed and efficiency of scientific research and drug discovery significantly.

The advantage of DNA microarray is that a single chip can reveal thousands of gene expression profiles. Therefore this technology is widely used in new gene discovery, DNA-DNA interactions, DNA-RNA interactions and DNA-protein interactions.

8.9.2 Theory and Principle

The basic principle of this technology is that, thousands of genes, oligonucleotides, ESTs or cDNA fragments are immobilized onto small slides or chips. These chips are then hybridized with complementary probes that can be generated from mRNA species.

DNA chips are commercially available, such as mouse array, human array etc., but these are quite costly. These arrays can also be produced in laboratory by spotting appropriate DNA, like cDNA, or PCR products or oligonucleotides, onto pre-coated glass slides using spotter (Figure 8.4). The spotted slides however should be compatible with commercial microarray scanner.

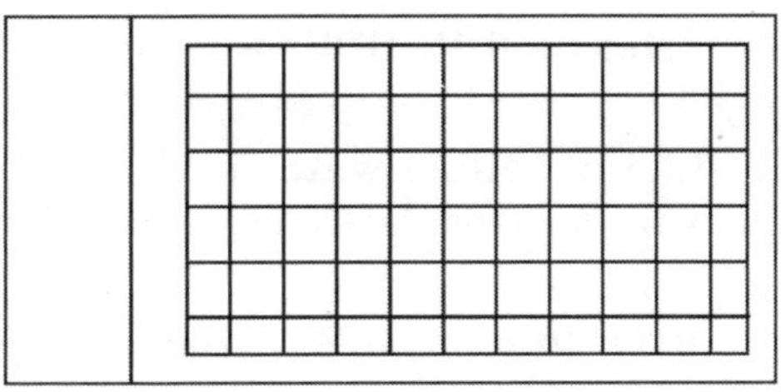

Fig. 8.4: DNA microarray slide

After hybridization, washing is done to remove any unhybridized probe and then these slides are scanned by commercially available scanners. This procedure is strikingly similar to southern or northern hybridization.

8.9.3 Protocol for DNA Microarray Generation

Materials required

Previously coated slides. These slides are available commercially. For example, Corning CMT-GAPS silane-coated slides.

- Spotting buffer: (pH=6.5)

DMSO	50%
Tris-HCl	20 mM

KCl	50 mM
UV cross-linker	

- Hybridization solution

Hybridization buffer	40 μl
Blocking solution	1 μl
Probe solution	10 μl

- Post-hybridization wash solution

2x SSC	
SDS	0.1%

Procedure

Pre-hybridization of gene microarrays

1. Mark the chip/microarray/slide with a marker pen.
2. Moisturize the slide with water vapour with array side facing down by holding over boiling water for 1 minute.
3. Quickly place the chip/slide in a UV cross-linker with the array side up.
4. Remoisturize the chip over steam as described in step 2.
5. Rinse the chip of slide in 0.1% SDS for 20 seconds and DDW for 20 seconds.
6. Heat the chip/slide in boiling water bath at 95ºC for 3 minutes. This is necessary for spotted dsDNA chips and optional for ssDNA chips.
7. Add 60 ml of pre-hybridization buffer to the chip and place a coverslip over the array.
8. Add 10 μl of DDW to each corner of the chamber to maintain humidity.
9. Place the chip/slide in the hybridization chamber and incubate in 50°C water bath for 1 hour.

Hybridization process

1. Remove the water from array slide.
2. Add 60 μl of hybridization solution over the array.
3. Place a new coverslip on the array and place the array in the hybridization chamber.
4. Perform hybridization at 50°C in a water bath, overnight.

Post-hybridization washing

1. After the hybridization is complete, place the slide in 50 to 100 ml wash solution. Carefully remove the coverslip from the array.
2. Shake the array gently for 10 minutes.

3. Replace 2x SSC with 100 ml of 0.2x SSC solution and shake the array for another 15 minutes.
4. Place the slide in 50 ml tube and spin for 5 minutes at 1000 rpm to dry it.
5. Store the array in dark place until scanning.

Scanning of DNA microarray

DNA microarray can be scanned using commercially available microarray scanner.

8.10 REVIEW QUESTIONS

1. What is nucleic acid hybridization? describe various commonly used solutions in nucleic acid hybridization processes.
2. What is probe? Describe in brief.
3. Give an account of southern blotting.
4. How northern hybridization differs from southern hybridization?
5. What is the difference between *in situ* hybridization and colony hybridization?
6. Describe colony hybridization.
7. What is FISH? How does it differ from *in situ* hybridization?
8. Give applications of FISH
9. What is multicolour FISH?
10. Which technique is used to visualize the sites of probe hybridization to target sequences on chromosomal DNA?
11. Name two commonly used reporter molecules in labelling of probes in the technique of FISH.
12. Describe what is DNA microarray technology. State the application of DNA microarray.

8.11 REFERENCES

Bentz, M., Dohner, H., Cabot, G. and Lichter, P. (1994). Fluorerescence *in situ* hybridization in leukemias: The FISH are spawing. *Leukemia,* **8:** 1447–1452.

Berger, C. (1986). *In situ* hybridization of immunoglobulin–specific RNA in single cells of the β–lymphocyte lineage with radiolabelled DNA probes. *EMBO J.,* **5:** 85–93.

Bishop, R. (2010). Applications of fluorescence *in situ* hybridization (FISH) in detecting genetic aberrations of medical significance. *Bioscience Horizons,* **3(1):** 85–95.

Emmerich, P., Loos, P., Jaunch, A., Hopman, A., Wiegant, J., Higgins, M., White, B., Van der Ploeg, M., Cremer, C. and Cremer, T. (1989). Double *in situ* hybridization in combination with digitized image analysis: A new approach to study interphase chromosome topography. *Exp. Cell Res.,* **181:** 126–140.

Forozan, F., Karhn, R., Kononen, J., Kallionemi, A. and Kallionemi, O. (1997). Genomic screening by comparative genomic hybridization. *Trends Genet.* **13:** 405–409.

Gozzetti, A. and Le Beau, M. (2000). Fluorescence *in situ* hybridization: Uses and limitations. *Seminars in Hematology,* **37(4):** 320–333.

Grunstein, M. and Hogness, D. (1975). Colony hybridization: a method for the isolation of cloned DNAs that contain a specific gene. *Proc. Natl. Acad. Sci.,* **72(10):** 3961–3965.

Haupl, T., Burmester, G. and Stuhlmuller, B. (2002). New aspects of molecular biology diagnosis: Array technology and expression profile for characterization of rheumatic diseases. *Z. Rheumatol.,* **61(4):** 396–404.

Hopman, A., Ramaekers, F., Raap, A., Beck, J., Devilee, P., Van der Ploeg, M. and Vooijs, G. (1988). *In situ* hybridization as a tool to study numerical chromosomal aberrations in solid bladder tumours. *Histochemistry,* **89:** 307–316.

Joos, S., Fink, T., Ratsch, A. and Lichter, P. (1994). Mapping and chromosome analysis: The potential of fluorescence *in situ* hybridization. *J. Biotechnol.,* **35:** 135–153.

Kearney, L. (2006). Multiplex FISH (M–FISH): Technique, developments and applications. *Cytogenet. Gen. Res.,* **114:** 189–198.

Khandjian, E. (1987). Optimized hybridization of DNA blotted and fixed to nitrocellulose and nylon membranes. *Biotechnology,* **5:** 165.

Langstrom, A. and Tefferi, A. (2006). Fluorescence *in situ* hybridization in the diagnosis, prognosis and treatment monitoring of chronic myeloid leukemia. *Leuk. Lymphoma,* **47:** 397–402.

Lawrence, J. and Singer, R. (1985). Quantitative analysis of *in situ* hybridization methods for the detection of actin gene expression. *Nucl. Aci. Res.,* **13:** 1777–1799.

Lopez *et al.* (2003). DNA microarray analysis of genes differently expressed in diet induced obese rats. *Obesity Res.,* **11(2):** 188–194.

Mitelman, F., Johansson, B. and Mertens, F. (2007). The impact of translocations and gene fusions on cancer causations. *Nat. Rev. Cancer,* **7:** 233–245.

Nam *et al.* (2002). Application of cDNA microarray technique to detection of gene expression in host cells infected with viruses. *Acta Virol.,* **46(3):** 141–146.

Nilda, H. and Nakanishi, M. (2006). DNA damage checkpoints in mammals. *Mutagenesis,* **21:** 3–9.

Reed, K. and Mann, D. (1985). Rapid transfer of DNA from agarose gels to nylon membranes. *Nucl. Aci. Res.,* **13:** 7207.

Sayler, G., Schields, M., Tedford, E., Breen, A., Hooper, S., Sirotkin, K. and Davis, J. (1985). Application of DNA–DNA colony hybridization to the detection of catabolic genotypes in environmental samples. *Appl. Environ. Microbiol.,* **49(5):** 1295–1303.

Scrock, E., Du Manoir, S., Veldman, T., Scoell, B., Wienberg, J., Ferguson–Smith, M., Ning, Y., Ledbetter, D., Bar–Am, I., Soenksen, D., Garini, Y. and Ried, T. (1996). Multicolour spectral karyotyping of human chromosomes. *Science,* **273:** 494–497.

Southern, E. (1975). Detection of specific sequences among DNA fragments separated by gel electrophoresis. *J. Mol. Biol.,* **98:** 503.

Speicher, M., Ballard, S. and Ward, C. (1996). Karyotyping human chromosomes by combanitorial multi–fluor FISH. *Nature Genet.,* **12:** 368–375.

Speicher, M., Ballard, S. and Ward, D. (1996). Karyotyping human chromosomes by combinatorial multi–fluor FISH. *Nat. Gen.,* **12:** 368–375.

Trembleau, A., Roche, D. and Calas, A. (1993). Combination of non–radioactive and radioactive *in situ* hybridization and immunohistochemistry: A new method allowing the simultaneous detection of two mRNAs and one antigen in the same brain tissue section. *J. Histochem. Cytochem.,* **41:** 489–498.

Wu, W. (1995). Electrophoresis, blotting and hybridization. *In*: Handbook of Molecular and Cellular Methods in Biology and Medicine. *Ed.:* Kaufman P, Wu W, Kim D and Leland C. CRC Press, FL. pp. 87–122.

9

Techniques of Chromatography

9.1 INTRODUCTION

Chromatography is a routine analytical technique used in laboratories for accomplishing a number of tasks such as separation of biomolecules from mixtures, preparative techniques, as well as purification and analytical techniques. Russian botanical scientist Semenovich Tswett first described the technique of chromatography in 20th century.

Chromatography is commonly used for the separation, isolation and purification of proteins from complex sample mixtures. In living cells, proteins are present associated with numerous other components such as lipids, carbohydrates and nucleic acids. In order to be analyzed, the proteins must be separated from all other cell components. Chromatography plays a crucial role in protein purification process.

9.2 COMMONLY USED TERMS IN CHROMATOGRAPHY

Retention time (t_R): Each component of an analyte elutes from the stationary phase at a specific time.

Chromatogram: The signal obtained by the detector when the components pass through chromatographic column is recorded and plotted on a graph. This is called chromatogram.

- The analyte is the substance to be separated during chromatography.
- The eluate is the mobile phase leaving the column.
- The eluent is the solvent that carries the analyte.

- The mobile phase is the phase that moves in a definite direction. It may be a liquid or gas chromatography.
- The retention time is the characteristic time it takes for a particular analyte to pass through the system (from the column inlet to the detector) under set conditions.
- The stationary phase is the material that is fixed in place for the chromatography procedure.

9.3 THEORY OF CHROMATOGRAPHY

Chromatography is basically a separation method where the analyte is contained within a liquid or gaseous mobile phase that is passed through a stationary phase. Depending on the polarity, components of the analyte interact differently with these two phases. This leads to separation of different components present in the sample.

The chromatographic methods can be broadly classified as gas chromatography (GC) and liquid chromatography (LC) depending upon the nature of mobile phase involved. Gas chromatography can be applied to only gaseous or volatile heat-stable substances. In GC, the mobile phase in usually an inert carrier gas such as nitrogen, hydrogen or helium. GC is not applied to large biomolecules like proteins, nucleic acids etc. as they are destroyed by high temperatures. But smaller molecules such as amino acids, fatty acids, and certain carbohydrates can be analyzed if they are modified chemically to increase their volatility. Volatile metabolites like aromatic oils produced by certain plants are readily analyzed by the technique of gas chromatography.

In liquid chromatography, the sample is dissolved and passed through a column containing stationary phase. Liquid chromatography is more versatile than gas chromatography as it is not restricted to volatile and heat-stable compounds.

The optimization of chromatographic technique is targeted towards complete separation of all components of the analyte in shortest time possible. This in turn depends upon composition of the mobile phase, stationary phase and flow rate.

In a typical chromatogram, the peak at t_0 or zero retention time, is represented by the substances that are not retarded by the stationary phase and are first to leave the column. Thereafter, the peaks formed, correspond to different substances that are retarded by the stationary phase, leave the chromatographic column at different times. The peak width is defined as the intersection of the tangents on each side of the peak with the baseline. The efficiency of chromatographic separation is dependent upon band broadening. If band broadening is large, peaks can overlap each other and resolution is lost.

The basic parameters such as retention time and peak width can be used to derive a number of other parameters that express the quality of the chromatographic separation.

Each type of compound spends different amount of time interacting with mobile and stationary phase.

$$k = \frac{t_R - t_0}{t_0}$$

The capacity factor 'k' describes the velocity of the analyte corresponding to the velocity of the mobile phase. The average velocity of a compound is dependent on how much time it spends in the mobile phase. If k is smaller than 1, analyte moves too quickly and elution time is so short that the exact determination of t_R is difficult. If the sample moves too slowly, the separation time is too high. The value of k should ideally fall between 1 and 5.

The selectivity factor α gives the relative velocities of analyte components A and B with respect to each other.

$$\alpha = \frac{k_B}{k_A} = \frac{t_R(B) - t_0}{t_R(A) - t_0}$$

The selectivity factor describes how well a chromatographic method can separate two analyte components.

9.4 PAPER CHROMATOGRAPHY

9.4.1 Introduction

Paper chromatography is a simple, yet extremely effective method for determining the presence of such small quantities of analyte. In this method strip of paper serves as the stationary phase. This is the simplest and least expensive form of chromatography technique.

Paper chromatography tells whether or not the sample is pure, and if it is not, how many different components are present. The positions of each spot can be compared with those from known compounds or standards to ascertain the presence or absence of that substance.

Although paper chromatography is simple by not requiring elaborate equipments or complex procedures, it still requires careful attention to detail and good technique to obtain satisfactory results.

9.4.2 Theory and Principle

Chromatography is used to separate mixtures of substances into their components. In paper chromatography, the stationary phase is a very uniform absorbent paper. The mobile phase is a suitable liquid solvent or mixture of solvents. The mobile phase flows through the stationary phase and carries the components of the mixture with it. Different components travel at different rates.

In paper chromatography, the sample mixture is applied to a piece of filter paper, the edge of the paper is immersed in a solvent, and the solvent moves up the paper by capillary action. Components of the mixture are carried along with the solvent up the paper to varying degrees, depending on the compound's preference to be adsorbed onto the paper versus being carried along with the solvent. The paper is composed of cellulose to which polar water molecules are adsorbed, while the solvent is less polar, usually consisting of a mixture of water and an organic liquid.

The choice of solvent in paper chromatography is crucial. The solvent must dissolve the various components in the mixture and there must be at least slight differences in solubility of each component. Therefore, there will be no separation. Also if the components are insoluble they will remain at the origin, unmoved. If all of the components are very soluble in the solvent system then each component of the mixture may travel along with the leading edge of the solvent *i.e.* the solvent front, unseparated. In practice it is found that one pure solvent will not give enough resolution and a mixture of solvents must be used for each sample to be analyzed.

Performing a chromatographic experiment is basically a three-step process:

- Application of the sample
- Developing the chromatogram by allowing the mobile phase to move up the paper
- Calculating Rf values and making conclusions

In order to obtain a measure of the extent of movement of a component in a paper chromatography experiment, we can calculate Rf value for each separated component in the developed chromatogram. The Rf value is a number that is defined as:

$$R_f = \frac{\text{Distance travelled by component from point of application}}{\text{Distance travelled by solvent from point of application}}$$

It is important to remember that distance traveled by the spot is measured to the middle of the spot (Fig. 9.1). It is clearly seen from Fig. 9.1

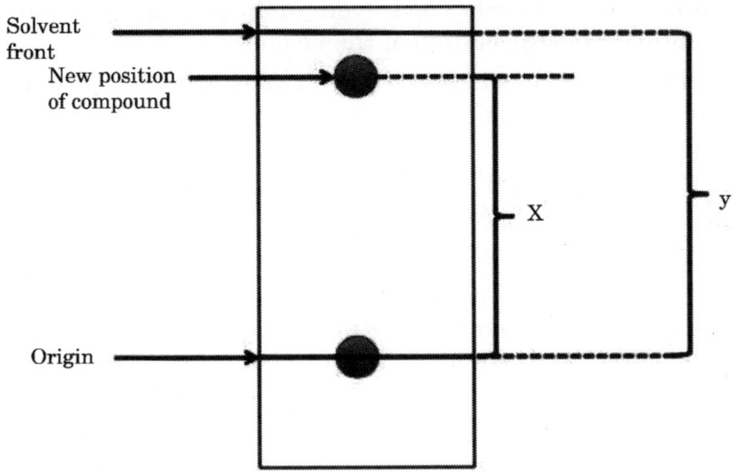

Fig. 9.1: Paper chromatographic separation of compounds. Rf values can be found by measuring the distance travelled by compound (x) and distance travelled by solvent (y)

that Rf=x/y. For a given substance, the distance it moves depends on the total time the experiment is carried out as well as on the physical and chemical properties of the system. However, the distance the compound moves relative to the distance the mobile phase moves is a characteristic of that compound and is known as the Rf value.

9.4.3 Procedure of Paper Chromatography

Materials required

- Chromatography paper
- Solvent- isopropanol: water (1:2) (v/v)
- Ink, pencil
- Beaker

Procedure

1. The chromatography paper is cut in about 4 × 10 cm strips. Along one of the shorter sides, draw a horizontal line in pencil about 1.5 cm from the edge of the strip. This will be the base line or the starting line where the samples will be spotted.
2. Apply a dot of ink from black felt tip pen on the baseline. Make sure dots are evenly spaced on the baseline and not too close to the edge. It is helpful to use a pencil to mark the spots before placing the ink on the paper.

3. Pour some of the developing solvent in the beaker, using only enough solvent to cover the bottom of the beaker.
4. Label each spot in pencil so that you know where each spot is. Place these strips in the clean glass beaker containing the mobile phase-solvent.
5. Stand the paper in the solution. Taping the top of the strips to the side of the beaker may be helpful.
6. Now cover the beaker with watch glass and allow the ink to migrate.
7. When the solvent has risen almost to the top of the paper strip, remove the strip from the chamber and immediately mark the level to which the solvent has risen and circle each spot present on the strip with a pencil.
8. Measure the distance each spot traveled and calculate Rf values for each spot.

9.4.4 Troubleshooting Guide

1. The marks of pencil do not move along with the mobile phase. So pencil marks are made to determine the original position of the samples.
2. The paper strips must be placed in such a way that the edge near the ink spot is submerged in the solvent. However, the ink spots must be above the solvent level.
3. While using more than one paper strips per beaker, the strips should not touch each other.
4. Avoid excessive handling of chromatographic paper.

9.5 THIN LAYER CHROMATOGRAPHY

9.5.1 Introduction

Thin layer chromatography commonly referred as TLC, is a mode of liquid chromatography in which the sample is applied as a small spot or streak at the origin of thin sorbent layer supported on a glass, plastic or metal plate.

The term 'thin layer chromatography' was coined by Egon Stahl in Germany in late 1950s. Stahl standardized several materials, procedures, and solvent systems for resolution of several compounds. Early workers in TLC had to make their own TLC plates, but today a variety of plates are available commercially.

Quantitative TLC was first described by Kirchner and coworkers in 1954 for the determination of biphenyl in citrus fruits.

TLC is a rapid, simple and inexpensive method for separation of a wide variety of substances.

9.5.2 Theory and Principle

Many of the principles and techniques of TLC are very similar to that of paper chromatography. Both paper chromatography and TLC are open systems in which multiple samples are applied to the planar stationary phase. The samples are resolved by mobile phase and solutes are detected statically after removal of mobile phase.

Ion exchangers, especially inorganic and cellulose-based exchangers, are used in thin layer chromatography. Chromatographic paper for this purpose is manufactured from finely ground resins and cellulose fibers. One use for this procedure is to filter small traces of metal ions from large volumes of solution.

In thin layer chromatography, a solid phase, the adsorbent, is coated onto a solid support as a thin layer (about 0.25 mm thick). In many cases, a small amount of a binder such as plaster of Paris is mixed with the absorbent to facilitate the coating. Many different solid supports are employed, including thin sheets of glass, plastic, and aluminium.

The Rf value:

The retention factor, or R_f, is defined as the distance traveled by the compound divided by the distance traveled by the solvent (Figure 9.1).

$$R_f = \frac{\text{distance travelled by the compound}}{\text{distance travelled by the solvent front}}$$

For example, if a compound travels 2.1 cm and the solvent front travels 2.8 cm, the R_f is 0.75.

The R_f for a compound is a constant from one experiment to the next only if the chromatography conditions below are also constant:

- Solvent system
- Adsorbent
- Amount of material spotted
- Temperature

Since these factors are difficult to keep constant from experiment to experiment, relative R_f values are generally considered. Relative R_f value means that the values are reported relative to a standard, or it means that you compare the R_f values of compounds run on the same plate at the same time.

The larger an R_f of a compound, the larger the distance it travels on the TLC plate. When comparing two different compounds run under identical chromatography conditions, the compound with the larger R_f is less polar because it interacts less strongly with the polar adsorbent on the TLC plate. Conversely, if you know the structures of the compounds in a mixture, you can predict that a compound of low polarity will have a larger R_f value than a polar compound run on the same plate.

The R_f can provide corroborative evidence as to the identity of a compound. If the identity of a compound is suspected but not yet proven, an authentic sample of the compound, or standard, is spotted and run on a TLC plate side by side (or on top of each other) with the compound in question. If two substances have the same R_f value, they are likely (but not necessarily) the same compound. If they have different R_f values, they are definitely different compounds. Note that this identity check must be performed on a single plate, because it is difficult to duplicate all the factors which influence R_f exactly from experiment to experiment.

9.5.3 Procedure of Thin Layer Chromatography

Materials required

- Acetone
- Agitator or vortex
- 25% ammonia solution
- Sodium bicarbonate
- Pirex jars or similars: 1 l, 0.5 l, 0.25 l.
- TLC plate (silica gel in an aluminium support)
- 2 big and 1 small glass containers used for TLC
- Funnel
- 1.5 ml eppendorfs
- Tweezers, spatulas, scissors, pencils and rulers.
- Latex gloves
- Ultraviolet lamp, of 24 nm, with protection screen
- Marquis reactive
- Methanol (MeOH).
- Capilary micropipetts
- Blotting paper
- Substance patterns, dissolved in methanol
- p-DMAB reactive
- 5 ml vials
- Small glass dishes (watch glass)

Procedure

Preparing the solvent

1. Use amphetamines and derivatives system (MeOH: NH_3) for the big container. Measure 29.25 ml in the test tube. Add 0.75 ml of ammonia. Mix it with the test tube well closed.
2. Add the solvent into the container, and make it soak the blotting paper.
3. The paper should be lying against the glass wall. Cover the container.
4. Leave it for 30 minutes, so the solvent vapours can saturate the glass container

Preparing the samples

1. Place 5 mg of the sample in an eppendorf. Add 0.5 ml of methanol in the eppendorf with the sample. Shake the sample manually or with an electric shaker.

Preparing the TLC plate

1. Cut the thin layer chromatography plate.
2. Draw with a pencil a parallel line at 1.5 cm above the lower border. Divide it with separate marks of 1 cm of distance between them and separate 1 cm of both sides of the plate.
3. Draw another line of 0.5 cm in the upper border scratching the silica with the spatula tip until the aluminium is seen.
4. Write with a pencil the initials of the sample patterns below of each mark.

Spotting the samples

1. Drop 3 µl of the sample on the TLC plate using a capillary micropipette. Drop the needed µl on the pattern or patterns with different capillary micropipettes.
2. Let the solvent evaporate.
3. Look at the stains that the samples and patterns have left in the TLC plate using the ultra violet developing camera. If the procedure has been done correctly and the filling has been enough, dark stains must be seen where puncture has been done. If they can not be seen, apply the samples again until the stains are visible.

Elution

1. Introduce the TLC plate, using the tweezers, into the glass container, and let the top of it lean against the dryer paper while the bottom is fixed in the small glass lump of the container.

2. Seal the glass container with its lid. Leave the TLC plate inside the container container. Wait till the solvent has gone up until reaching the upper line.
3. Take the TLC plate out of the container and let it dry in a ventilated place.

Analyzing the stains / spots

Introduce the dry TLC plate into the ultra violet developing camera. Compare the sample stains with the patterns. Circle with a pencil the top of the stains.

9.5.4 Troubleshooting Guide

1. If the compound runs as a streak rather than a spot: The reason could be that sample was overloaded. Run the TLC again after diluting your sample. Otherwise sample might just contain many components, creating many spots, which run together and appear as a streak.
2. If the sample runs as a smear or a upward crescent: Compounds which possess strongly acidic or basic groups like amines or carboxylic acids sometimes show up on a TLC plate with this behavior. Add a few drops of ammonium hydroxide amines or acetic acid carboxylic acids to the eluting solvent to obtain clearer plates.
3. If the sample runs as a downward crescent: It is likely, the adsorbent was disturbed during the spotting, causing the crescent shape.
4. If many random spots are seen on the plate: Make sure that you do not accidentally drop any organic compound on the plate.
5. The plate solvent front runs faulty: Either the adsorbent has flaked off the sides of the plate or the sides of the plate are touching the sides of the container or the paper used to saturate the container as the plate develops. Crooked plates make it harder to measure R_f values accurately.
6. If you see a blur of blue spots on the plate as it develops, then perhaps you used an ink pen instead of a pencil to mark the origin.
7. Absence of spots on the plate: You might not have spotted enough compound, perhaps because the solution of the compound is too dilute. Try concentrating the solution, or spot it several times in one place, allowing the solvent to dry between applications.
8. Some compounds do not show up under UV light; Try another method of visualizing the plate such as staining or exposing to iodine vapor.
9. If the solvent level in the developing jar is deeper than the original spotting line of the TLC plate, the solvent will dissolve the compounds into the solvent reservoir instead of allowing them to move up the plate by capillary action. Thus, you will not see spots after the plate is developed.

9.6 AFFINITY CHROMATOGRAPHY

9.6.1 Introduction

Affinity chromatography has highest specificity and selectivity as compared to all other types of chromatographic techniques. This technique is a powerful method for the separation and purification of biomolecules that are present in trace amounts. This method is advantageous where the target molecules needs to be isolated and detected from complex mixtures like blood or serum.

Target molecules can be purified from complex biological mixtures, native forms can be separated from denatured forms of the same substance and small amounts of biological material can be purified from high levels of contaminating substances.

Affinity chromatography exploits highly specific biorecognition phenomenon and is highly suited for purification of biomolecules. In 1910, Emil Starkenstein published a report regarding influence of chloride ions on the enzymatic activity of liver α-amylase. This was the first description of biospecific adsorption of enzyme onto starch molecules used as a solid substrate.

In 1951, Campbell and coworkers used the affinity principle to isolate rabbit anti-bovine serum albumin antibodies on specific immuno-adsorbent column containing bovine serum albumin coupled to π-aminobenzyl-cellulose. Later on many other scientists purified several enzymes and nucleotides by employing affinity chromatography principle.

The term 'affinity chromatography' however, was coined around 1968. Group specific ligands were introduced around 1970s. The advantage of ligands with broad spectrum of specificity is that it is not necessary to devise a new strategy for purification of similar biomolecules.

9.6.2 Theory and Principle

Affinity chromatography separates proteins on the basis of a reversible interaction between a protein (or group of proteins) and a specific ligand coupled to a chromatography matrix. The technique is ideal for a capture or intermediate step in a purification protocol and can be used whenever a suitable ligand is available for the protein(s) of interest.

Affinity chromatography uses highly specific molecular recognition of certain biomolecules. By attachment of a special ligand such as antigen to the stationary phase material, the matching antibody can be specifically and reversibly adsorbed. This molecular recognition does not only occur between antigens and antibodies. Other binding partners such as enzyme-coenzyme, receptor protein-hormone and single stranded oligonucleotides-its matching strand, can also be employed (Refer Table 9.1).

Table 9.1: The commonly used binding partners in affinity chromatography

Ligands	***Adsorbent/target molecule***
Antibodies	Antigens
Hormones	Hormone receptors
Nucleic acids	Complementary sequences
Enzyme inhibitors	Enzymes
Protein A	IgG
Protein G	IgG
Biotin	Streptavidin
Heparin	Coagulation factors
Lectins	Polysaccharides, glycoproteins

The coupled ligand must retain its specific binding affinity for the target molecules and, after washing away unbound material, the binding between the ligand and target molecule must be reversible to allow the target molecules to be removed in an active form.

The ligands used for the affinity chromatography are of two types, mono-specific and group-specific. Mono-specific ligands show affinity for only one analyte while group-specific ligands bind to similar molecules such as proteins belonging to same class. The examples of group specific ligands are lectins, coenzymes, metal ions, protein A etc.

Biological interactions between ligand and target molecule can be a result of electrostatic or hydrophobic interactions, van der Waals' forces and/or hydrogen bonding.

Nickel-affinity chromatography uses the ability of histidine to bind nickel. Six histidine amino acids present at the end of a protein (either N or C terminus) is known as a 6X His tag. In this technique, nickel is bound to agarose beads by chelation using nitroloacetic acid (NTA) beads. These NTA beads can be procured from commercial suppliers. The general method is to batch absorb the protein onto the column, by mixing the beads with the sample, then pouring the slurry of NTA beads and protein into a column. The low concentrations of phosphate and imidazole are used to remove low affinity bound proteins. If needed, the imidazole concentration can be increased to 20 mM before most His-tagged proteins are eluted. Finally, higher concentrations of imidazole is used to elute the protein from the NTA-beads.

The process of affinity chromatography is divided into following steps;

- Sample introduction
- Adsorption

- Washing/elution of impurities
- Desorption and collection of target molecule

The chromatographic column used in this technique is packed with agarose or cellulose beads making the stationary phase, on which the ligands are covalently attached. After addition of sample mixture, the molecules that have affinity for the ligands are adsorbed and retained. All other substances that have no affinity for the ligand are eluted out. Washing is done to ensure removal of non-specifically bound components. In the next step, adsorbed molecules are removed by desorption. To achieve desorption, various strategies are employed such as changing pH, increasing ionic strength, addition of denaturing agents like urea or addition of some other organic solvent.

The desorption methods are usually non-specific and they remove any bound molecule. On the other hand, specific desorption can be achieved by introducing another type of molecule that binds more strongly to the ligands. To elute the target molecule from the affinity medium the interaction can be reversed, either specifically using a competitive ligand, or non-specifically, by changing the pH, ionic strength or polarity.

9.6.3 Procedure of Affinity Chromatography

Materials required

- 6X Histidine Wash Buffer (25 ml)

Phosphate Buffer pH 7.0	50 mM
NaCl	300 mM
Imidazole	1 mM

  Adjust pH to 7.0
- 6X Histidine Elution Buffer (10 ml)

Phosphate Buffer pH 7.0	50 mM
NaCl	300 mM
Imidazole	150 mM

  Adjust pH to 7.0
- NTA agarose beads
- Plastic chromatography column
- Vials for collection of fractions

Procedure

1. While preparing protein samples, ensure your protein solution is clear and not cloudy. Cloudy solution forms from precipitated protein.

Remove the precipitation by centrifugation or filter the solution through a 0.45 mm syringe filter. Save a sample of the lysate for later analysis.

2. Resuspend the resin by inverting the bottle several times. The resin comes in a 50% slurry in 20% ethanol.
3. Pipette 2-4 ml of slurry (1-2 ml of resin) into the column and wash 10 ml of equilibration buffer through the resin. The resin has about a 3 mg of protein/ml of resin binding capacity.
4. Transfer the beads and load solution to a 50 ml tube. Incubate at room temperature for 30–60 minutes with gentle shaking to keep the beads in suspension.
5. Transfer the beads and solution to the chromatographic column with the bottom closed and let the beads settle.
6. Drain the remaining buffer until it reaches the top of the resin bed. Do not let the resin bed dry out. Save the flow through as one fraction.
7. Wash the column once with 5 column volumes of wash buffer. The wash buffer should have a low concentration of imidazole, which helps wash off non-specifically bound proteins. Collect one ml fractions.
8. Elute the protein with 5 column volumes of elution buffer. Collect one ml fractions. Most of the protein will elute in the second and third fraction.
9. Analyze each tube for the protein for total protein concentration by Bradford assay and fluorescence Assay.
10. Prepare a chromatograph after taking the readings of each fraction.

9.6.4 Troubleshooting Guide

1. Increasing the pH of the equilibration buffer from 7 to 8 will increase the binding of protein to the NTA beads, but will also increase non-specific binding of other proteins. Start using a pH=7.0 buffer unless you are having a difficulty in getting the protein to bind to the beads.
2. Adding imidazole to the buffer will change the pH of the solution. Check the pH of the solution after adding imidazole.
3. If there is a high level of contaminant, the imidazole in the equilibration buffer can be increased to 50-75 mM.
4. Imidazole has a higher affinity for the metal than does histidine. By looking at the structure of histidine and imidazole, it can be observed that they basically have the same functional group.
5. Free imidazole has been known to denature protein when freezing and thawing. Therefore, it is important to continue to the next chromatography before freezing sample.
6. The beads can be regenerated with 10 column volumes of the Buffer wash at pH 5.0, or wash with water, or 20% ethanol. Store the beads in 20% ethanol.

7. Imidazole may interfere with the protein detection assays giving a false-positive reading.

9.7 ION EXCHANGE CHROMATOGRAPHY

9.7.1 Introduction

Ion-exchange chromatography as the name suggests, depends on the ionic interaction of target molecules with the molecules of stationary phase. When a mixture of two kinds of ions that are held by the exchanger with differing strengths is introduced at the top of a column, the mixture of ions separates as it moves down the column, with the result that the original single band of ions is resolved into two separate bands. This process is called ion exchange chromatography. Ion-exchange chromatography is an important tool in chemical analysis because it permits separation of materials that are very difficult to separate by other means. It can be applied to organic and inorganic ions and even to substances that are not ionic. It is often used to separate mixtures of many components.

The two requirements for ion exchange, fixed ionic charges on a supporting material and permeability of the material to a solution are met in a large number of materials. The fixed charges may be negative, or positive. The mobile ions must be of opposite charge to the fixed ions. Materials with fixed negative charges exchange positive ions, or cations, and the process is called cation exchange. Those having fixed positive charges correspondingly exchange negative charges, or anions, and are said to undergo anion exchange.

A major breakthrough in ion-exchange technology came in 1935, when the first ion-exchange resins were discovered by the English chemists Basil Albert Adams and Eric Leighton Holmes. These resins were chemical relatives of the plastic Bakelite and were made by condensing polyhydric phenols or phenolsulfonic acids with formaldehyde.

Styrene and divinylbenzene are liquids and are polymerized as spherical droplets, with the result that the resins have the form of beads that are almost perfect spheres. The beads swell when placed in water, and though they look smooth and impermeable, they are actually very permeable to water and small ions. They may have diameters ranging from a few microns to one to two millimeters.

Applications of ion exchange chromatography

Ion exchange is used for both analytical and preparative purposes in the laboratory, however analytical uses are the more common. An important use of ion-exchange chromatography is in the routine analysis of amino acid mixtures.

Columns of cation-exchange resins are commonly used, and the solutions are maintained sufficiently acidic so that the amino acids are at least partly in their cationic forms. The 20 principal amino acids from blood serum or from the hydrolysis of proteins can be separated in a few hours, and their concentrations are determined automatically by light absorption methods. Such analysis is used in clinical diagnosis.

In a highly important application of ion-exchange chromatography the products of hydrolysis of nucleic acids are analyzed. In this way, information is gained about the structure of these molecules and how it relates to their biological function as carriers of hereditary information. Because of their use in analyzing the structures of complex biological materials, ion-exchange chromatographic procedures have been of great importance in the development of modern molecular biology the explanation of biological processes in terms of the interactions of molecules.

Anion exchange in hydrochloric acids is an effective way to separate metal ions. Most metals form negatively charged chloride complexes that can be held by anion-exchange resins carrying quaternary ammonium groups. These complexes differ greatly in their stabilities in solution and in their affinities for the resin. The distribution of metal ions between the solution and the resin depends on the hydrochloric acid concentration and the identity of the metal ion. The separations of metal ions can be achieved by manipulating the hydrochloric acid concentration.

Miscellaneous analytical uses of ion exchange include the dissolving of sparingly soluble salts like calcium sulphate, the determination of total dissolved salts in natural waters by passing them through hydrogen-loaded, cation-exchange resins and titrating the acid formed, and the identification of minute traces of ions by absorbing them onto a single resin bead along with a colour-producing reagent.

Ion exchange finds its major industrial application in the treatment of water. Hard water is caused by the presence of calcium and magnesium ions, which form insoluble precipitates with soaps, is softened by exchanging its calcium and magnesium ions with sodium ions. To accomplish this, the hard water is passed through a column of cation exchanger containing sodium ions.

In other use of ion exchange, salts are removed from sugar juices to raise the yield of crystallized sugar. Deionization also can improve the flavour and storage time of pineapple juice and wine. In these and other beverage applications, ion exchange removes traces of heavy metals, which not only taste bad but also catalyze oxidation.

9.7.2 Theory and Principle

Ion exchange chromatography separates analytes according to their overall charge. This technique can be used for the separation of almost any type of molecule such as large proteins, nucleotides and amino acids. It is commonly used for protein purification.

The principle of ion exchange chromatography is based on the competitive interaction between charged molecules of analyte, and salt ions of mobile phase with charged functional groups of the stationary phase. The stationary phase of the ion exchange chromatography is usually a gel consisting of agarose or cellulose beads with covalently attached charged groups. Anion exchangers are positively charged functional groups while cation exchangers are negatively charged surface groups. Cellulose-based exchangers are especially useful in biochemistry.

The ion-exchanger is packed into a tube or column through which the solution is made to flow. Commonly used ion exchangers are diethyl aminoethyl (DEAE) and carboxymethyl (CM) cellulose. The charge and capacity of the ion exchangers depends upon the pH of the mobile phase. For example, anion exchanger DEAE will be deprotonated and neutralized at high pH. The CM and DEAE cellulose work efficiently at pH between 4 and 8.

Care is needed in preparing columns for the laboratory. Dry resin must be stirred with water to let it swell before it is poured into the column. Air bubbles must not be allowed to form in the resin bed, for they interfere with the liquid flow. It is desirable to backwash the column that is, to pass liquid upward to expand the resin bed in order to release air bubbles and to segregate the resin particles according to size before the column is used. The aim of the preparation procedure is to assure even packing and even flow.

The column arrangement forces the ion-exchange reaction, which is intrinsically reversible, to go to completion in the desired manner. The solution flowing down the column continually meets fresh exchanger, and a reaction that goes half way in the first centimeter of the column may be three-quarters completed in the second centimeter, seven-eighths in the third, and so on. In a short time, the exchangeable ions that entered the column have been adsorbed and become undetectable analytically. When the exchangeable ions do start to emerge from the end of the column, however, the column has become completely saturated with them. It may be restored to its original condition, or regenerated, by passing through it a solution of the ions that it originally contained.

Separation of proteins

Proteins are amphoteric molecules, *i.e.* having both basic amino group and acidic carboxyl groups. The overall charge of proteins will be sum of all charges on all the component amino acids. Therefore, depending on the pH of the solvent/mobile phase, proteins will have either net positive or negative charge. The pH at which the net charge on a protein is zero, is called isoelectric point (pI). Adsorption of proteins to a stationary phase will be minimum when pH of the mobile phase is close to the pI. But if pH differs considerably from a protein's pI, then the protein will have a significant net charge and will interact strongly with functional groups of the stationary phase.

In ion exchange chromatography, the column is usually packed with porous particles having positively charged functional groups on their surfaces. After adding the sample, negatively charged components are adsorbed to the positively charged functional groups of the stationary phase. The positively charged and neutral components are not retained and are eluted out.

The negatively charged analytes are then desorbed and eluted gradually from the column by increasing the salt concentration of the mobile phase or by changing the pH.

For example to adsorb a protein onto cation exchanger CM-cellulose, the protein should have net positive charge. To accomplish this, the pH of the mobile phase must be lower than the protein's pI. In the same way, to adsorb a protein onto DEAE-cellulose anion exchanger, the protein should have net negative charge. Therefore, the pH of the mobile phase must be higher than the pI of the protein.

Also, to minimize the competition between buffer ions and sample components for the binding sites on the stationary phase, the buffer concentration should be kept very low at adsorption step.

After adsorption step, when positively charged and neutral components have been eluted out, the immobilized analytes are removed by gradual desorption. This is achieved by continuous increase in the ionic strength or change in pH of the mobile phase. Salt gradients, usually employing NaCl, are commonly used for desorption process in cation as well as anion exchanger systems. The salt concentration is gradually increased form 0 to 1 M or higher. At this stage, the salt ions compete with the proteins for the charged binding sits of stationary phase.

As a result, weakly bound proteins having weaker charge are eluted first, whereas strongly charged proteins are retained and are eluted at very high ionic strength. This method allows bound proteins to be desorbed gradually according to their net charge and eluted from the column at different rates.

9.7.3 Procedure of Ion-Exchange Chromatography

The following procedure takes advantage of lysozyme's remarkably high pI (over 11). Even well above neutral pH, lysozyme has a net positive charge, resulting in its retention by a cation exchange matrix when most other proteins would elute out through due to a net negative charge. Cation exchange matrices that might be used include phosphocellulose and carboxymethyl or sulfopropyl-substituted cellulose.

Materials required

- Preparation of Sample: Break an egg, collect the egg white separately. Prepare an egg white mixture by diluting 6 ml of egg white with 6 ml of equilibration buffer. Mix it for 10 minutes. Centrifuge it at 10,000 rpm for 10 minutes. Collect the supernatant.
- Preparation of 1M NaCl: Take 2 ml of 5M NaCl in a 20 ml beaker and add 8 ml of distilled water. Mix well.
- Equilibration buffer (pH=5.5)
- Elution buffer (pH=10.0)
- CM-cellulose
- Chromatography column

Procedure

1. Wash the empty column with distilled water. Fix the column to the stand.
2. Remove the top cap of the column and pack the column with 2.5 ml of CM-cellulose.
3. Remove the bottom cap and equilibrate the column with 30 ml of Equilibration buffer.
4. Add 10 ml of pre-prepared sample to the CM-cellulose and allow to equilibrate for 1 hour at 25°C in an orbital shaker at 80 rpm.
5. After incubation, fix the column to the stand. The unbound egg white solution should be removed after the column matrix is settled at the bottom.
6. Wash the column with 45-50 ml of Equilibration buffer.
7. Take 2 ml 15 microfuge tubes and label them from 1 to 15.
8. Elute lysozyme from matrix by using 15 ml of Elution buffer. Start collecting the eluate in microfuge tubes sequentially with 2 ml fractions in each tube.
9. Take equilibration buffer to set blank. Monitor O.D. at A_{280} nm and pool the fractions that show A_{280} as 0.5 and above. Label this as eluate.
10. Wash the column with 10 ml of 1M NaCl. Replace the top and bottom caps, respectively. Store at 4°C for next use.

9.7.4 Troubleshooting Guide

1. If poor separation is observed, it could be due to the reason that column got dried up. Improper washing can also result in poor separation. Store the column in a refrigerator. Make sure it does not get dried. Wash the column with 50 ml of equilibration buffer before elution.
2. If the column is clogged, it is possibly due to microbial growth in the column. Microbial growth rarely occurs in columns during use, but steps should always be taken to prevent infection of packed columns, buffers and matrix suspensions. Store matrix in the presence of 20% ethanol or an antimicrobial agent.
3. If reduced or poor flow is observed through the column, then repacking the column may be necessary.

9.8 SIZE EXCLUSION CHROMATOGRAPHY

9.8.1 Introduction

Gel filtration chromatography or size exclusion chromatography is a very gentle method, as no harsh pH or ionic strength environments are required. One popular application of this technique is the separation of proteins from low molecular weight compounds such as peptides and amino acids. Another application is separation of biomolecules such nucleotides according to their molecular sizes.

9.8.2 Theory and Principle

In size exclusion chromatography, sample components are separated according to their sizes, which are closely related to their molecular weights. When this method is applied to molecules in non-aqueous solutions, it is referred to as gel permeation chromatography. When the technique is used for the separation of biomolecules in aqueous solutions the method is called as gel filtration chromatography.

In this technique, the chromatographic column is filled with porous material such as agarose beads with diameters ranging between 10-40 μ. The separation occurs depending upon the pore size of the gel beads. The large molecules cannot enter the pores of gel beads. Therefore they are not retained in the stationary phase and are eluted first. Smaller molecules enter the pores and have retention times depending upon the size and shape of the molecule.

Differentiation and separation occurs over a certain range of molecular sizes between 2 kDa to 200 kDa. This range can be increased to 1000 kDa by use of specialized gels. This depends on the pore size and pore size distribution in the matrix of gel.

In size exclusion chromatography, retention volumes V_R are measured instead of retention times. The total volume of separation column V_t is sum of the volume of gel particles V_g, volume of solvent inside the pores, also called intrinsic volume V_i and volume of the free solvent outside the pores or inter-particle volume V_o

$$V_t = V_g + V_i + V_o$$

All analyte components are eluted between V_o and V_o+V_i. The logarithm of the molecular weight is plotted against the retention volume V_R. Unlike other methods, in size exclusion chromatography, the mobile phase acts just as a solvent. Its physical properties do not influence the separation process. Solvent gradients do not alter the retention volume. As all compounds leave the column between V_o and V_o+V_i no sample is lost on the stationary phase. The mobile phase usually consists of an aqueous buffer with ionic strength of 50-100 mM. Typical flow rates are in the order of 0.1-1 ml/min.

9.8.3 Procedure of Size Exclusion Chromatography

Materials required

- Sephadex G100 gel beads, test tubes, 10 ml measuring cylinders, pipettes, glass rod, glasswool, glass column, spectrophotometer at 620 nm and 750 nm, cuvettes etc.
- 0.01 M phosphate buffer (pH + 7.5)
- 10mg/ml solutions of BSA,trypsin and blue dextran
- Alkaline sodium carbonate solution (I) : 2 % Na_2CO_3 in 0.1N NaOH
- Copper sulphate –sodium potassium Tartarate solution (II): 0.5% $CuSO_4$ + 1% Na-K tartarate
- Reagent (III): 50 ml (I) + 1 ml (II). Prepare fresh before use.
- Folin Ciocalteaus Reagent; Dilute the commercially available phenol reagent 1:1 with D/W. Prepare fresh.

Procedure

1. A known weight of the gel is allowed to swell in distilled water.
2. Packing of the column: The tapering end of the glass column is plugged with glass wool. The column is packed with the gel filling it with about 80 ml of the swollen gel. Care is taken to avoid incorporation of air bubbles in the gel column. The gel is allowed to settle completely and then washed with 0.01 M phosphate buffer (pH=7.5). A thin layer of the buffer is always retained on the column to avoid drying of the gel.

3. The flow rate in ml/hr of the column is determined. The flow rate is adjusted to about 35-40 ml/hour.
4. Determination of void volume (V_o): The void volume is determined using a completely excluded compound (blue dextran). Blue dextran is a high molecular weight polysaccharide covalently bound to blue dye.
5. The blue dextran is eluted from the gel using phosphate buffer and its elution volume is determined. The elution volume of blue dextran is the void volume of the column. Void volume usually constitutes about 30-35% of the volume of the column
6. Application of the sample: The sample is first dissolved in the buffer before loading. The concentrated sample is carefully pippetted onto the surface. After the sample entered the column buffer was added to facilitate elution.
7. Fractions of 2 ml each are collected and analyzed by Folin –Lowry protein estimation and measured at 750 nm. Blue dextran is a coloured compound, which can be read directly at 620 nm.

9.8.4 Troubleshooting Guide

1. Packing of the column is very critical stage. Poorly packed column will give uneven flow, peak broadening, and loss of resolution.
2. Use well degassed buffers and equilibrate all materials to the temperature at which the separation is to be performed.
3. Use freshly prepared calibration standards. Standards should be selected so that the expected molecular weight values are covered by the entire calibration range.
4. Avoid trapping of air bubbles and blockage of the column.

9.9 GAS CHROMATOGRAPHY

9.9.1 Introduction

The gas chromatography method is often used in the determination of alcohol in blood or urine. In these cases, high sensitivity is required since 0.1% blood alcohol content is considered to be legally intoxicated in most states most cases.

Gas chromatography (GC) is a powerful and widely used tool for the separation, identification and quantitation of components in a mixture. In this technique, a sample is converted to the vapor state and a flowing stream of carrier gas, which is often helium or nitrogen, sweeps the sample into a thermally-controlled column, where the sample is analyzed.

9.9.2 Theory and Principle

In the case of gas-liquid chromatography, the column is usually packed with solid particles that are coated with a non-volatile liquid, referred to as the stationary phase. As the sample mixture moves through the column, sample components that interact strongly with the stationary phase spend more time in the stationary phase vs. the moving gas phase and thus require more time to move through the column.

Retention time

The time taken for a particular compound to travel through the column to the detector is known as its retention time. This time is measured from the time at which the sample is injected to the point at which the display shows a maximum peak height for that compound. Retention time is defined as the time from injection of the sample to the time a specific sample component is detected. While polarity is usually the major factor governing separation, the boiling points of components of the sample also play a role in determining retention time. Components with higher volatility *i.e.* lower boiling points, tend to spend more time in the moving gas phase and therefore tend to have shorter retention times. After exiting the column the separated components are detected and a detector response is recorded.

Different compounds have different retention times. For a particular compound, the retention time will vary depending on:

- The boiling point of the compound. A compound which boils at a temperature higher than the column temperature will spend nearly all of its time condensed as a liquid at the beginning of the column. So high boiling point means a long retention time.
- The solubility in the liquid phase. The more soluble a compound is in the liquid phase, the less time it will spend being carried along by the gas. High solubility in the liquid phase means a high retention time.
- The temperature of the column: A higher temperature will tend to excite molecules into the gas phase either because they evaporate more readily, or because they are so energetic that the attractions of the liquid no longer hold them. A high column temperature shortens retention times for everything in the column.

The components of gas chromatography instrument are as follows:

Carrier gas

The carrier gas must be chemically inert. Commonly used gases include nitrogen, helium, argon, and carbon dioxide. The choice of carrier gas is often dependant upon the type of detector, which is used. The carrier gas system also contains a molecular sieve to remove water and other impurities.

Sample injection port

For optimum column efficiency, the sample volume should not be too large, and should be introduced onto the column at once. Slow injection of large samples causes band broadening and loss of resolution.

The most common injection method is where a microsyringe is used to inject sample through a rubber septum into a flash vaporizer port at the head of the column. The temperature of the sample port is usually about 50°C higher than the boiling point of the least volatile component of the sample.

The injector can be used in one of two modes, split or splitless. The injector contains a heated chamber containing a glass liner into which the sample is injected through the septum. The carrier gas enters the chamber and can leave by three routes when the injector is in split mode. The sample vaporizes to form a mixture of carrier gas, vaporized solvent and vaporized solutes. A proportion of this mixture passes onto the column, but most of it exits through the split outlet. The septum purge outlet prevents septum bleed components from entering the column.

Columns

There are two general types of column, packed and capillary (open tubular). Packed columns contain a finely divided, inert, solid support material commonly based on diatomaceous earth, coated with liquid stationary phase. Most packed columns are 1.5–10 m in length and have an internal diameter of 2–4 mm.

Wall-coated columns consist of a capillary tube whose walls are coated with liquid stationary phase. In support-coated columns, the inner wall of the capillary is lined with a thin layer of support material such as diatomaceous earth, onto which the stationary phase has been adsorbed.

Column temperature

For precise work, column temperature must be controlled carefully. The optimum column temperature is dependant upon the boiling point of the sample. As a rule of thumb, a temperature slightly above the average boiling point of the sample results in an elution time of 2-30 minutes. Minimal temperatures give good resolution, but increase elution times. If the sample components have a wide boiling range, then temperature programming can be useful. The column temperature is increased either continuously or in steps as separation proceeds.

Detectors

There are many detectors, which can be used in gas chromatography. Different detectors will give different types of selectivity. A non-selective detector responds to all compounds except the carrier gas, a selective detector responds to a range of compounds with a common physical or chemical property and a specific detector responds to a single chemical compound. Detectors can also be grouped into concentration dependant detectors and mass flow dependant detectors. The signal from a concentration dependant detector is related to the concentration of solute in the detector, and does not usually destroy the sample. Dilution of sample with make-up gas will lower the detectors response. Mass flow dependant detectors usually destroy the sample, and the signal is related to the rate at which solute molecules enter the detector. The response of a mass flow dependant detector is unaffected by make-up gas.

During the process of gas chromatography, the effluent from the column is mixed with hydrogen and air, and ignited. Organic compounds burning in the flame produce ions and electrons which can conduct electricity through the flame. A large electrical potential is applied at the burner tip, and a collector electrode is located above the flame. The current resulting from the pyrolysis of any organic compounds is measured. Flame ionization detectors are mass sensitive rather than concentration sensitive; this gives the advantage that changes in mobile phase flow rate do not affect the detector's response.

The FID is a useful general detector for the analysis of organic compounds. It has high sensitivity, a large linear response range, and low noise. It is also robust and easy to use, but unfortunately, it destroys the sample.

9.9.3 Procedure of Gas Chromatography

The process of GC is in fact automated. The temperature and other parameters are set and then the sample is injected in the GC instrument. The instrument analyzes sample and produces chromatogram.

9.10 HIGH PRESSURE LIQUID CHROMATOGRAPHY

9.10.1 Introduction

High performance (or pressure) liquid chromatography is basically a highly improved form of column chromatography. Instead of a solvent being allowed to drip through a column under gravity, it is forced through under high pressures of up to 400 atmospheres. This makes the process faster, and allows use of very much smaller particle size for the column packing material, which gives a much greater surface area for interactions between the stationary phase and the molecules flowing past it. This allows a much better separation of the components of the mixture. This method is highly automated and extremely sensitive.

HPLC has a high degree of versatility not found in other chromatographic systems and it has the ability to easily separate a wide variety of chemical mixtures.

9.10.2 Theory and Principle

HPLC utilizes a liquid mobile phase to separate the components of a mixture. The stationary phase can be a liquid or a solid phase. These components are first dissolved in a solvent, and then forced to flow through a chromatographic column under a high pressure. In the column, the mixture separates into its components. The amount of resolution is important, and is dependent upon the extent of interaction between the solute components and the stationary phase.

In HPLC chromatography, separation columns are tightly packed with small particles of 1-5 μ in diameter. To achieve ambient flow rates in these columns, high pressure of up to 300-400 bars needs to be generated. The computer controlled pumps force the mobile phase through the system.

HPLC is a dynamic adsorption process. Analyte molecules, while moving through the porous packing beads, tend to interact with the surface adsorption sites. Depending on the HPLC mode, the different types of the adsorption forces may be included in the retention process.

Retention time

The time taken for a particular compound to travel through the column to the detector is known as its retention time. This time is measured from the time at which the sample is injected to the point at which the display shows a maximum peak height for that compound.

Different compounds have different retention times. For a particular compound, the retention time will vary depending on:

- The pressure used because that affects the flow rate of the solvent
- The nature of the stationary phase (not only what material it is made of, but also particle size)
- The exact composition of the solvent
- The temperature of the column

This means that conditions have to be carefully controlled if you are using retention times as a way of identifying compounds.

Types of HPLC

There are many ways to classify HPLC. If this classification is based on the nature of the stationary phase and the separation process, following modes can be specified.

- *Adsorption chromatography*: In this, the stationary phase is an adsorbent like silica gel or any other silica based packing and the separation is based on repeated adsorption-desorption steps.
- *Ion-exchange chromatography*: The stationary bed has charged surface of opposite charge to the sample ions. This technique is used almost exclusively with ionic or ionizable samples. The stronger the charge on the sample, the stronger it will be attracted to the ionic surface and thus, the longer it will take to elute. The mobile phase is an aqueous buffer, where both pH and ionic strength are used to control elution time. This technique is well suited for the separation of inorganic and organic anions and cations in aqueous solutions. Ionic dyes, amino acids, and proteins can be separated by ion exchange HPLC.
- *Size exclusion chromatography*: In this type, the column is filled with material having precisely controlled pore sizes, and the sample is simply screened or filtered according to its solvated molecular size. Larger molecules are rapidly washed through the column; smaller molecules penetrate inside the porous of the packing particles and elute later. This technique is also called gel filtration or gel permeation chromatography.

Concerning the first type, two modes are defined depending on the relative polarity of the two phases: normal and reversed-phase chromatography.

- In normal phase chromatography, the stationary bed is strongly polar in nature *e.g.* silica gel and the mobile phase is nonpolar such as n-hexane, iso-ocatane, ethyl acetate etc. Polar samples are thus retained on the polar surface of the column packing for longer than less polar materials. This technique is useful for water-sensitive compounds, geometric isomers, cis-trans isomers and chiral compounds.
- Reversed-phase chromatography is the inverse of this. The stationary bed is nonpolar in nature, while the mobile phase is a polar liquid, such as mixtures of water and methanol or acetonitrile. Here the more nonpolar the material is, the longer it will be retained. This technique can be used for non-polar, polar, ionizable and non-ionizable molecules. Reverse phase chromatography is used for almost 90% of all chromatographic applications.

There are many detection systems used to detect the compounds eluting from an HPLC column. The most common are:

- Spectroscopic Detection
- Refractive Index Detection
- Fluorescence Detection

HPLC instrument

HPLC instrumentation includes a pump, injector, column, detector and data system. The heart of the system is the column where separation occurs. Since the stationary phase is composed of micrometre size porous particles, a high pressure pump is required to move the mobile phase through the column. The chromatographic process begins by injecting the solute onto the top of the column. Separation of components occurs as the analytes and mobile phase are pumped through the column. Eventually, each component elutes from the column as a narrow band (or peak) on the recorder.

Mobile phase reservoir

The most common type of solvent reservoir is a glass bottle. Most of the manufacturers supply these bottles with special caps, Teflon tubing and filters to connect to the pump inlet and to the helium gas that is used to remove dissolved air. Helium purging and storage of the solvent under helium is not sufficient for degassing aqueous solvents. It is useful to apply a vacuum for 5-10 min. and then keep the solvent under a helium atmosphere.

Pump

High pressure pumps are needed to force solvents through packed stationary phase beds. Smaller bed particles require higher pressures. There are many advantages of using smaller particles, but they may not be essential for all separations. The most important advantages are: higher resolution, faster analyses, and increased sample load capacity. However, only the most demanding separations require these advances in significant amounts. Many separation problems can be resolved with larger particle packings that require less pressure.

Injector

Sample introduction can be accomplished in various ways. The simplest method is to use an injection valve. In more sophisticated LC systems, automatic sampling devices are incorporated where the sample is introduced with the help of autosamplers and microprocessors. In liquid chromatography, liquid samples may be injected directly and solid samples need only be dissolved in an appropriate solvent. The solvent need not be the mobile phase, but frequently it is chosen to avoid detector interference, column/component interference, loss in efficiency or all of these.

It is always best to remove particles from the sample by filtering over a 5 μm filter, or centrifuging, since continuous injections of particulate material will eventually cause blockages in injection devices or columns.

Chromatographic column

Typical HPLC columns are 5, 10, 15 and 25 cm in length and are filled with small diameter particles of 3, 5 or 10 μm. The internal diameter of the columns is usually 4.6 mm; this is considered the best option for sample capacity, mobile phase consumption, speed and resolution. However, if pure substances are to be collected on the preparative scale, then larger diameter columns may be needed.

Detector

Today, optical detectors are used most frequently in liquid chromatographic systems. These detectors pass a beam of light through the flowing column effluent as it passes through a low volume (~10 μl) flow cell. The variations in light intensity caused by UV absorption, fluorescence emission or change in refractive index, from the sample components passing through the cell, are monitored as changes in the output voltage. These voltage changes are recorded on a strip chart recorder and frequently are fed into a computer to provide retention time and peak area data. The most commonly used detector in HPLC is the ultraviolet absorption detector. A variable wavelength detector of this type, capable of monitoring from 190 to 400 nm, is suitable for the detection of the majority samples. Other detectors in common use include: Photo Diode Array UV detector (PAD), refractive index (RI), fluorescence (FLU), electrochemical (EC). The RI detector is universal but is also the less sensitive one. FLU and EC detectors are quite sensitive.

Data system

Since the detector signal is electronic, using modern data collection techniques can aid the signal analysis. In addition, some systems can store data in a retrievable form for highly sophisticated computer analysis at a later time. The main goal in using electronic data systems is to increase analysis accuracy and precision, while reducing operator attention. There are several types of data systems, each differing in terms of available features.

9.10.3 Procedure of HPLC Chromatography

The following procedure is for the determination of caffeine content of beverages by using HPLC. A series of caffeine standards that cover the range of the unknown sample concentration will be measured to construct a calibration curve. A comparison of the caffeine peak area in the soft drink sample compared to those for the standards permits a quantitative determination of the caffeine content.

Materials Required

Equipment

- HPLC chromatography instrument with UV-VIS detector.
- Tandem reversed-phase cartridge columns, Pecosphere, 4.6 × 33 mm, 3 m particles coated with octadecylsilane groups.
- A 25-µl flat nose syringe.

Solutions

- Methanol/water mobile phase for elution (filtered)
- Caffeine standard solution (1000 µg/ml, filtered)
- Unknown caffeinated beverage (de-gassed and filtered)
- Caffeinated beverage of choice

Procedure

Calibration of standards

1. Make a series of dilutions to prepare 4 calibration standards with concentrations from 20 to 100 µg/ml starting with the 1000 µg/ml caffeine standard.
2. Use only 10 ml and 25 ml volumetric flasks and HPLC grade water for the dilutions. Store each solution tightly capped in a labeled 15 ml vial.

Preparation of caffeinated beverage

1. Obtain a sample for which you wish to measure the caffeine content. If you choose a carbonated beverage, your sample must be degassed before injection.
2. Now dilute the sample so that its concentration is in the range of the standards. Filter a few millilitres of the diluted sample using a syringe and submicron filter disk.

Analysis of standards

1. Use the software to adjust components of HPLC system. This will run the system with the methanol/water solvent at a flow rate of 0.80 ml/min and a total run time of 2.5 minutes.

2. Now start the software. When the system gives message for injection of the samples, load the injection port with at least 3-4 syringe volumes of the highest concentration caffeine standard. When the injection loop (20 µl) is properly flushed and filled, the sample can be injected.
3. Run the HPLC analysis.
4. When the run is finished, confirm that the baseline drawn for the caffeine peak is appropriate, and, if necessary, draw a manual baseline for the caffeine peak before printing out the results. Print out a copy of the HPLC results.
5. Note the retention time (t_R) and peak area for caffeine.
6. Repeat the above procedure for other standards.

Analysis of unknown beverage

1. The analysis of unknown sample is done in the same way as analysis of standards.
2. On the chromatogram display locate the caffeine peak. Verify that is it adequately resolved from the other peaks and that its peak area is in the range of the values found for your standards.
3. The presence of other peaks in the sample chromatogram affects the baseline of the caffeine peak. Draw a manual baseline, as you did for the standard peaks, for the caffeine peak in each chromatogram of the unknown solution and beverage before printing out the results.
4. Make two additional injections of the unknown solution and print out the chromatogram and report for each run.

Preparation of calibration curve

1. Prepare a plot of peak area versus caffeine concentration using the data obtained for your standard solutions.
2. Take into account the actual volume delivered by your pipette when determining the standard concentrations.

Analysis of unknown and caffeinated beverage

From the equation for the least-squares line and the peak area for each unknown and beverage analysis, calculate the caffeine concentration (µg/ml) in your samples.

9.10.4 Troubleshooting Guide

1. Temperature control in HPLC system is crucial because Retention in HPLC is temperature-dependent. If temperature varies, then it is

difficult to assign peaks to specific compounds in the chromatogram and the peak areas/heights may vary.

2. Certain chemical compounds may have low solubility in the HPLC mobile phase. If they are injected into the flow stream they may precipitate or other difficulties may arise.
3. Certain chemical compounds, especially biological compounds such as enzymes or proteins, may not be stable at room temperature or higher.

9.11 REVIEW QUESTIONS

1. What is chromatography? Describe the principle underlaying chromatography.
2. What is paper chromatography? Describe the process of paper chromatography.
3. Compare paper chromatography with thin layer chromatography.
4. What is Rf value? Describe in brief.
5. Give principle of affinity chromatography.
6. Describe various ligands that are used in affinity chromatography.
7. Describe the principle of ion-exchange chromatography.
8. What are the applications of ion-exchange chromatography?
9. Describe principle and theory of gas chromatography.
10. Describe different components of gas chromatography system.
11. What is size exclusion chromatography? Describe the procedure of size-exclusion chromatography.
12. What is HPLC? How does it differ from other forms of chromatography?
13. Describe different HPLC systems.
14. Describe different components of HPLC instrument.

9.12 REFERENCES

Ahmed, S., Sheraz, M., Aminuddin, M., Ahmed, I., Mahmood, K. and Iqbal, K. (2010). Analysis of amino acids by paper chromatography. *International Journal of Chemical and Analytical Science,* **1(8):** 174–176.

Bailon, P., Ehrlich, J., Fung, W. and Berthold, W. (2000). Affinity chromatography: Methods and Protocols. Humana Press.

Cochin, J. and Daly, J. (1963). The use of thin layer chromatography for the analysis of drugs. Isolation and identification of barbiturates and nonbarbiturates hypnotics from urine, blood and tissues. *Journal of Pharmacological and experimental therapeutics,* **139(2):** 154–159.

Hossu, A., Maria, M., Radulescu, C., Ilie, M. and Mageaaru, V. (2009). TLC applications on separation and quantification of fat soluble vitamins. *Romanian Biotechnological Letters,* **14(5):** 4615–4619.

Mayer, C. *et al.* (2009). Size exclusion chromatography can identify faster-associating protein complexes and evaluate design strategies. *BMC Research Notes,* **2:** 135.

McNair, H. and Miller, J. (2011). Basic Gas Chromatography. 2nd Edition. John Wiley and Sons, Inc.

Miller, J. (2005). Chromatography: Concepts and Contrasts. 2nd Edition. John Wiley and Sons, Inc.

Rutter, L. (1949). Streaming potential in paper chromatography. *Nature,* **163:** 487.

Sherma, J. (1986). Thin layer and paper chromatography. *Anal. Chem.,* **58(5):** 69–81.

Snyder, L., Kirkland, J. and Dolan, J. (2011). Introduction to Modern Liquid Chromatography. 3rd Edition. John Wiley and Sons, Inc.

Vromen, H. and Cohen, C. (1967). Separation of sterol acetates by column and thin layer argentation chromatography. *Journal of Lipid Research,* **8:** 150–152.

10

Polymerase Chain Reaction (PCR)

10.1 INTRODUCTION

PCR is a technique that is used for *in vitro* amplification of DNA or RNA of an organism or gene defect without the requirement of culturing or physically multiplying the organism. PCR takes advantage of an enzyme that is known as thermostable Taq Polymerase.

The PCR, like recombinant DNA technology, had an enormous impact on both basic and diagnostic aspects of molecular biology because it can produce large amounts of a specific DNA fragment from small amounts of a complex template. PCR represents a form of *in vitro* cloning that can generate, as well as modify, DNA fragments of defined length and sequence in a simple automated reaction.

Kary Mullis conceived the method of PCR in 1983, at the side of a California mountain road while he was resting. This technique was first presented in 1986 at the 51st Cold Spring Harbour Laboratory Symposium. In 1993, Kary Mullis was awarded the Nobel Prize in Chemistry for his invention of PCR. With PCR, minute quantities of nucleic acids can be analyzed. Even a single DNA molecule is enough for amplification.

Limitations of PCR

PCR is an extremely sensitive technique but is prone to contamination from extraneous DNA, leading to false positive results. Another potential problem is due to cross-contamination between samples. It is for this reason that sample preparation, running PCR and post-amplification detection must be carried out in separate rooms. Concentration of Mg is very crucial as low Mg^{+2} leads to low yields (or no yield) and high Mg^{+2} leads to accumulation of nonspecific products. Non-specific binding of primers and primer-primer dimmer formation are other possible reasons for unexpected results. Also, reagents and equipments are costly, hence can't be afforded by small laboratories.

10.2 THEORY AND PRINCIPLE

The polymerase chain reaction (PCR) is capable of synthesizing millions of copies of a specific DNA sequence in a simple, rapid and automated reaction. PCR is *in vitro* enzymatic synthesis of specific DNA sequences. In this technique, oligonucleotide primers are used commonly. Twenty cycles of PCR assuming 100% efficiency would give about a million fold of DNA amplification.

The PCR process requires a repetitive series of the three fundamental steps that defines one PCR cycle:

First, denaturation of double-stranded DNA template occurs. The DNA sequence which is to be amplified by PCR, is known as the template. The double-stranded DNA template must be denatured into two complementary single strands of DNA before the reaction can commence. DNA undergoes rapid denaturation at 94°C. Five minutes are allowed for each denaturation step in a 30 cycle PCR, the enzyme will be subjected to 94°C for a total time of 2.5 hours. In practice, 30 seconds to 2 minutes are allowed for the denaturation step, with 1 minute being the optimum choice for most templates.

Annealing of two oligonucleotide primers to the single-stranded template is the second step. After denaturation, reaction mixture is quickly cooled, preventing immediate reannealing of long DNA strands. Due to their small size, oligonucleotides rapidly anneal to the single strands of DNA at positions containing specified template sequence. In these positions, they act as primers for Taq polymerase. Formation of the specific primer-template complex is highly temperature dependent, so for annealing to take place, the temperature of the reaction must be lowered to a preset level calculated to maximize primer-template interaction.

Enzymatic extension of the primers to produce copies can serve as templates in subsequent cycles. This is also known as polymerization, this is the final step of the PCR cycle in which the temperature of the reaction is adjusted to the optimum for Taq polymerase activity, which is between 75°C and 80°C. During this step, the polymerase enzyme incorporates nucleotides into the nascent DNA strand, producing a complementary copy of the DNA template in the region specified by the annealed primer. The new temperature is invariably above that at which annealing occurs, but does not lead to denaturation of primer-template complex, presumably because the enzyme is already active at the annealing temperature and significantly increases primer length during the annealing step, thus raising its denaturation temperature above that of the polymerization step.

10.3 PRE-REQUISITES OF PCR- ENZYMES, dNTPS, PRIMERS

Enzymes

DNA polymerases are enzymes that can synthesize a new strand of DNA from deoxyribonucleotide triphosphates using a template DNA strand and a primer with free 3' –OH group for initiation of polymerization. In the initial stages, PCR was done using Klenow Fragment of *E. coli* DNA polymerase I. However this enzyme has temperature limitation that produced hindrance in the PCR reaction. The Klenow fragment is active only at 37°C at which it carries out polymerization process. However this enzyme is inactivated at high temperatures that are required to melt the template DNA strands. This thermolability of Klenow Fragment requires the enzyme to be added during each cycle of PCR. This increased the cost of experiment tremendously as well as gave poor results due to repeated opening of PCR tubes, resulting in temperature fluctuations.

The introduction of thermostable Taq DNA polymerase in the PCR eliminated need for enzyme addition after denaturation. Taq Polymerase also simply termed as 'Taq', is a thermostable DNA polymerase used in polymerase chain reaction (PCR) to catalyze the DNA replication reaction in the PCR cycle. Taq was first isolated from *Thermus aquaticus*, a bacterium that lives in hot springs and hydrothermal vents. Taq was the first polymerase that was able to withstand the denaturing conditions (over 90°C) required during PCR cycling.

One of Taq polymerases' major disadvantages is its low replication fidelity. This is because Taq does not have 3' to 5' exonuclease proof reading mechanism to replace an accidental mismatch in the newly synthesized DNA strand.

Today many thermostable DNA polymerases other than Taq polymerase have been identified in various other organisms (extremophiles) that retain polymerase activity high temperatures. These other modern polymerases include Tth, Pwo and Pfu. Out of these Pwo and Pfu exhibit proof reading activity in addition to polymerase activity. This proof reading activity improves the fidelity of reaction.

Primers

All DNA polymerases are primer dependent *i.e.* they require a primer to initiate new DNA strand synthesis using a template. To make PCR a specific, efficient and cost effective tool for researchers and clinicians the most important aspect is oligonucleotide primer. The standard primer concentration is 100-900 nM. Too much concentration of primers can increase the formation of nonspecific product.

Length of a primer is a critical parameter. The thumb rule is to use a primer with a minimal length that ensures a denaturation at temperature of 55-56°C. This greatly enhances specificity and efficiency. For general studies, primers of typically 17-34 nucleotides in length are the best. Primer >16 nucleotides in length are not generally annealed specifically to non-target DNA sequence. For example, a short primer sequence, such as a 12 bp oligonucleotide, binds to 200 specific annealing sites in the human genome.

In contrast, a 20-mer sequence is expected to randomly exist only once every 420 nucleotides and as such, has only a 1 in 400 probability of existing by chance in the human genome. Primers, 18-24 mers are accepted as best in being sequence specific if the annealing temperature of the PCR reactions is set within 5°C of the primer T_d (dissociation temperature of the primer/template duplex). Primers work exceptionally well for the sequence with least intra-strand secondary structure. This is because secondary structure impedes primer annealing and extension.

Longer primers are required only to discriminate homologous genes within different species or when a perfect complementary sequence to all the templates is not expected. They could also be used when extra sequence information *e.g.* a motif-binding site, restriction endonuclease site or GC clamp is attached to 5' end. Such extensions do not generally alter annealing to the sequence specific portion of the primer.

Although the following formula is generally used for determining melting temperature (Tm):

$$Tm = 4\,(G+C) + 2(A+T)$$

Primers are about 10-30 nucleotides long that share sequences with template DNA. The primers are chosen such that they hybridize with 3' ends of template DNA. Two distinctive primers are used for PCR amplification- forward and reverse primer. Each of the primer hybridizes to one of the two strands of double stranded DNA molecule that is to be amplified. DNA synthesis always proceeds in the direction from 5' to 3'. Within the primer there should not be complementary sequences that might lead to primer dimers. For example, the primer with sequence 5'–CCAAATTTGG-3' can lead to pairing of bases even in the single stranded conformation due to the presence of complementary sequences within the primer. This leads to problems in primer annealing to the template DNA and results in poor amplification. While designing primers, unusual sequences such as long stretches of polypurines or polypyrimidines and repetitive sequences should be avoided.

Deoxyribunucleotide Triphosphates (dNTPs)

The four nucleotide triphosphates, dATP, dCTP, dGTP and dTTP are the building bocks of DNA that are required for the synthesis of new DNA strand. The reaction must contain excess of dNTPs as they are depleted during DNA synthesis. It is important that concentrations of all dNTPs should be equal. Too much nucleotide concentration reduces specificity by increasing the error rate of the polymerase and also chelates magnesium ions changing the effective magnesium ion concentration. The standard concentration of each nucleotide in the final reaction mixture is ~200 µM for synthesizing 12.5 µg of DNA when half of the dNTPs are incorporated.

PCR Buffer

The buffer pH and ionic strength is chosen according to the polymerase used. The ionic strength is crucial in influencing the specificity of PCR. The buffer impacts the amplification of target DNA by maintaining the pH range, minimizing the effect of inhibitors, protecting enzymes from premature loss of activity, and stabilizing template. As polymerase enzymes have a narrow optimum pH range, a slight fluctuation in the pH will reduce the PCR yield.

Magnesium Chloride

The Mg^{+2} ions form a soluble complex with DNA and polymerase enzyme. Their role is to bring the enzyme and DNA in close proximity and to balance the negative charge on DNA molecules. They also stimulate polymerase activity.

The concentration of magnesium ions is related to both specificity and yield of PCR reaction. At low magnesium ion concentration, the enzymatic activity of polymerase enzyme decreases. However, excess of magnesium ions results in poor denaturation of DNA as these ions stabilize the DNA structure.

Additives

A number of additives are used in PCR buffer to stabilize the enzyme or for better primer annealing. The commonly used additives are BSA, gelatin, glycerin, DMSO (Dimethyl sulfoxide), PEG etc.

10.4 WHAT IS PRIMER? COMPARING PRIMER, PROBE AND GENE

A primer is simply a strand of nucleic acid that serves as a starting point for DNA synthesis. Primer is required for DNA replication *in vivo* because the enzymes that catalyze this process known as DNA polymerases, can only add new nucleotides to an existing strand of DNA. The PCR primers on the other hand are short chemically synthesized oligonucleotides having about 10-20 basepairs. PCR primers are invariably DNA primers.

The probe, in molecular biology is commonly referred as hybridization probe. A probe is a fragment of DNA or RNA, 100-1000 basepairs long. Probes are used to hybridize with the complementary sequences on the sample DNA for detection of those sequences. To detect hybridization of the probe to its complementary sequence, probes are usually labelled with radioactive or fluorescent molecules. The probes can be synthesized chemically or multiplied and labelled in PCR.

A gene is simply the unit of heredity. Gene is a DNA molecule that is capable of producing RNA transcripts by the process of transcription. Gene is normally very large molecule as compared to primers or probes.

In case of genetic engineering experiments, the gene of interest is inserted into the target cells. To detect the successful incorporation of gene in the genome of the target organism, that particular gene can be used as a primer in PCR reaction. Successful primer annealing and amplification indicates presence of the gene of interest in the target organism. Alternatively, the same gene can be used in hybridization experiments like southern hybridization. The successful hybridization of probe with sample DNA and detection of hybridized probe by radioactive or non-radioactive means indicates presence of gene of interest.

10.5 COMPARISON OF PCR REACTION WITH *IN VIVO* DNA REPLICATION PROCESS

The aim of both PCR and *in vivo* DNA replication process is the amplification of DNA. However, this amplification is a controlled process inside cells where various enzymes govern the number of copies of DNA that are to be produced. The PCR reaction on the other hand is an uncontrolled process that continues to amplify DNA until the dNTPs are exhausted.

The main difference between the PCR and DNA replication is that in case of PCR, heat is applied for melting of template strand and in DNA replication parental DNA duplex is melted by enzyme helicase. For this reason, in PCR, thermostable enzymes are required. Today, most commonly used enzyme in PCR is Taq DNA polymerase.

Another difference between the two processes is that in PCR, DNA primer is used for the initiation of DNA amplification, while RNA primer is used for initiation of DNA replication process *in vivo*.

Apart from these differences, both the processes require a pool of free dNTPs for the synthesis of new strand. The PCR reaction is inspired from *in vivo* replication process. Therefore PCR can be said to be DNA replication process *in vitro*.

10.6 APPLICATIONS OF PCR

PCR has been applied to a large number of medical procedures. The first application of PCR can be said for genetic testing where a sample of DNA is analyzed for the presence of genetic disorder or mutations. Prospective parents can be tested for being carriers of genetic disorders or their children might be tested for actually being affected by a particular disease. PCR can also be used as a part of sensitive test for tissue typing that is vital to organ transplantations. PCR technique is also used in cancer research. Many forms of cancers involve alterations to oncogenes. By using PCR-based tests to study these mutations, therapy regimens can sometimes be individually customized to a patient.

Characterization and detection of infectious disease organisms have been revolutionized by PCR. The Human Immunodeficiency Virus (HIV), responsible for AIDS, is difficult to diagnose and eradicate. The earliest tests for infection relied on the presence of antibodies to the virus circulating in the bloodstream. However, antibodies don't appear until many weeks after infection, maternal antibodies mask the infection of a newborn, and therapeutic agents to fight the infection don't affect the antibodies. PCR tests have been developed that can detect as little as one viral genome among the DNA of over 50,000 host cells. It can, for example, detect the AIDS virus sooner during the first few weeks after infection than the standard ELISA test. With the help of PCR, infections can be detected earlier, donated blood can be screened directly for the virus, newborns can be immediately tested for infection, and the effects of antiviral treatments can be quantified.

Some disease organisms, such as tuberculosis pathogen, are difficult to sample from patients and slow to be grown in the laboratory. PCR-based tests have allowed detection of small numbers of disease organisms either live or dead, in convenient samples. Detailed genetic analysis can also be used to detect antibiotic resistance, allowing immediate and effective therapy. The effects of therapy can also be immediately evaluated.

The development of PCR-based DNA fingerprinting protocols has been a widespread application in forensic sciences. DNA fingerprinting can uniquely discriminate any one person from the entire population of the world. Minute samples of DNA can be isolated from a crime scene, and compared to that from suspects, or from a genetic database of earlier evidence or convicts. Simpler versions of these tests are often used to rapidly rule out suspects during a criminal investigation. Evidence from decades-old crimes can be tested, confirming the people originally convicted.

DNA fingerprinting can also help in parental testing, where an individual is matched with their close relatives. DNA from unidentified human remains can be tested, and compared with that from possible parents, siblings, or children. Similar testing can be used to confirm the biological parents of an adopted or kidnapped child.

PCR has been applied to many areas of research in molecular biology and genetics. PCR allows rapid production of short pieces of DNA, even when nothing more than the sequence of the two primers is known. This ability of PCR facilitates many methods, such as generating hybridization probes for southern or northern hybridization. PCR supplies these techniques with large amounts of pure DNA, sometimes as a single strand, enabling analysis even from very small amounts of starting material.

The task of DNA sequencing can also be assisted by PCR. PCR has numerous applications to the more traditional process of DNA cloning. It can extract segments for insertion into a vector from a larger genome, which may be only available in small quantities. Using a single set of vector primers, it can also analyze or extract fragments that have already been inserted into vectors.

Another application of PCR is the phylogenic analysis of DNA from ancient artifacts, such as that found in the recovered bones of pre-historic man, or from frozen tissues of extinct animals like mammoths. In some cases the highly degraded DNA from these sources might be reassembled during the early stages of amplification.

A common application of PCR is the study of patterns of gene expression. Tissues or individual cells can be analyzed at different stages to see which genes have become active, or which have been switched off.

PCR can also be used to detect chromosomal aberrations. Similarly, unusual deletions, insertions, translocations, or inversions can be analyzed.

PCR is also important in answering basic scientific questions. In the field of evolutionary biology, PCR has been used to establish relationships among species. In anthropology, it has been used to understand ancient human migration patterns.

In archaeology, it has been used to help identify ancient human remains. Paleontologists have used PCR to amplify DNA from extinct insects preserved in amber for 20 million years. The Human Genome Project, which had a goal of determining the sequence of the 3 billion base pairs in the human genome, relied heavily on PCR. The genes responsible for a variety of human diseases have been identified using PCR.

By amplifying and analyzing DNA from bone and mummified soft tissue, scientists have also found that moas, a group of large New Zealand birds that were hunted to extinction, are not related to the still-extant New Zealand kiwi, despite the fact that both bird species could not fly. Leaping far back in time, researchers have suggested, however, that termites imprisoned in amber 40 million years ago differed little from the termites of today.

PCR is the most sensitive and specific test for *Helicobacter pylori*, the disease organism now known to cause almost all stomach ulcers. Unlike previous tests, PCR can detect three different sexually transmitted disease organisms on a single swab namely, herpes, papillomaviruses, and chlamydia and can even distinguish the particular strain of papillomavirus that predisposes to cancer, while other tests cannot.

With PCR analysis of cells shed into feces, for example, doctors have already demonstrated premalignant changes in the gastrointestinal tract, such as mutations in genes that protect against tumors. This can help them select high-risk candidates for colon cancer tests. Researchers have also detected potentially metastatic cells in the circulation of patients with newly diagnosed tumors.

DNA sequencing reveals crucial variations in the nucleotides that constitute genes. These mutational changes produce disease and even death by forcing the genes to produce abnormal proteins, or sometimes no proteins at all. DNA sequencing involves first isolating and duplicating DNA segments for nucleotide analysis. Thus PCR is an essential tool for the Human Genome Project because it can quickly and easily generate an unlimited amount of any piece of DNA for this kind of study.

Researchers have used the technique to aid in reducing illegal trade in endangered species, and products made from them. Because PCR is a relatively low-cost and portable technology, it is adaptable for field studies of all kinds in the developing countries. It is also a tool for monitoring the release of genetically engineered organisms into the environment.

10.7 NATIVE PCR

10.7.1 Introduction

Native PCR is the basic form of PCR that is very much the same that was invented by Kary Mullis. PCR is a DNA photocopier in a test tube. PCR is really controlled by heating and cooling of reaction mixtures at specific temperatures.

10.7.2 Principle and Working

The amplification of nucleic acids is done in a single PCR tube, placed in the thermal cycler or PCR machine. This PCR tube contains primer, dNTPs, enzyme that is usually Taq polymerase, sample to be amplified and PCR buffer. In addition to these, a magnesium ion source such as $MgCl_2$ is added to assist the activity of enzyme. The crucial factor of the PCR reaction is temperature control. The reaction takes place in three basic steps (Fig. 10.1),

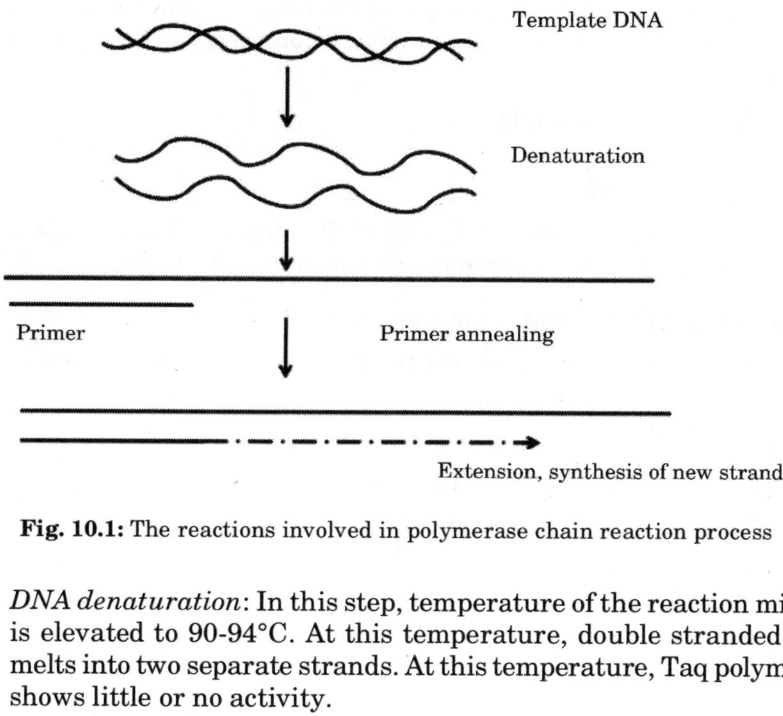

Fig. 10.1: The reactions involved in polymerase chain reaction process

1. *DNA denaturation*: In this step, temperature of the reaction mixture is elevated to 90-94°C. At this temperature, double stranded DNA melts into two separate strands. At this temperature, Taq polymerase shows little or no activity.
2. *Primer annealing*: In this step, temperature is lowered to 50-60°C. At this temperature, due to excess of primer molecules, primers anneal to the complementary sequences on template single stranded DNA. This happens before the two separated DNA strands are able to hybridize again.
3. *Primer extension*: At this stage, actual synthesis of DNA takes place. The synthesis is catalyzed by DNA polymerase by continuous incorporation of complementary nucleotides into the chain. The primer extension occurs from 3' to 5' end of template strand.

These three stages form one cycle of PCR. About 20 to 30 cycles of PCR are usually sufficient to amplify sample DNA in most of the applications. This process of repeated heating and cooling of the reaction mixture is called as thermal cycling.

On excessive cycling, DNA synthesis will become less sufficient as primers and dNTPs are used up. Therefore, after the last cycle, the enzyme is allowed to finish any incomplete synthesis by the final extension cycle of 10-15 minutes at 72°C. After completion of the program, the samples are held at 4°C to preserve the reaction products, especially during overnight runs.

In practice, PCR conditions have to be optimized for each type of template DNA and primer combinations used. The factors that influence the outcome of PCR reaction are annealing temperature, template concentration, primer concentration, $MgCl_2$ concentration, extension time and number of cycles.

10.7.3 Procedure of PCR

Materials required

Usually the reagents such as primers, PCR buffer and Taq enzyme etc. are directly procured from commercial manufacturers by buying a PCR kit.

- Purified DNA samples to be amplified
- Oligonucleotide primers- upstream and downstream primers
- Sterile PCR grade water
- 10x PCR buffer

Tris-HCl (pH=8.3)	100 mM
KCl	500 mM
$MgCl_2$	15 mM
Gelatin	0.1% (w/v)

- dNTP stock 10 mM (working 1.25 mM)
- Light mineral oil
- Ammonium acetate 7.5 M (store at room temperature)
- Ethanol 100%
- Taq DNA polymerase (store at 4°C)

Procedure

1. Prepare a master mix of PCR reagents containing,

Sterile water	172.50 μl
PCR buffer	25.00 μl
$MgCl_2$ stock	5.00 μl

dNTPs	10.00 µl
Forward primer	3.75 µl
Reverse primer	3.75 µl
Taq polymerase	2.50 µl

2. Mix the reagents gently and dispense 23 µl of reaction mixture in each of the sterile PCR tubes.
3. Add 2 µl of purified DNA sample to each of the tube to make final volume of 25 µl.
4. On top of the reaction mixture, add few drops of light mineral oil to prevent evaporation.
5. Perform the PCR reaction as given in Table 10.1

Table 10.1: Reactions of PCR with specified temperatures and time

Stage	*Temperature*	*Time*	*Cycles*
Initial denaturation	94°C	5 minutes	1
Denaturation	94°C	30 seconds	–
Annealing	55°C	30 seconds	–
Extension	72°C	1 minute	40
Final extension Hold at 4°C	72°C	10 minutes	1

6. After completion of PCR reaction, remove the samples and add 50 µl of chloroform to each tube.
7. Due to addition of chloroform, mineral oil sinks to the bottom of the tube. Without disturbing the tube, place tubes in centrifuge. Spin the tubes at top speed for about 30 seconds.
8. Transfer the upper phase to clean microfuge tubes. Add 15 µl of ammonium acetate. Mix well. Ammonium acetate removes primers and unincorporated nucleotides.
9. Add 100 µl of 100% ethanol and mix the contents. Keep on ice for 30 minutes.
10. Centrifuge at top speed in tabletop microcentrifuge 15 minutes.
11. Decant the supernatant. Air-dry the pellet. Dissolve in sterile water and store at –20°C till further use.
12. The results can be seen by performing agarose gel electrophoresis of the PCR reaction products.

10.7.4 Troubleshooting Guide

1. Gloves should always be worn while handling PCR reagents.
2. If there is low yield of PCR products, it could be due to low template quality or quantity, suboptimal thermal conditions, and improper primer.

3. Non-specific reaction products may result from reaction set-up at room temperature, or suboptimal reaction conditions or suboptimal thermal cycling conditions. Set-up the reaction on ice.
4. Non-specific amplification can occur from overabundance of primer or template. In this case, decrease primer/template concentration.

10.8 RT- PCR

10.8.1 Introduction

The RT-PCR technique involves amplification via reverse transcription polymerase chain reaction in which a very low copy number of RNA molecules can be detected. RT-PCR is widely used in the diagnosis of genetic diseases and in the determination of the abundance of specific different RNA molecules within a cell or tissue as a measure of gene expression. Northern blot analysis is used to study the RNA's gene expression further.

Many cellular functions are regulated by changes in gene expression. Thus, quantification of transcription levels of genes plays a central role in the understanding of gene function and of abnormal alterations in regulation that may result in a diseased state. The innovation of the real-time polymerase chain reaction (PCR) technique played a crucial role in molecular medicine and clinical diagnostics.

RT-PCR is commonly used in studying the genomes of viruses whose genomes are composed of RNA, such as influenza virus and retroviruses.

Reverse transcriptase polymerase chain reaction (RT-PCR) is becoming a widely used method to quantify cytokines from cells, tissues, or tissue biopsies.

Examples of applications of RT-PCR are the quantitation of relative gene expression, detection of minimal residual disease, cancer diagnostics, and pathogen detection. Other applications include detection of genetically modified organisms in food samples, measurements of DNA or transgene copy number, and allelic discrimination.

In the field of immunology, many studies have shown that chronic inflammatory disorders, autoimmune diseases, and transplant rejections are closely associated with specific changes in the balance between pro- and anti-inflammatory cytokines in the affected cells or tissues. Quantification of the cytokines involved is essential for gaining more insight into the immune processes involved. Because tissue samples available for analysis are often too small to allow quantification of cytokines at the protein level, analysis of messenger RNA (mRNA) is widely used to investigate the cytokine levels at sites of immune infiltration or inflammation.

RT-PCR (reverse transcription-polymerase chain reaction) is a sensitive method for the detection of mRNA expression levels.

10.8.2 Principle and Working

When RNA is to be analyzed instead of DNA, RT-PCR is used. However, PCR is not capable of using RNA, as a template, and therefore amplification can not take place. In order to analyze RNA, PCR is combined with another enzymatic reaction called reverse transcription (RT). In reverse transcription, enzyme RNA-directed DNA polymerase or reverse transcriptase is used. This enzyme synthesizes DNA from RNA templates, a process that is reverse of the normal transcription phenomenon. When reverse transcriptase produces DNA that is complementary to the mRNA, it is called cDNA or complementary DNA.

Traditionally RT-PCR involves two steps- the RT reaction and PCR amplification. RNA is first reverse transcribed into cDNA using a reverse transcriptase as described here, the resulting cDNA is used as templates for subsequent PCR amplification using primers specific for one or more genes.

The reverse transcription can be performed in the same reaction tube that is to be used for further PCR. Initially, the reaction mixture contains RNA template, reverse transcriptase enzyme, a DNA primer directed to the targeted RNA region and dNTPs. The RNA template is first denatured at 72°C to break any random basepairs present. The reaction mixture is then cooled to 40°C to allow the primers to anneal to complementary sequences on RNA template. Further reverse transcriptase enzyme catalyzes primer extension in 5' to 3' direction, forming cDNA strand.

The reaction mixture is then heated to 94°C. At this temperature, reverse transcriptase enzyme is inactivated. Thereafter PCR reaction components are added to the mixture and reaction is carried out in the thermal cycler.

RT-PCR can also be carried out as one-step RT-PCR in which all reaction components are mixed in one tube prior to starting the reactions. Although one-step RT-PCR offers simplicity and convenience and minimizes the possibility for contamination, the resulting cDNA cannot be repeatedly used as in two step RT-PCR.

It is important to include a negative control in the experiment. This negative control is set up just like the cDNA synthesis reaction, except that RT enzyme is not added to it. Absence of PCR products when using the negative control reaction indicates that the PCR products did not arise from genomic DNA contamination of the RNA sample.

10.8.3 Procedure of RT- PCR

Materials required

Reverse Transcription reaction mixture

10 X RT Buffer

500 mM Tris-HCl, pH 8.3

750 mM KCl

30 mM $MgCl_2$

50 mM DTT

Random primers (50 μM)

dNTP (2.5 mM each dNTP)

RNase Inhibitor (10 units/μl)

PCR reaction components

10X PCR Buffer

Forward primer	5 μM
Reverse primer	5 μM
dNTP mix	2.5 mM
Taq Polymerase	5 U/μl

Procedure

First step of RT-PCR

1. Place RNase Inhibitor and Reverse Transcriptase on ice.
2. Thaw 10x reaction buffer, random primers, and dNTP mix quickly and place on ice.
3. Use small 0.25 ml PCR tubes. Assemble your reaction as given in Table 10.2 on ice. Add the enzyme last.
4. Mix the components of reaction mixture gently and spin briefly.

Table 10.2: Components of the reaction mixture for reverse transcription reaction

Component	*Stock*	*Final concentration*	*Sample (+RT)*	*Control (-RT)*
Total RNA	–	–	~1-2 μg	~1-2 μg
Random primers	50 μM	5 μM	2 μl	2 μl
10X RT Buffer	10 X	1 X	2 μl	2 μl
dNTP mix	2.5 mM	0.5 mM	4 μl	4 μl
RNase Inhibitor	10 U/μl	10 U	1 μl	1 μl
Reverse Transcriptase	100 U/μl	100 U	1 μl	0 μl
Nuclease-free water	–	–	Make final volume 20 μl	Make final volume 20 μl

5. Incubate in the thermal cycler at,
- 44°C for 1 hr.
- 92°C for 10 min to inactivate the reverse transcriptase.
6. Store reaction mixture at –20°C or proceed to the PCR.

Second step of RT-PCR

1. Mix the components of the PCR reaction mixture thoroughly in properly labelled PCR tubes (Refer Table 10.3).

Table 10.3: Components of reaction mixture for PCR reaction

Components	*Final concentration*	*Sample*	*Control*
RT reaction mixture	–	1-2 µl	0 ml
PCR Buffer	1 X	2.5 µl	2.5 µl
Forward primer	0.25 µM	1.25 µl	1.25 µl
Reverse primer	0.25 µM	1.25 µl	1.25 µl
dNTP mix	0.125mM	1.25 µl	1.25 µl
Taq Polymerase	1U	0.2 µl	0.2 µl
Sterile water	–	Make final volume 25 µl	Make final volume 25 µl

2. On top of the reaction mixture, add few drops of light mineral oil to prevent evaporation.
3. Perform the PCR reaction as given in Table 10.4

Table 10.4: Reactions of PCR

Stage	*Temperature*	*Time*	*Cycles*
Initial denaturation	94°C	5 minutes	1
Denaturation	94°C	30 seconds	–
Annealing	55°C	30 seconds	–
Extension	72°C	1 minute	40
Final extension Hold at 4°C	72°C	10 minutes	1

4. After completion of PCR reaction, remove the sample tubes and add 50 µl of chloroform to each tube.
5. Due to addition of chloroform, mineral oil sinks to the bottom of the tube. Without disturbing the tube, place tubes in centrifuge. Spin the tubes at top speed for about 30 seconds.
6. Transfer the upper phase to clean microfuge tubes. Add 15 µl of ammonium acetate. Mix well. Ammonium acetate removes primers and unincorporated nucleotides.
7. Add 100 µl of 100% ethanol and mix the contents. Keep on ice for 30 minutes.

8. Centrifuge at top speed in tabletop microcentrifuge for 15 minutes.
9. Decant the supernatant. Air-dry the pellet. Dissolve in sterile water and store at –20°C till further use.
10. The results can be seen by performing agarose gel electrophoresis of the PCR reaction products.

10.8.4 Troubleshooting Guide

1. If there is low or no yield of RT-PCR products, it could be due to low template quality or quantity, suboptimal thermal conditions, and improper primer.
2. Non-specific reaction products may result from reaction set-up at room temperature, or suboptimal reaction conditions or suboptimal thermal cycling conditions.
3. Non-specific amplification can occur from overabundance of primer or template. In this case, decrease primer/template concentration.

10.9 REVIEW QUESTIONS

1. Give principle of PCR reaction.
2. Compare PCR reaction with DNA replication.
3. Compare native PCR and RT-PCR.
4. What are the pre-requisites of PCR?

10.10 REFERENCES

Blaschke, V., Reich, K., Blaschke, S., Zipprich, S. and Neumaan, C. (2000). Rapid quantitation of proinflammatory and chemoattractant cytokine expression in small tissue samples and monocyte-derived dendritic cells: Validation of a new real-time RT-PCR technology. *J. Immunol. Methods,* **246:** 79–90.

Ehrlich, H. (1989). Polymerase chain reaction. *Journal of Clinical Immunology,* **9(6):** 437–447.

Overbergh, L., Giulietti, A., Valckx, D., Decallonne, B., Bouillon, R. and Mathieu, C. (2003). The use of real-time reverse transcriptase PCR for the quantification of cytokine gene expression. *J. Biomol. Tech.,* **14(1):** 33–43.

Singh, V. and Kumar, A. (2001). PCR primer design. *Molecular Biology Today,* **2(2):** 27–32.

White, B. (1993). PCR Protocols: Current Methods and Applications. Humana Press. pp. 1–41.

Wu, D., Ugozzoli, B., Qian Pal, J., and Wallace, R. (1991). The effect of temperature and oligonucleotide primer length on specificity and efficiency of amplification by the polymerase chain reaction. *DNA Cell Biol.,* **10:** 233–238.

11

Molecular Marker Techniques

11.1 INTRODUCTION

A molecular marker is the site of heterozygosity for some type of DNA variation not associated with any other measurable phenotypic variation. Such a DNA locus can be used in genetic mapping analysis just like a conventional heterozygous allele pair can be used. As molecular markers can be easily detected and are numerous in genome of an organism; they are usually mapped by linkage analysis. In genetic mapping, the biological significance of the DNA marker is not important in itself; as the heterozygous site is merely a convenient reference point that will be useful in finding one's way around the chromosomes.

The development of molecular markers has provided the study of dispersal with a new, potentially powerful tool. Nowadays many types of molecular marker techniques are available, the most widely used include RAPD (random amplified polymorphic DNA), RFLP (restriction fragment length polymorphism), AFLP (amplified fragment length polymorphism), SNP (single nucleotide polymorphism) and microsatellites or SSR (simple sequence repeats) markers. These markers differ in the type and amount of variability they express, in their suitability for each particular trait and in the ease and costs of their development and application.

Most molecular markers may represent variation in non-coding DNA regions. Although there may be cases where markers linked are to quantitative trait loci (QTL) or microsatellites associated with neurological disorders in humans, where the number of repeats is correlated to the severity of the symptoms of the disease; it is generally assumed that variation expressed by these markers is essentially selectively neutral. By applying a random set of markers a significant influence of selection or bias on the results is cancelled out.

Molecular markers vary, displaying different amounts of variation and different modes of inheritance. They may be either dominant or co-dominant, and may or may not be subjected to genetic recombination. Use of molecular markers has inspired the development of molecular techniques that take the evolutionary history of alleles into account while estimating gene flow.

Markers can be classified as co-dominant, meaning that banding patterns of homozygotes can be distinguished from the patterns of heterozygotes. In case of dominant markers, the banding patterns of heterozygotes are identical to the patterns of one of the homozygotes and thus homo- and heterozygotes cannot be distinguished. Co-dominant markers such as allozymes, microsatellites and most RFLP allow easy estimation of allele frequencies in populations. Dominant markers like RAPD, and AFLP on the other hand, allow estimation of genotype but not of allele frequencies.

Highly variable molecular markers, especially microsatellites, have facilitated a direct genetic approach to measuring gene flow based on parental analysis. Minisatellite fingerprints are codominant but, because of the large numbers of bands that typically result from this technique, it is often very difficult to calculate allele frequencies. Dominant markers, like RAPD and AFLP, do not allow estimation of allele frequencies.

11.2 IMPORTANCE AND APPLICATIONS OF MOLECULAR MARKER TECHNIQUES

Molecular markers should not be considered as normal genes, as they usually do not have any biological effect, and instead can be thought of as constant landmarks in the genome. They are identifiable DNA sequences, found at specific locations on the genome, and are transmitted by the standard laws of inheritance from one generation to the next. The existence of various molecular techniques and differences in their principles and methodologies require careful consideration in choosing one or more of such marker types.

Multi-locus marker: Marker that screens many loci in the genome, as in the random amplified polymorphic DNA-polymerase chain reaction (RAPD–PCR), amplified fragment length polymorphism–PCR (AFLP–PCR) and minisatellite DNA fingerprinting is called multi-locus marker.

Single-locus marker: Marker that derives from a single locus in the genome, such as allozymes, most RFLPs and the typical microsatellite marker.

Dominant marker: Marker that is scored as present or absent (null) and thus does not allow identification of homologous alleles is called dominant marker.

Co-dominant marker: Marker that allows identification of homologous alleles and scoring of homozygote and heterozygote alleles is called codominant marker. For many experiments of population genetics, co-dominant markers are clearly superior to dominant markers because they allow estimation of allele frequencies. For a given level of analytical power, co-dominant markers require smaller sample sizes than dominant markers.

The development and use of molecular markers for the detection and exploitation of DNA polymorphism is one of the most significant developments in the field of molecular genetics. As scientists learn where markers occur on a chromosome, and how close they are to specific genes, they can create a genetic linkage map. Such genetic maps serve several purposes, including detailed analysis of associations between economically important traits and genes or quantitative trait loci (QTLs) and facilitate the introgression of desirable genes or QTLs through marker-assisted selection (MAS).

In case of ornamentals, vegetables or any other agronomically important crops, the conventional testing of varieties is a time consuming and erroneous process. In such cases, molecular markers provide advantages both in accuracy and speed.

Molecular markers have proven to be powerful tools in the assessment of genetic variation and in the elucidation of genetic relationships within and among species. Several molecular markers viz. RFLP, RAPD, SSRs, ISSRs, AFLP and SNPs are presently available to assess the variability and diversity at molecular level (Refer Table 11.1). Information regarding genetic variability at molecular level could be used to help, identify and develop genetically unique germplasm that compliments existing species and varieties.

The various PCR-based techniques are of two types depending on the primers used for amplification:

Table 11.1: Comparisons of different molecular marker systems

Feature	***RFLP***	***RAPD***	***AFLP***	***SSR***	***SNP***
PCR based	No	Yes	Yes	Yes	Yes
Amount of DNA required	High	Low	Medium	Low	Low
DNA quality	High	High	Moderate	Moderate	High
Number of loci analyzed	1.0-3.0	10-50	20-100	1.0-3.0	1.0
Ease of use	Difficult	Easy	Easy	Easy	Easy
Reproducibility	High	Low	High	High	High
Type of probes/primers	Low copy genomic DNA or cDNA clones	Usually 10 bp random nucleotides	Specific sequence	Specific repeat DNA sequence	–

- Arbitrary or semi-arbitrary primed PCR techniques that have been developed without prior sequence information (*e.g.* RAPD, AFLP, ISSR).
- Site-targeted PCR techniques that developed from known DNA sequences (*e.g.*, EST, SSR, SCAR, STS).

SSRs are now the marker of choice in most areas of molecular genetics as they are highly polymorphic even between closely related lines, require low amount of DNA, can be easily automated for high throughput screening, can be exchanged between laboratories, and are highly transferable between populations

The analysis of DNA sequence variation is of major importance in genetic studies. In this context, molecular markers are a useful tool for assaying genetic variation, and have greatly enhanced the genetic analysis of crop plants. A variety of molecular markers, including restriction fragment length polymorphisms (RFLPs), random amplification of polymorphic DNAs (RAPDs), amplified fragment length polymorphisms (AFLPs) and microsatellites or simple sequence repeats (SSRs), have been developed in different crop plants.

11.3 RESTRICTION FRAGMENT LENGTH POLYMORPHISM (RFLP)

11.3.1 Introduction

In the last few years, molecular tools like restriction endonucleases have opened a door for laboratory research in genetics.

RFLP markers were first used in 1975 to identify DNA sequence polymorphisms for genetic mapping of a temperature-sensitive mutation of adenovirus serotypes. It was then used for human genome mapping, and later adopted for plant genomes.

Restriction Fragment Length Polymorphism (RFLP) is a difference in homologous DNA sequences that can be detected by the presence of fragments of different lengths after digestion of the DNA samples in question with specific restriction endonuclease enzymes. RFLP, as a molecular marker, is specific to a single clone or restriction enzyme combination.

RFLP markers have become useful in all areas of molecular biology. The restriction endonuclease enzymes recognize a specific nucleotide sequence and cleave both strands of duplex DNA at or near the recognition site. Meselson and Yuan in 1968, reported presence of a peculiar enzyme in *E. coli* bacteria that was able to digest foreign DNA sequence of bacteriophage. This enzyme was later named as 'restriction endonuclease'. Later on similar enzymes were found in other bacteria.

In 1980, Botstein and coworkers developed the theoretical framework for identifying polymorphic single basepairs by RFLP and using them as marker loci to map human genomes.

Advantages of RFLP

- RFLP is highly robust methodology with good reproducibility between various protocols and laboratory conditions for same restriction enzyme.
- This molecular marker is codominantly inherited and as such, can estimate heterozygosity.
- No sequence information is required for development of RFLP markers.
- Because this marker is based on sequence homology, it is highly recommended for phylogenetic analysis between related species, as well as is suited for constructing genetic linkage maps.
- This is a locus-specific marker, which allows synteny studies.
- Given the availability of suitable probes, the technique can readily be applied to any organism. This simplicity of the RFLP technique provides a great advantage over other molecular markers.

Limitations of RFLP

- Large amounts of DNA are required for RFLP analysis. Not only that, but it requires substantially larger sample sizes than other forms of analysis. This is a considerable limitation especially when the samples to be analyzed are extremely rare and valuable.
- RFLP analysis is a slow and more tedious process compared to some of the newer DNA analysis techniques.
- Low levels of polymorphism are detected in some species by RFLP markers.
- Few loci are detected per assay
- This technique is time consuming, especially with single-copy probes. The duration of the process itself is a lengthy one as well, taking up to a month to complete.
- This technique can be very costly due to high price of reagents, restriction enzymes and probes.

11.3.2 Theory and Principle

In RFLP analysis, organisms are differentiated on the basis of lengths of fragments generated by cleaving their genomic DNA with restriction enzymes. The polymorphism is the result of variation in the location of restriction site on the genome or intervening sequences in the genome of different organisms that yield DNA fragments of varying lengths.

If two organisms differ in the distance between sites of cleavage of a particular restriction endonuclease, the length of the fragments produced will differ when the DNA is digested with a restriction enzyme. Although two individuals of the same species have almost identical genomes, they will always differ at a few nucleotides due to one or more of the following causes: point mutation, insertion, deletion, translocation, inversion and duplication. Some of the differences in DNA sequences at the restriction sites can result in the gain, loss, or relocation of a restriction site.

Restriction Endonucleases

Restriction endonucleases are enzymes that cleave DNA molecules at specific nucleotide sequences depending on the particular enzyme used. Enzyme recognition sites are usually 4 to 6 base pairs in length. Generally, the shorter the recognition sequence, the greater the number of fragments generated. If molecules differ in nucleotide sequence, fragments of different sizes may be generated.

Restriction enzymes are isolated from a wide variety of bacterial genera and are thought to be part of the cell's defenses against invading bacterial viruses. These enzymes are named by using the first letter of the genus, the first two letters of the species, and the order of discovery (Refer Table 11.2).

Table 11.2: Various restriction enzymes that are used in RFLP system

Restriction enzyme	*Source organism*	*Restriction site (5'→3')*
Bam HI	*Bacillus amyloliquifaciens*	G↓GATCC
*Bgl*II	*Bacillus gliobigli*	A↓ATCT
*Hae*III	*Haemophillus aegypticus*	GG↓CC
*Hha*I	*Haemophilus haemolyticus*	GCG↓C
*Pst*I	*Providencia stuartii*	CTGCA↓G
*Sma*I	*Serratia marcescence*	CCC↓GGG
*Eco*RI	*E. coli* RY 13	G↓AATTC
*Eco*RII	*E. coli* R245	↓CCTGG

The greatest resolution is obtained by using restriction enzymes recognizing a four base pair sequence because there are many such sites in the genome. The fragments produced will be relatively small, which provides a better chance of identifying single base alterations. Conversely, use of enzymes that recognizes an eight basepairs sequence will require fewer probes, because larger fragments of the genome are analyzed at one time. As a result of this, only large alterations of the DNA will be visualized and complex electrophoresis system must be used to resolve the fragments into discrete bands.

Most RFLP markers are co-dominant, which means that both alleles in heterozygous sample will be detected and are highly locus-specific.

An RFLP probe is a labeled DNA sequence that hybridizes with one or more fragments of the digested DNA sample after they were separated by gel electrophoresis, thus revealing a unique blotting pattern characteristic to a specific genotype at a specific locus. Short, single- or low-copy genomic DNA or cDNA clones are typically used as RFLP probes.

RFLP is a type of length polymorphism in the genomes of homologous segments of DNA of different individuals that have been digested with restriction enzymes. The digested DNA is resolved on agarose gels and specific probes are used to identify regions of the DNA that exhibit polymorphism (Fig. 11.1).

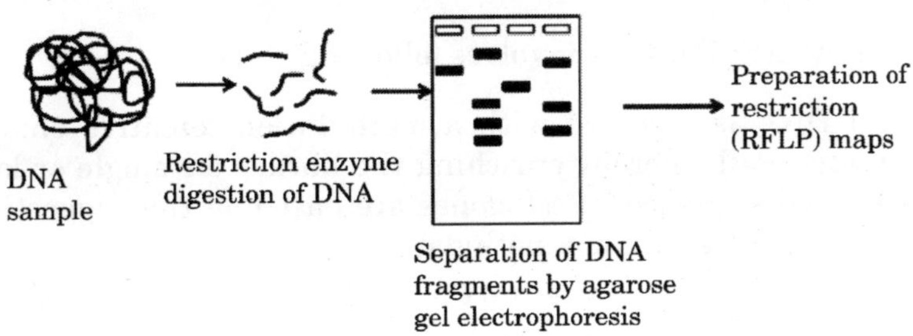

Fig. 11.1: Restriction fragment length polymorphism

The steps involved in the RFLP analysis of DNA are,

1. Isolation of DNA
2. Digestion of DNA with restriction enzymes.
3. Separation of restriction fragments by agarose gel electrophoresis

4. Transferring restriction fragments to membrane by southern blotting
5. Detection of particular sequence/restriction fragment with probe

After digestion by restriction enzymes, the DNA is present as a mixture of linear double-stranded molecules of various lengths. These are then separated by electrophoresis through agarose or polyacrylamide gels. The choice between agarose and polyacrylamide is based on the restriction enzymes chosen.

The fragments of DNA cut by the restriction enzymes contain genes and non-coding DNA. The non-coding DNA includes regions consisting of direct repeats of the same sequence of bases, referred to as tandem repeats. The number of repeats of the sequence is genetically determined and, provided that the sequence is long enough and is repeated a sufficient number of times, will affect the length of the restriction fragment. These regions are called variable number of tandem repeats (VNTRs) loci.

The RFLP probes are frequently used in genome mapping and in variation analysis such as genotyping, forensics, paternity tests, hereditary disease diagnostics, etc. The probes used for hybridization are preferably single locus and mostly species-specific probes of about 500 to 3000 bp in size. Probes used to identify specific DNA fragments by hybridization are of two types: genomic clones giving fragments of nuclear DNA and cDNA clones that are DNA copies of mRNA molecules.

Developing RFLP probes involves following steps:

- Total DNA is digested with a methylation-sensitive enzyme for example, *Pst*I, thereby enriching the library for single or low-copy expressed sequences. *Pst*I clones are based on the suggestion that expressed genes are not methylated.
- The digested DNA is size-fractionated on a preparative agarose gel, and fragments ranging from 500 to 2000 bp are excised, eluted and cloned into a plasmid vector for example, pUC18.
- Digested plasmids are screened to check for DNA inserts.
- Southern blots of the DNA inserts can be probed with total sheared DNA to select clones that hybridize to single and low-copy sequences.
- The probes are screened for RFLPs using genomic DNA of different genotypes digested with restriction endonucleases. Typically, in species with moderate to high polymorphism rates, two to four restriction endonucleases are used such as *Eco*RI, *Eco*RV, and *Hind*III. In species with low polymorphism rates, additional restriction endonucleases can be tested to increase the chance of finding polymorphism.

The genetic status of RFLP loci on single chromosome is called as haplotype. The determination of alleles of an RFLP locus in an individual is called haplotyping or DNA typing. During 1980s, identification of thousands of RFLP loci expanded the number of alleles that were used for genetic studies.

The information obtained with the RFLP technique depends upon both the number of probes and restriction enzymes used. Each different probe hybridizes with a different set of genomic DNA fragments and each enzyme excises a segment of genomic DNA at different points. The major strength of RFLP markers are high reproducibility, codominant inheritance, good transferability between laboratories, provide locus specific markers that allow synteny studies which target conserved order of genes between related organisms and no sequence information required.

11.3.3 Applications of RFLP

The use of RFLP to help map genes involved in human diseases, was first suggested by David Botstein. This technique is now widely used to map genes that contribute to diseases.

In an extension of mapping analysis, RFLP alleles can be used as diagnostic tools. For example, in a family with a record of a certain disease, if it can be established that the people who have the disease also carry a specific allele of an RFLP, then this fact suggests not only that the RFLP locus is linked to the disease gene locus, but furthermore that the specific RFLP allele is linked with the disease allele. Hence the RFLP allele becomes a diagnostic marker for the disease, and this information can be used in genetic diagnostics.

RFLPs can be used to measure genetic divergence between different populations or related species. The restriction-site difference is effectively a difference in DNA sequence, so a measure of the total number of RFLP differences represents a measure of genetic difference. Hence RFLPs are important in studies of evolution and population genetics.

DNA fingerprinting is a technique that is used to identify patterns that occur in DNA. No two organisms have identical DNA so this procedure can be used to identify if a sample of DNA came from a particular individual. The technique of DNA fingerprinting requires that the DNA be cut up into small fragments. Restriction enzymes are used to perform this digestion. In RFLP analysis, the DNA of an organism is cut up into fragments using restriction enzymes. A large number of short fragments of DNA will be produced.

Restriction enzymes always cut at the same base sequence. Because no two individuals have identical DNA, no two individuals will have the same length fragments. For example, the enzyme *Eco*RI always cuts DNA at the sequence GAATTC. Different people are going to have different locations of this particular sequence on their genomes and will therefore have different fragment lengths. Thus RFLP is an excellent tool in DNA fingerprinting.

RFLPs can be used in many different settings to accomplish different objectives.

RFLPs can be used in forensic applications such as paternity cases or criminal cases to determine the source of a DNA sample. RFLPs can be used determine the disease status of an individual. It can be used in the detection of genetic mutations. RFLPs can be used to measure recombination rates, which can lead to a genetic map while distance between RFLP loci can be measured and compared with genetic database.

In forensic science, RFLP has become very important tool. Development of the first DNA fingerprinting techniques by Alec Jeffreys in 1984. It plays an important role in allowing scientists to map the human genome as well as provide information on genetic diseases. By using RFLP analysis, scientists are then able to determine others who might be at risk for the disease or a carrier of the mutated gene.

RFLP analysis is frequently used to characterize whole genomes or PCR amplification products of bacterial isolates.

DNA typing was introduced into forensic science in the mid 1980s, arising from discoveries made in biomedical research. In 1980s, White and colleagues described the first polymorphic RFLP marker and proposed methods for mapping the human genome based on RFLP technology.

11.4 RANDOM AMPLIFIED POLYMORPHIC DNA (RAPD)

11.4.1 Introduction

In the last decade, the random amplified polymorphic DNA (RAPD) technique based on the polymerase chain reaction (PCR) has been one of the most commonly used molecular techniques to develop DNA markers. RAPD markers are amplification products of anonymous DNA sequences using single, short and arbitrary oligonucleotide primers, and thus do not require prior knowledge of a DNA sequence. Low expense, efficiency in developing a large number of DNA markers in a short time and requirement for less sophisticated equipment has made the RAPD technique valuable.

RAPD is a popular molecular marker system because it is technically easy to implement, relatively cheap and produces results quickly. RAPD markers were developed and applied in plants in 1990. The number of RAPD loci used for the analyses of plant genomes varies between several tens to hundreds.

Limitations of RAPD

- Nearly all RAPD markers are dominant, *i.e.* it is not possible to distinguish whether a DNA segment is amplified from a locus that is heterozygous or homozygous. Codominant RAPD markers, observed as different-sized DNA segments amplified from the same locus, are detected only rarely.
- PCR is an enzymatic reaction therefore the quality and concentration of template DNA, concentrations of PCR components, and the PCR cycling conditions may greatly influence the outcome. Thus, the RAPD technique is laboratory dependent and needs carefully developed laboratory protocols to become reproducible.
- Mismatches between the primer and the template may result in the total absence of PCR product as well as in a merely decreased amount of the product. Thus, the RAPD results can be difficult to interpret.

11.4.2 Theory and Principle

RAPD are multilocus markers and their mode of inheritance is dominant. The genotyping using RAPD is very simple. Therefore it has become popular in many genetic analytical laboratories.

RAPD markers are decamer (10 nucleotide length) DNA fragments. RAPD does not require any specific knowledge of the DNA sequence of the target organism. RAPD reactions are PCR reactions, but they amplify segments of DNA, which are essentially random.

The standard RAPD technology utilizes short synthetic oligonucleotides (10-12 bases long) of random sequences as primers to amplify nanogram amounts of total genomic DNA under low annealing temperatures by PCR. Amplification products are generally separated on agarose gels and stained with ethidium bromide (Fig. 11.2). Decamer primers are commercially available from various sources.

Unlike the conventional PCR targeted on the identified sequence, RAPD reaction uses a single type of primer and not a pair of different primers. If two neighbouring primers are in correct position in the proper orientation, and not too far from one another, *i.e.* less than 5kbp, then PCR generates a product of amplification. RAPD can be used to quickly obtain various fragments of genomic DNA that are to be analyzed.

Polymorphism in RAPD results due to presence or absence in the primer binding sites. Amplification products represent one allele per locus.

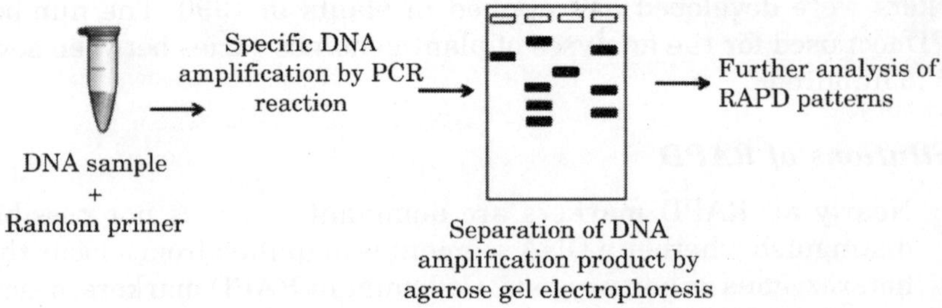

Fig. 11.2: Random amplified polymorphic DNA (RAPD) technique

At an appropriate annealing temperature during the thermal cycle, oligonucleotide primers of random sequence bind several priming sites on the complementary sequences in the template genomic DNA and produce discrete DNA products. Nucleotide variation between different sets of template DNAs will result in the presence or absence of bands because of changes in the priming sites. The conditions of sample preparation and PCR cycles must be optimized for each application. It is essential to standardize amplification conditions and run all samples for comparative analysis using the same protocol and same PCR equipment.

Most RAPD fragments result from the amplification of one locus, and two kinds of polymorphism occur: the band may be present or absent, and the brightness (intensity) of the band may be different.

Reproducibility of RAPD Markers

Although the RAPD method is relatively fast, cheap and easy to perform in comparison with other methods that have been used as DNA markers, the issue of reproducibility has been of much concern since the invention of the technique. In fact, ordinary PCR is also sensitive to changes in reaction conditions, but the RAPD reaction is far more sensitive than conventional PCR because of the length of a single and arbitrary primer used to amplify anonymous regions of a given genome.

Several factors have been reported to influence the reproducibility of RAPD reactions: quality and quantity of template DNA, PCR buffer, concentration of magnesium chloride, primer to template ratio, annealing temperature, Taq DNA polymerase brand or source, and thermal cycler brand.

This reproducibility problem is usually the case for bands with lower intensity. The reason for bands with high or lower intensity is still not known. Perhaps some primers do not perfectly match the priming sequence, amplification in some cycles might not occur, and therefore bands remain fainter. The chance of these kinds of bands being sensitive to reaction conditions of course would be higher than those with higher intensity amplified with primers perfectly matching the priming sites. The most important factor for reproducibility of the RAPD profile has been found to be the result of inadequately prepared template DNA.

Differences between the template DNA concentrations of two individual DNA samples may result in the loss or gain of some bands.

Since RAPD amplification is directed with a single, arbitrary and short oligonucleotide primer, DNA from virtually all sources is amenable to amplification. Therefore, DNA from the genome in question may include contaminant DNA from infections and parasites in the material from which the DNA has been isolated. Special care is needed for keeping out the DNA to be amplified from other sources of DNA.

Although the sequences of RAPD primers are arbitrarily chosen, two basic criteria must be met while selecting a primer. A minimum of 40% GC content and the absence of palindromic sequences. Palindromic sequence is a base sequence that reads exactly the same from right to left as from left to right. Because G-C bond consists of three hydrogen bridges and the A-T bond of only two, a primer-DNA hybrid with less than 50% GC will probably not withstand the 72°C temperature at which DNA elongation takes place by DNA polymerase.

11.4.3 Applications of RAPD

Because of the simplicity and low cost of the RAPD technique, it has found a wide range of applications in many areas of molecular biology. Some of the areas where the technique is used are described below:

Genetic Mapping

The necessity for sequence information for PCR was circumvented using short primers of arbitrary sequences to amplify DNA segments, namely RAPD. The speed and efficiency of RAPD analysis encouraged scientists to perform high-density genetic mapping in many plant species such as alfalfa, apple, pomegranate etc. in a relatively short time. Over 250 RAPD markers in *Arabidopsis* have been mapped.

The RAPD approach has also been used to create saturated genetic maps in fish species. About 401 polymorphic DNA markers have been mapped in zebrafish. RAPD markers are also being used to construct genetic maps of tilapia species, *O. niloticus* and *O. aureus* and rainbow trout, *Oncorhynchus*.

One disadvantage of RAPD markers is that they are dominant markers, hence the statistical information generated is less per marker in F_2 populations. Therefore, when mapping with dominant markers, it is necessary to use backcross or recombinant inbred populations, haploid or gametophytic tissue, or alternatively an F_2 population where only RAPD markers amplified from a single parent are mapped.

The RAPD technique can provide a ready source of hybridization probes for southern blot analysis.

Developing genetic markers linked to target trait

One of the most widely used applications of the RAPD technique is the identification of markers linked to traits of interest without the necessity for mapping the entire genome. This approach based on near-isogenic lines (NILs) is accomplished by repeatedly backcrossing a line carrying a gene of interest or donor parent to a cultivated line having otherwise desirable characteristics recurrent parent. The introgression of the target gene produces a line with a small segment of donor parent genome in a genetic background, which is almost exclusively from the recurrent parent. Thus markers that show polymorphisms between these 2 lines are likely to be linked to the gene of interest.

RAPD analysis has been successful in identifying markers linked to disease resistance genes in tomato (*Lycopersicon* sp.), lettuce (*Lactuca* sp.) and in common bean (*Phaseolus vulgaris*).

Another RAPD-based approach is bulked segregant analysis. BSA involves bulking DNA samples of individuals segregating in a single population. Bulked DNA samples from individuals that have the target trait or gene are compared to bulked DNA samples of individuals lacking the trait. Markers that are polymorphic between the pools will be genetically linked to loci determining the trait used to construct the pool. The technique has been successfully used to identify RAPD markers linked to disease resistance genes in crops like lettuce and barley.

This strategy has also been useful in the identification of sex-linked RAPD markers in dioecioius plants like angiosperm *Silene latifola*. RAPD markers that are linked to a particular chromosome or specific region of a chromosome can be developed using the pooled DNA method. RAPD analysis of pooled DNA samples further increases the efficiency of gene tracking.

Population and Evolutionary genetics

Application of DNA-based approaches to population genetic studies has been limited, probably due to the need for large samples of individuals from each population to provide an accurate estimate of allele and genotype frequencies.

The RAPD technique has received a great deal of attention from population geneticists because of its simplicity and rapidity in revealing DNA-level genetic variation.

A major drawback of RAPD markers in population genetic studies of outbreeding organisms is that they are dominant. Thus gene frequency estimates for such loci are less accurate than those obtained with codominant markers such as allozymes and RFLPs. The assumption of homology between bands of apparently the same molecular weight from the same primer is potentially another problem for RAPD surveys. Homology between co-migrating bands in different individuals is a good assumption when individuals are from the same population. However, this may not be true when individuals belong to different species or widely divergent populations. The RAPD analysis gives more accurate estimates between closely related populations and less accurate estimates for distantly related populations. RAPD data have been used for phylogenetic studies.

RAPD markers are more suitable for screening of clonal organisms than sexually reproducing organisms. As they breed asexually, a polymorphic fragment among individuals can be used to determine the clonal identity.

As stated above, the RAPD technique enables the development of large numbers of genetic markers more efficiently than RFLP based methods that have been used to construct saturated mapping of complex genomes. These markers can be used in monitoring these loci during introgression and selection programs.

The ability of the RAPD technique to reveal intra-specific variation can be used in screening for the degree of inbreeding in commercial plant and animal species to prevent an increase in the frequency of deleterious recessive alleles in populations. Polymorphic RAPD markers transformed to SCAR markers can be more advantageous in commercial breeding programs if a quick assay can be developed to detect the presence or absence of the product.

11.5 AMPLIFIED FRAGMENT LENGTH POLYMORPHISM (AFLP)

11.5.1 Introduction

Amplified fragment length polymorphisms (AFLPs) are polymerase chain reaction (PCR)-based markers for the rapid screening of genetic diversity. AFLP methods rapidly generate hundreds of highly replicable markers from DNA of any organism; thus they allow high-resolution genotyping of fingerprinting quality. The time and cost efficiency, reproducibility and resolution of AFLPs are superior or equal to those of other markers.

AFLP-PCR was first described by researcher Pieter Vos and his colleagues in 1995.

This PCR (polymerase chain reaction) based marker method requires no prior sequence knowledge and provided some diversity exists at the molecular level, *i.e.* plants are not highly inbred, and a reasonable number of polymorphic markers will be obtained from a series of primer combinations. The other main advantage of using AFLP is that it incorporates a series of molecular techniques that are relevant to many other molecular marker types, *e.g.* RFLP (Restriction Fragment Length Polymorphism), SSAP (Sequence Specific Amplified Polymorphism), SNP (Single Nucleotide Polymorphism), SSR (Simple Sequence Repeats) etc.

In short, the AFLP method reduces the whole genome into amplifiable fragments of less than 800 base pairs in length so that a subset of them can be examined on denaturing polyacrylamide gels. The series of procedures uses a range of DNA modifying enzymes and finally relies on fragment amplification by PCR.

The AFLP technique provides a novel and very powerful DNA fingerprinting technique for DNAs of any origin or complexity.

AFLP analysis is not as easy to perform as RAPDs, but is more efficient than using RFLPs. The advantages of AFLP include:

- It is highly reliable, and
- It does not require any DNA sequence information from the organism under study.
- It is information-rich due to its ability to analyze a large number of polymorphic loci simultaneously with a single primer combination on a single gel as compared to RAPDs, RFLPs and microsatellites
- Co-migrating AFLP amplification products are mostly homologous and locus specific with exceptions in polyploidy species.

The limitations of AFLP include:

- It requires template DNA free of inhibitor compounds that interferes with the restriction enzyme.
- It involves additional cost to purchase, both restriction and ligation enzymes as well as adapters.
- Like RAPD, most AFLP loci are dominant, which does not differentiate dominant homozygotes from heterozygotes. This reduces the accuracy of AFLP markers in population genetic analysis, genetic mapping, and marker assisted selection.

11.5.2 Theory and Principle

AFLP technique combines the power of RFLP with the flexibility of PCR-based technology by ligating primer recognition sequences to the restricted DNA.

AFLPs are DNA fragments that are between 80-500 base pairs in length. The AFLP technique is based on the selective PCR amplification of restriction fragments from a total digest of genomic DNA. These are obtained by digesting DNA using restriction enzymes and then ligating oligonucleotide adaptors to the digested fragments and finally amplifying these sequences by PCR.

Primers used in this technique are semi-specific, consisting of a core adaptor sequence, a restriction enzyme-specific sequence and a tail of 1-5 other nucleotides. Depending on genome size, restriction and ligation generates thousands of adapted fragments. For visualization after electrophoresis, only a subset of these fragments is amplified. To achieve selective amplification of a subset of these fragments, primers are extended into the unknown part of the fragments usually one to three arbitrarily chosen bases beyond the restriction site

The tail seems to be responsible for specificity. Therefore higher the number of other nucleotides in the tail, lower the number of bands obtained in PCR.

In AFLP technique, two cycles of amplification are normally carried out, using more specific primers in the second round. The selective amplification is achieved by the use of primers that extend into the restriction fragments, amplifying only those fragments in which the primer extensions match the nucleotides flanking the restriction sites.

The polymorphism results from the variation in the restriction sites and intervening regions. Using this method, sets of restriction fragments may be visualized by PCR without knowledge of nucleotide sequence. The method allows the specific co-amplification of high numbers of restriction fragments.

AFLP markers can be generated for DNA of any organism without initial investment in primer/probe development and sequence analysis. Both good quality and partially degraded DNA can be used for digestion but the DNA should be free of restriction enzyme and PCR inhibitors.

AFLP-PCR reactions are carried out under stringent conditions, permitting only the selective amplification of those genomic DNA fragments that are perfectly complementary to the 3' ends of the primer sequences. The stringent PCR conditions employed in AFLP-PCR reactions lead to highly reproducible results that are readily comparable among different samples.

The technique involves three steps:

- Restriction of the DNA and ligation of oligonucleotide adapters
- Selective amplification of sets of restriction fragments
- Gel electrophoretic analysis of the amplified fragments.

Typically 50-100 restriction fragments are amplified and detected on denaturing polyacrylamide gels. Polyacrylamide gel electrophoresis provides maximum resolution of AFLP banding patterns to the level of single-nucleotide length differences, whereas fragment length differences of less than ten nucleotides are difficult to score on agarose gels.

AFLP markers can be generated for DNAs of any organism, and no initial investment in primer/probe development or sequence analysis is required. Partially degraded DNA can be used, but DNA should be highly purified and free of polymerase chain reaction (PCR) inhibitors. Extremely small amounts of DNA (about 50 ng) are digested with two restriction enzymes, and AFLP adaptors are ligated to these ends. Adaptor ligations are performed in the presence of restriction enzymes such that any fragment to fragment ligations are immediately recleaved by the restriction enzymes. The adaptor is designed so that ligation of a fragment to an adaptor does not reconstitute the restriction site. The end sequences of each adapted fragment now consist of the adaptor sequence and the remaining part of the restriction sequence. These known end sequences serve as priming sites in the subsequent AFLP–PCR.

In AFLP technique, PCR amplification will only occur where the primers are able to anneal to fragments, which have the adaptor sequence plus the complementary base pairs to the additional nucleotides called selective nucleotides.

The technique of AFLP can be described in brief as follows:

Preparation of the AFLP template

AFLP analysis starts with a sample of genomic DNA isolated and purified from a single individual. Techniques for the isolation of genomic DNA are widely available and can be easily adapted for any organism of choice. After the genomic DNA is obtained, it is digested with a pair of restriction enzymes, such as MseI and EcoRI.

MseI recognizes 5' -TTAA-3' and cleaves after the first 5' -T, whereas EcoRI recognizes 5' -GAATTC-3' and cleaves after the 5' -G. The genomic

DNA is typically incubated with these enzymes for a period long enough to permit complete digestion of the DNA into restriction fragments. MseI and EcoRI generate DNA fragments with 5' overhangs (5'-TA-3' and 5'-AATT-3', respectively) that are distinct from each other and are non-complementary.

Ligation Reaction with restriction fragments and adaptors

The MseI adaptor and an EcoRI adaptor, is a short, double-stranded DNA molecule with a 5' overhang that is complementary to the 5' overhang generated by MseI (5' -TA-3') and EcoRI (5' -AATT-3'), respectively.

Rather than containing a 3' -A after the 5' -TA-3' overhang, the MseI adaptor contains a 3' -C, which destroys the MseI recognition site once ligated. Similarly, rather than containing a 3' -C after the 5' -AATT-3' overhang, the EcoRI adaptor contains a 3' -G, which destroys the EcoRI recognition site once ligated. The MseI and EcoRI adaptors each contain their own specific DNA sequence, typically between 19 and 22 base pairs in length, that is located upstream of their respective 5' overhangs.

This ligation reaction involves incubation of the MseI/EcoRI digested genomic DNA together with the two adaptors and with DNA ligase. DNA ligase is an enzyme that covalently links the adaptors with their respective complementary 5' ends. MseI and EcoRI restriction enzymes can also be added to this ligation reaction, as they will prevent re-ligation between EcoRI-EcoRI and MseI-MseI genomic fragments and will not recognize the ligated adaptor-genomic DNA fragments.

Selective PCR amplification

Now each unique genomic DNA fragment is sandwiched between the MseI adaptor and the EcoRI adaptor and their associated unique DNA sequences.

Although the MseI and EcoRI adaptor-ligated ends of each DNA fragment will be identical, the sequences of the genomic DNA fragments, which begin just after the original MseI and EcoRI sites, will differ.

If PCR reactions are carried out using primers corresponding to the MseI and EcoRI adaptor sequences, it would amplify every single genomic DNA fragment, and one would therefore be faced with an indecipherable set of DNA fragments. Thus, in order to selectively amplify a smaller number of genomic DNA fragments, it is advisable to use primer sets that are complementary to the MseI or EcoRI adaptor sequences starting at their 5' ends but that add up to three unique nucleotides following the end of the original MseI or EcoRI recognition site. The addition of only a small number of base pairs, called selective nucleotides, at the end of each primer permits the amplification of a subset of genomic DNA fragments.

As the number of nucleotides added to the 3' end of each primer is increased, the selectivity of the primer also increases. After all, more genomic DNA fragments will be amplified by a primer that contains a single unique base at its 3' end compared to a primer that contains three unique bases at its 3' end.

Primer design does not require any previous knowledge of the genome under study. Rather, researchers simply set up PCR reactions using a variety of primer sets, each of which consists of one MseI-associated primer and one EcoRI-associated primer.

Electrophoretic separation of amplified DNA fragments

In order to determine the AFLP-PCR-derived DNA amplifications, one must be able to identify and characterize the set of DNA fragments produced in a given PCR reaction. Thus, in most cases, one of the two PCR primers is radioactively or fluorescently labeled (typically the EcoRI primer), which leads to the production of a labeled PCR product that can be easily detected. One can also use high-resolution electrophoresis to separate the DNA fragments based on their size and overall negative charge, large fragments will migrate more slowly, and small fragments will migrate more quickly when exposed to an electrical field.

Analysis

Following electrophoretic separation of AFLP-PCR products, one must analyze the resulting DNA banding patterns to determine the DNA fingerprint associated with a given sample. Depending on the complexity of the banding patterns, one can manually determine band sizes, or they can turn to automated approaches. Because AFLP-PCR can only detect dominant genetic markers, it cannot report whether an individual is homozygous for a given marker. However, this technique does permit the simultaneous amplification of multiple genomic DNA fragments often between 50 and 100 fragments within a single PCR reaction with high specificity and reproducibility in the absence of any genome sequence data.

11.5.3 Applications of AFLP

Because of their high reproducibility and ease of use, AFLP markers have emerged as a major new type of genetic marker with broad application in systematics, pathotyping, population genetics, DNA fingerprinting and quantitative trait loci (QTL) mapping.

The key feature of AFLP–PCR is its capacity for the simultaneous screening of many different DNA regions distributed randomly throughout the genome.

AFLP markers have proved useful for assessing genetic differences among individuals, populations and independently evolving lineages, such as species. For a wide range of taxa, including plants, fungi, animals and bacteria, AFLP markers have been used to uncover cryptic genetic variation of strains, or closely related species, that had been impossible to resolve with morphological or other molecular systematic characters.

AFLP markers have also been used to infer phylogenetic relationships based on measures of genetic distance. These markers have found the widest application in analyses of genetic variation below the species level, particularly in investigations of population structure and differentiation. AFLP markers have been applied to evaluate gene flow and dispersal, outcrossing, introgression and cases of hybridization.

Because AFLP methods can generate many genome-wide polymorphic markers without prior sequence knowledge, AFLPs are a powerful tool for generating linkage maps. AFLP markers have been used extensively for constructing linkage maps for QTL analyses of agronomic plant traits, such as disease resistance and salt tolerance.

In short, DNA fingerprinting involves the generation of a set of distinct DNA fragments from a single DNA sample; these fragments are then used as a source of genotypic information. A wide range of techniques can be used to generate DNA fingerprinting data. Today, researchers are increasingly turning to AFLP-PCR in their DNA fingerprinting efforts. AFLP-PCR uses many of the same steps as RFLP, SSR, and RAPD, including restriction enzyme digestion of genomic DNA, PCR, and electrophoretic separation of DNA fragments. However, this method includes additional steps that permit high-resolution interrogation of the entire genome, and it yields highly specific, reproducible genotypic data. Although AFLP can only be used to study dominant genetic markers, it does not rely on any previous knowledge of the genome sequence.

This technique can be especially useful in addressing questions related to organisms whose genome sequence has not yet been determined, including many plants, fungi, and bacteria. AFLP can help indicate whether two organisms are members of the same species. Furthermore, it can be used to assess genetic variation within a species or among closely related species. Population geneticists also use AFLP approaches to determine genetic variation across different populations. Indeed, this method has allowed researchers to refine the taxonomic classification of organisms based on AFLP associated genetic markers.

Of course, AFLP can also be used to study organisms whose genome has been sequenced. For example, AFLP has been used to study human DNA samples in criminal investigations and in paternity tests.

In the medical backgound, AFLP has been used to track pathogenic outbreaks to determine whether an outbreak is associated with the spread of a single strain or multiple distinct strains. In this type of analysis, the pathogenic organism is isolated from infected patients and subjected to AFLP analysis. By comparing AFLP data collected from different patients, researchers can readily determine whether the disease outbreak stems from one or from many different strains.

11.6 SIMPLE SEQUENCE REPEATS (SSR)

11.6.1 Introduction

SSR markers detect variation in the number of short repeat sequences, usually two or three base-repeats. The numbers of such repeat units have been found to change at a high frequency and allow the detection of multiple alleles. Simple sequence repeats or microsatellite markers have been developed recently for crop plants and this marker system is predicted to lead to even more rapid advances in marker development and breeding programs.

The use of molecular markers to track loci and genome regions in crop plants is now routinely applied in many breeding programs. Improvements in marker screening techniques have also been important in facilitating the tracking of genes. For markers to be effective, they must be closely linked to the target locus and be able to detect polymorphisms in material likely to be used in a breeding program.

Simple sequence repeats (SSRs), when resolved using agarose gels, may be a viable and cost effective alternative to RFLPs and isozymes.

11.6.2 Theory and Principle

The genomes of higher organisms contain three types of multiple copies of simple repetitive DNA sequences like satellite DNAs, minisatellites, and microsatellites; arranged in arrays of vastly differing sizes. Microsatellites are also known as simple sequence repeats.

SSRs are stretches of tandemly repeated nucleotide motifs with one to six nucleotides per motif and are usually highly polymorphic due to a variable number of repeats. Insertions and deletions of repeat units are the result of DNA polymerase error that may occur during replication or unequal recombination.

Investigators often prefer to work with loci containing tri and tetra-nucleotide repeat arrays rather than dinucleotide arrays because the former frequently give fewer stutter bands. The stutter bands are multiple near-identical ladders of PCR products, which are one or two nucleotides shorter or longer than the full length product.

Microsatellites are born from regions in which variants of simple repetitive DNA sequence motifs are already over represented. The repeated sequence is often simple, consisting of two, three or four nucleotides. One common example of a microsatellite is a dinucleotide repeat (CA)n, where n refers to the total number of repeats that ranges between 10 and 100. These markers often present high levels of inter and intra-specific polymorphism, particularly when tandem repeats number is ten or greater.

A simple sequence repeat consisting of two to six, but usually two or three, nucleotides that are repeated many times in tandem -CACACA...- show high variation in repeat number between individuals. By developing PCR primers for the regions flanking a microsatellite repeat, microsatellite allele variation at this site can be screened through high-resolution electrophoresis of microsatellite PCR products.

SSR primers are developed by cloning random segments of DNA from the target species. These are inserted into a cloning vector, which is in turn, implanted into *Escherichia coli* bacteria for replication. Colonies are then developed, and screened with single or mixed simple sequence oligonucleotide probes that will hybridize to a microsatellite repeat, if present on the DNA segment. If positive clones for microsatellite are obtained from this procedure, the DNA is sequenced and PCR primers are chosen from sequences flanking such regions to determine a specific locus. This process involves significant trial and error on the part of researchers.

Microsatellite primers developed for one species can rarely be used beyond the very closest relatives; therefore, microsatellite primers need to be developed de novo for each new species. The analytical strengths of microsatellite markers are co-dominance and hypervariability.

PCR reactions for SSRs are run in the presence of forward and reverse primers that anneal at the 5' and 3' ends of the template DNA, respectively. PCR fragments are usually separated on polyacrylamide gels in combination with $AgNO_3$ staining, autoradiography or fluorescent detection systems.

The development of microsatellite markers involves several distinct steps from obtaining the library to developing a working set of primers that can amplify polymorphic microsatellite loci.

1) Microsatellite library construction.
2) Identification of unique microsatellite loci.
3) Identifying a suitable area for primer design.
4) Obtaining a PCR product.
5) Evaluation and interpretation of banding patterns.
6) Assessing PCR products for polymorphism

Several methods have been used for the development of microsatellite markers. SSR-containing sequences can be found by simply searching public sequence databases, like GenBank. On the other hand, genomic libraries can be constructed and screened for SSR-containing clones. This can be laborious and many colonies have to be screened, because the proportion of microsatellites to the entire genome is generally quite low.

The advantages of microsatellite markers over the RAPD markers are their high specificity, high polymorphism, codominance, unambiguous scorability, and high reproducibility.

When PCR products are separated on agarose gels (or acrylamide gels) extra bands (or peaks) could be observed. These extra bands are called stutter bands. Stutter bands are artifacts and are usually shorter and have weaker signal intensity than the actual microsatellite alleles. Due to this technical limitation, when the microsatellite bands or peaks are scored these extra bands have to be eliminated. Improperly scored bands or peaks will affect the data analysis and final results.

Null alleles: A microsatellite null allele is defined as any allele at a microsatellite locus that consistently fails to amplify due to primer template mismatch or elimination of the gene product. This can occur in normal SSR applications. The primary sources of microsatellite null alleles are poor primer annealing due to nucleotide sequence divergence like mutation in one or both primers, or differential amplification of size-variant alleles and fragment rearrangements such as insertions or inversions.

11.6.3 Applications of SSR

Microsatellite or simple sequence repeat (SSR) markers have been successfully used for genomic mapping, DNA fingerprinting, and marker-assisted selection in many plant species.

Microsatellite loci are more common in some organisms than in others, and screening may produce few useful loci in some species. The efficiency of microsatellite marker development depends on the abundance of repeats in the target species and the ease with which these repeats can be developed into informative markers

SSRs are mostly codominant markers, and are excellent for studies of population genetics and mapping

To better understand the population biology of organisms microsatellite (simple sequence repeat) markers are more informative markers. Microsatellite markers have several advantages over RFLPs or RAPDs: they are technically simple to be handled, small amounts of DNA are needed for PCR amplification, they are unambiguously scorable, and results are highly reproducible. In addition, microsatellite markers are codominant and, therefore, widely used in genome mapping as well as in population genetics.

SSR markers have been useful for integrating the genetic, physical and sequence-based physical maps in plant species, and simultaneously have provided breeders and geneticists with an efficient tool to link phenotypic and genotypic variation.

Microsatellites as DNA markers have been widely used in many living organisms including *E. coli*, humans, mice, cows and sheep, plants and insects for different kinds of basic genetic research.

As a diagnostic technique, microsatellite markers have been developed for identifying these disorders. In plants and insects, microsatellite markers have been developed for cultivar and insect identification, marker-assisted selection (MAS), quantitative trait loci (QTL) analysis, exploring insect resistance, linkage maps and population and evolutionary genetics studies.

These markers have proved useful in disease research. There are more than 40 neurological, neurodegenerative, and neuromuscular disorders associated with human diseases, which are ultimately caused by tandem repeat instability.

SSR markers can be used as anchor markers for joining large pieces of overlapped DNA fragments such as bacterial artificial chromosomes (BACs). Physical maps will give us a real physical distance between markers or genes in bp (base pair) or kbp. SSR markers have been used to construct a whole genome physical map of model crop species.

Since SSR markers are abundant and highly polymorphic, they have been used in the identification of *E. coli* strains and yeast strains.

SSR based marker-assisted selection (MAS) can also greatly enhance the efficiency of plant breeding programs. SSR markers used for selection can be classified into flanking SSR markers closely linked to the locus for a trait and targeted gene SSR markers developed within the targeted gene itself. Therefore SSR marker-assisted selections have been used in many breeding programs.

Phylogenetic relationships reflect the relatedness of a group of species based on a calculated genetic distance such as sequence conservation or diversification in their evolutionary history. SSR markers often make a powerful system for revealing interspecific or intraspecific phylogenetic relationships.

Microsatellite markers can be used to determine the population structure within and among natural populations and/or identify the potential progenitors.

Chloroplast SSR markers have been used to survey the variation among barley populations. SSR marker-assisted selection and cultivar identification have been used in plant breeding programs and cultivar commercialization.

11.7 SINGLE NUCLEOTIDE POLYMORPHISM (SNP)

11.7.1 Introduction

Over the past few years, a large number of different SNP genotyping methods have been developed based on various methods of allelic discrimination and detection platforms.

When any two human genomes are compared side by side, they are 99.9% identical. However, with a 3.2 billion base pair genome, each person harbors some 3.2 million differences in his/her diploid genome. Most of the differences are due to single base substitution polymorphisms, popularly known as single nucleotide polymorphisms (SNPs). While the majority of the SNPs are of no biological consequence, a fraction of the substitutions have functional significance and are the basis for the diversity found among humans.

11.7.2 Theory and Principle

A single-nucleotide polymorphism is a variation occurring when a single nucleotide- A, T, G or C, in the genome that differs between members of a biological species or paired chromosomes in the same individual.

A SNP marker is just a single base change in a DNA sequence, with a usual alternative of two possible nucleotides at a given position. In this technique, allele discrimination can not be based on size differences on a gel.

Two sequenced DNA fragments from different individuals, AAGCCTA to AAGCTTA, contain a difference in a single nucleotide. In this case we say that there are two alleles, C and T. Similarly, almost all common SNPs have only two alleles.

The majority of SNP genotyping assays to one of four groups are based on molecular mechanisms like allele specific hybridization, primer extension, oligonucleotide ligation and invasive cleavage. Allele specific hybridization, also known as allele specific oligonucleotide hybridization, is based on distinguishing between two DNA targets differing at one nucleotide position by hybridization. Two allele-specific probes are designed, usually with the polymorphic base in a central position in the probe sequence. Under optimized assay conditions, only the perfectly matched probe-target hybrids are stable, and hybrids with one-base mismatch are unstable.

Allele specific oligonucleotide ligation is a method for SNP typing based on the ability of ligase to covalently join two oligonucleotides when they hybridize next to one another on a DNA template. The invader assay is based on the specificity of recognition, and cleavage, by a flap endonuclease, of the three-dimensional structure formed when two overlapping oligonucleotides hybridize perfectly to a target DNA.

Different aspects should be taken into account to determine the best suitable technology in terms of sensitivity, reproducibility, accuracy, capability of multiplexing for high throughput analysis, cost effectiveness in terms of initial investment for equipment and cost per data-point, flexibility of the technology for uses other than SNP discovery, and time-consumption for analysis.

Analytical methods to detect novel SNPs and known SNPs include:

- DNA sequencing
- Single strand conformation polymorphism
- Electrochemical analysis
- RFLP analysis
- Hybridization analysis

Pyrosequencing is a sequencing-by-synthesis method in which a cascade of enzymatic reactions yields detectable light, which is proportional to incorporated nucleotides. One feature of typing SNPs with pyrosequencing is that each allelic variant will give a unique sequence compared to the two other variants. These variants can easily be distinguished by a pattern recognition software. The software displays the allelic alternatives and allows for direct comparison with the pyrosequencing raw data. For optimal determination of SNPs, various protocols of nucleotide dispensing order should be investigated .

SNPs do not necessarily produce any adverse effects. This is because SNPs within a coding sequence do not necessarily change the amino acid sequence of the protein that is produced, due to degeneracy of genetic code. SNPs that are not in protein-coding regions may still affect gene splicing, transcription factor binding, messenger RNA degradation, or the sequence of non-coding RNA.

In addition to pharmacogenomic, diagnostic, and biomedical research implications, SNP maps are helping to identify thousands of additional markers in the genome.

The genomic distribution of SNPs is not homogenous, SNPs usually occur in non-coding regions more frequently than in coding regions. Single-nucleotide polymorphisms may fall within coding sequences of genes, non-coding regions, or in the intergeneric regions.

The examples of non-coding SNPs are

- 5' UTR
- 3' UTR
- Introns
- Intergenic regions
- Pseudogenes
- Splicing sites
- Transcriptional regulation (promoter & TF binding sites)
- Translational regulation (initiation or termination)
- Regulatory miRNA target sites

The examples of coding SNPs are

- Synonymous SNPs (third position variation)
- Replacement SNPs (changing amino acid)
- Functional SNPs (with acceptable amino acid replacement)
- Non-functional SNPs

Besides natural selection, other factors like recombination and mutation rate can also determine SNP frequency.

11.7.3 Applications of SNP

Single nucleotide polymorphism (SNP) detection technologies are used to scan for new polymorphisms and to determine the allele(s) of a known polymorphism in target sequences. The abundance of these polymorphisms in plant genomes makes the SNP marker system an attractive tool for mapping, marker-assisted breeding and map-based cloning.

Scientists believe that SNP maps will help in identifying the multiple genes associated with complex ailments such as cancer, diabetes, vascular disease, and some forms of mental illness.

Although more than 99% of human DNA sequences are the same, variations in DNA sequence can have a major impact on how humans respond to disease; environmental factors such as bacteria, viruses, toxins, and chemicals; and drugs and other therapies. This makes SNPs valuable for biomedical research and for developing pharmaceutical products or medical diagnostics. SNPs are also evolutionarily stable *i.e.* do not change much from generation to generation. This makes them easier to use in population studies.

As genetic markers, SNPs can be used to follow the inheritance patterns of chromosomal regions from generation to generation and are powerful tools in the study of genetic factors associated with human diseases.

SNPs normally do not cause disease, but they can help determine the likelihood that someone will develop a particular illness. One of the genes associated with Alzheimer's disease, apolipoprotein E or *ApoE*, is a good example of how SNPs affect disease development. *ApoE* contains two SNPs that result in three possible alleles for this gene: E2, E3, and E4. Each allele differs by one DNA base, and the protein product of each gene differs by one amino acid.

Each individual inherits one maternal copy of *ApoE* and one paternal copy of *ApoE*. Research has shown that a person who inherits at least one E4 allele will have a greater chance of developing Alzheimer's disease. Apparently, the change of one amino acid in the E4 protein alters its structure and function enough to make disease development more likely. Inheriting the E2 allele, on the other hand, seems to indicate that a person is less likely to develop Alzheimer's.

SNP detection encompasses two broad areas: scanning DNA sequences for previously unknown polymorphisms and genotyping individuals for known polymorphisms. SNP detection technologies are applicable not only to humans, but to any living organism on earth.

The study of SNPs is also important in crop and livestock breeding programs. The nature and organization of single nucleotide polymorphisms, between genomes of individuals are of great interest, because these variations can be associated with phenotypic traits, including disease susceptibility in agronomically important crops.

11.8 SEQUENCE CHARACTERIZED AMPLIFIED REGION (SCAR)

11.8.1 Introduction

SCARs are DNA fragments amplified by the Polymerase Chain Reaction (PCR) using specific 15-30 bp primers, designed from nucleotide sequences established in cloned RAPD (Random Amplified Polymorphic DNA) fragments linked to a trait of interest. By using longer PCR primers, SCARs avoid the problem of low reproducibility generally encountered with RAPDs. Obtaining a codominant marker may be an additional advantage of converting RAPDs into SCARs.

11.8.2 Theory and Principle

To overcome the problems associated with RAPD markers, they are converted into simpler and more robust marker known as SCAR. SCAR markers are developed by end sequencing of RAPD fragments and using longer primers. SCAR markers are codominant in expression.

A SCAR marker is a genomic DNA fragment that is identified by PCR amplification using a pair of specific oligonucleotide primers. SCARs are derived by cloning and sequencing the two ends of RAPD markers that appeared to be diagnostic for specific purposes, for example, a RAPD band present in disease resistant lines but absent in susceptible lines. SCARs are advantageous over RAPD markers as they detect only a single locus, however their amplification is less sensitive to reaction conditions, and they can potentially be converted into co-dominant markers.

This procedure increases reproducibility of the markers. It also avoids occurrence of non-homologous markers of same molecular weight. SCAR markers are obtained by introducing polymorphic RAPD bands into markers, which are then sequenced and specific primers are designed usually by expanding the original decamer primer sequence with 10-15 bases so that only the polymorphic band of interest is amplified.

Based on the sequence primers can be designed that can amplify a specific region of DNA. If a given set of genotypes displays length variation between the amplified fragments, the marker is called as SCAR marker.

11.8.3 Applications of SCAR

Randomly Amplified Polymorphic DNA (RAPD) is easy to develop and simple molecular marker, but lack of reproducibility makes it less reliable for authentication of herbal drugs. Therefore, it is a better option to improve the reproducibility of RAPD by converting RAPD amplicons into Sequence Characterized Amplified Region (SCAR) markers. SCAR markers are easy, reliable and reproducible thus, have an advantage over RAPD markers for authentication of medicinal herbs used in the preparation of traditional medicines. These markers however, have been developed for only a few medicinal herbs.

There are many applications of SCAR markers in insect taxonomy and population biology. SCARs have been used as genetic markers to distinguish Asian gypsy moths from the established North American populations. SCAR markers have also been used for identification of different species and biotypes of Asian gall midge, a major insect pest of rice and for the specific identification of six *Trypanosoma cruzi* lineages.

The potential application of SCAR markers in plant studies is by development of markers linked to a specific trait of interest. For example, SCAR markers linked to disease resistance, insect tolerance, or some other qualitative trait can definitely be helpful in marker assisted breeding in plants. Identification of markers linked to genes conferring economically important traits, can provide a rapid and non-destructive method for plant breeders. These markers can also provide a starting point for a map-based cloning of these genes. The identification of genes governing important traits enables scientists to predict gene functions and conduct genetic engineering experiments.

11.9 REVIEW QUESTIONS

1. What is importance and application of molecular markers?
2. What is RAPD marker? Describe in brief.
3. What is restriction fragment length polymorphism?
4. Give applications of RFLP.
5. Compare RFLP and AFLP.
6. Give applications of AFLP.
7. What are SSR markers? Give their applications.
8. What is SNP? Give its significance.
9. Compare RAPD and SCAR markers.

11.10 REFERENCES

Ahmadian, A., Gharizadeh, B., Gustafsson, A., Sterky, F., Nyrén, P., Uhlén, M. and Lundeberg, J. (2000). Single-nucleotide polymorphism analysis by pyrosequencing. *Analytical Biochemistry,* **280(1):** 103–110.

Ashley, C. and Warren, S. (1995). Tri-nucleotide repeat expansion and human disease. *Annual Review of Genetics,* **29:** 703–728.

Bardakci, F. and Skibinski, D. (1999). A polymorphic SCAR-RAPD marker between species of *Tilapia*. *Animal Genetics,* **30:** 78–79.

Botstein, D., White, R., Skolnick, M. and Davis, R. (1980). Construction of a genetic linkage map in man using restriction fragment length polymorphisms. *Am. J. Hum. Genet.,* **32:** 314–331.

Botstein, D., White, R., Skolnick, M. and Davis, R. (1980). Construction of a genetic linkage map in man using restriction fragment length polymorphisms. *Amer. J. Human Genet.,* **32:** 314–331.

De Vicente, M., Truco, M., Truco, J., Egea, J., Burgos, L. and Arús, P. (1998). RFLP variability in apricot (*Prunus armeniaca* L.). *Plant Breed.*, **117:** 153–158.

Dubreuil, P. and Charcosset, A. (1999). Relationships among maize inbred lines and populations from European and North American origins as estimated using RFLP markers. *Theor. Appl. Genet.*, **99:** 473–480.

Erlich, H.A. (1989). PCR Technology Principles and Applications for DNA Amplification. Stockton Press, New York.

Klein-Lankhorst, R., Vermut, A., Weide, R., Liharska, T. and Zabel, P. (1991). Isolation of molecular markers for tomato (*L. esculentum*) using random amplified polymorphic DNA (RAPD). *Theor. Appl. Genet.*, **83:** 108–114.

Mackill, D.J. *et al.* (1996). Level of polymorphism and genetic mapping of AFLP markers in rice. *Genome,* **39:** 969–977.

Maughan, P. *et al.* (1996). Amplified fragment length polymorphism (AFLP) in soybean: Species diversity, inheritance, and near-isogenic line analysis. *Theor. Appl. Genet.,* **93:** 392–401.

Paran, I. and Michelmore, R. (1993). Development of reliable PCR-based markers linked to downy mildew resistance genes in lettuce. *Theor. Appl. Genet.*, **85:** 985–993.

Queller, D., Strassman, J. and Hughes, C. (1993). Microsatellites and Kinship. *Trends Ecol. Evol.,* **8:** 285–288.

Scott, M., Haymes, K. and Williams, S. (1992). Parentage analysis using RAPD-PCR. *Nucleic Acid Res.*, **20:** 5493.

Senior, M., Murphy, J., Goodman, M. and Stuber, C. (1998). Utility of SSRs for determining genetic similarities and relationships in maize using an agarose gel system. *Crop Science,* **38(4):** 1088–1098.

Tanksley, S. (1993). Mapping polygenes. *Annual Review of Genetics,* **27:** 205–233.

White, R., Leppert, M., Bishop, D., Barker, D., Berkowitz, J., Brown, C., Callahan, P., Holm, T. and Jerominski, L. (1985). Construction of linkage maps with DNA markers for human chromosomes. *Nature,* **313:** 101–105

Williams, J., Kubelik, A., Livak, K., Rafalski, J. and Tingey, S. (1990). DNA polymorphisms amplified by arbitrary primers are useful as genetic markers. *Nucleic Acids Res.*, **18:** 6531–6535.

Wolff, K., Schoen, E. and Peters-Van, J. (1993). Optimizing the generation of random amplified polymorphic DNA in chrysanthemum. *Theor. Appl. Genet.,* **86:** 1033–1037.

Vos, P., Hogers, R., Bleeker, M., Reijans, M., Van De Lee, T., Hornes, M., Frijters, A., Pot, J., Peleman, J., Kuiper, M. and Zabeau, M. (1995). AFLP: A new technique for DNA fingerprinting. *Nucleic Acids Res.* **23:** 4407–4414.

Subject Index

H

I

J

L

M

N

S

T

U

V

W

X

Z